# Geoscience and Geosystems

# Geoscience and Geosystems

Edited by Joe Carry

SYRAWOOD
PUBLISHING HOUSE
New York

Published by Syrawood Publishing House,
750 Third Avenue, 9th Floor,
New York, NY 10017, USA
www.syrawoodpublishinghouse.com

**Geoscience and Geosystems**
Edited by Joe Carry

International Standard Book Number: 978-1-68286-657-3 (Hardback)

**Cataloging-in-Publication Data**

Geoscience and geosystems / edited by Joe Carry.
        p. cm.
Includes bibliographical references and index.
ISBN 978-1-68286-657-3
1. Geology. 2. Earth sciences. 3. Physical geography. I. Carry, Joe.
QE26.3 .G46 2019
550--dc23

# TABLE OF CONTENTS

# PREFACE

This book aims to highlight the current researches and provides a platform to further the scope of innovations in this area. This book is a product of the combined efforts of many researchers and scientists, after going through thorough studies and analysis from different parts of the world. The objective of this book is to provide the readers with the latest information of the field.

Geoscience refers to the study of the planet Earth. It encompasses all phenomena and physical characteristics of Earth. It uses other fields of science as research tools for developing scientific models related to the study of large scale geosystems. Geoscience delves into a detailed analysis of the lithosphere, atmosphere, biosphere and hydrosphere. This book traces the progress of this field and highlights some of its key concepts and applications. The extensive content herein also presents researches that have transformed this discipline and aided its advancement. Scientists and students actively engaged in this field will find this book full of crucial and unexplored concepts. For all readers who are interested in geoscience and geosystems, the case studies included in this book will serve as an excellent guide to develop a comprehensive understanding.

I would like to express my sincere thanks to the authors for their dedicated efforts in the completion of this book. I acknowledge the efforts of the publisher for providing constant support. Lastly, I would like to thank my family for their support in all academic endeavors.

**Editor**

# Optical laboratory facilities at the Finnish Meteorological Institute –Arctic Research Centre

**Kaisa Lakkala**[1], **Hanne Suokanerva**[1], **Juha Matti Karhu**[1], **Antti Aarva**[2], **Antti Poikonen**[2], **Tomi Karppinen**[1], **Markku Ahponen**[1], **Henna-Reetta Hannula**[1], **Anna Kontu**[1], **and Esko Kyrö**[1]

[1]Finnish Meteorological Institute – Arctic Research Centre, Tähteläntie 62, 99600 Sodankylä, Finland
[2]Finnish Meteorological Institute, Observation Services, Helsinki, Finland

*Correspondence to:* Kaisa Lakkala (kaisa.lakkala@fmi.fi)

**Abstract.** This paper describes the laboratory facilities at the Finnish Meteorological Institute – Arctic Research Centre (FMI-ARC, http://fmiarc.fmi.fi). They comprise an optical laboratory, a facility for biological studies, and an office. A dark room has been built, in which an optical table and a fixed lamp test system are set up, and the electronics allow high-precision adjustment of the current. The Brewer spectroradiometer, NILU-UV multifilter radiometer, and Analytical Spectral Devices (ASD) spectroradiometer of the FMI-ARC are regularly calibrated or checked for stability in the laboratory. The facilities are ideal for responding to the needs of international multidisciplinary research, giving the possibility to calibrate and characterize the research instruments as well as handle and store samples.

## 1 Introduction

The location of the Finnish Meteorological Institute – Arctic Research Centre (FMI-ARC) (67.367° N, 26.629° E) is ideal for atmospheric and environmental research in the boreal and sub-Arctic zone. Numerous international projects have been conducted in the FMI-ARC, and the need for multidisciplinary laboratory facilities have been obvious. In this paper, we present the optical laboratory facilities of the FMI-ARC and focus on the measurements of optical instruments used for stratospheric and climate research at the FMI-ARC.

The Brewer spectroradiometer (hereafter referred to simply as Brewer) measurements (Bais et al., 1993) started in 1988 at Sodankylä. First the focus was on total ozone measurements, as Sodankylä is affected by the springtime Arctic ozone loss in the stratosphere. Since 1990, the Brewer is also used to measure spectral solar ultraviolet (UV) irradiances. The Brewer is a single monochromator with a wavelength range from 290 to 325 nm. The wavelength step of the recorded spectrum is 0.5 nm. The slit function of the Brewer is 0.56 nm at full width at half maximum (FWHM). In 2012, a second Brewer was purchased to measure next to the old one on the roof of the sounding station of the observatory. The second Brewer is a double monochromator with a wavelength range from 290 to 365 nm. The Brewer time series is one of the longest homogenized time series measured in the Arctic (Lakkala et al., 2003, 2008).

The long-term solar UV radiation is also measured with multichannel NILU-UV radiometers (Høiskar et al., 2003) located on the roof of the sounding station since 2007. In addition, the NILU-UV instruments have been set up to measure in the peatland field experiment and forest experiment of the Finnish Ultraviolet International Research Centre (FU-VIRC) during the summers of 2002–2011 (Lakkala et al., 2016). The NILU-UV monitors the UV-B, UV-A, erythemally weighted (McKinlay and Diffey, 1987) UV radiation, photosynthetically active radiation (PAR), and total ozone column, and it provides information on cloudiness. The radiometer is a filter instrument with five UV channels, with central wavelengths around 305, 312, 320, 340, and 380 nm and bandwidths of around 10 nm at FWHM. The sixth channel measures the PAR in the 400–700 nm wavelength region. The radiometer has a Teflon diffuser, silicon detectors, and high-quality bandpass filters and is temperature-stabilized to 40 °C. One-minute averages of measured irradiances and detector temperature are recorded.

The FieldSpec Pro JR Full Range spectroradiometer, manufactured by Analytical Spectral Devices (ASD), Inc., now known as PANalytical, measures solar UV spectrum in the wavelength region from 350 to 2500 nm, covering the longer wavelengths of the UV part of the solar spectrum as well as the visible and near-infrared part of the solar spectrum. The measurements at FMI-ARC are used for validation of satellite measurements and algorithm development (e.g. Heinilä et al., 2014; Pulliainen et al., 2014; Niemi et al., 2012). The measurements started in 2006 and are located on a 30 m high tower (Sukuvaara et al., 2007), from which reflected radiation from both a forested and open area was measured until 2013. Currently only measurements over the forest are performed.

A common thing for these optical measurements is that the instruments' measurement capacity tends to change as a function of time. For example, the sensitivity of the channels of the multifilter radiometer tends to drift over time (Lakkala et al., 2005). In order to obtain reliable and homogenized measurements, the instruments need to be well characterized and regularly calibrated (Webb et al., 1998, 2003; Seckmeyer et al., 2001, 2010). The quality control and quality assurance of the measurements require monitoring of the stability of the instruments using regular lamp tests, which need to be performed in an appropriate optical laboratory. This work describes the characteristics of the optical laboratory facilities at the FMI-ARC and shows typical measurement protocols for the above-mentioned instruments.

## 2  Laboratory facilities

### 2.1  Optical laboratory

The optical laboratory was initially built in 1998 and moved to its present location in November 2002. It is a duplication of the optical laboratory at the FMI Jokioinen observatory. The laboratory comprises two adjacent rooms: the control room and the dark room. The temperature in the rooms is monitored using PT100 sensors, and both rooms are equipped with adjustable air conditioning. The floor is covered with a black plastic membrane, and the walls together with the ceiling lamps of the dark room are painted with antireflection black paint. Ceiling lamps are turned away from the measurement system. Lockers are covered with antireflection black cloths. An opening has been made on one wall of the dark room to serve as a lead-out for cables or installations which need outdoor air.

An optical table, from the manufacturer Melles Griot, is placed in the dark room. The average height from the floor is 91.5 cm, and its dimensions are $100 \times 150$ cm. The lamp holders and needed sensors can be fixed to the table with high precision. A lamp holder is set up to fulfil the needs of the calibration of the Brewer spectroradiometer. Also UV-B (BN-9102-147 UVB XB03) and UV-A (BN-9102-130

UVA XB05) sensors are set up in order to monitor the calibration lamp. The sensors are temperature-stabilized using circulating water.

The electronics of the laboratory include a $0.1\,\Omega$ shunt resistor (Burster-1282-0.1), a high-precision digital multimeter (Hewlett Packard 3458A), a high-precision power supply (Hewlett Packard 6675A system DC power supply 0-120V/0-18A), a voltage limiter (from 150 to 10 VDC), a voltage-to-voltage converter with galvanic isolation (Nokeval Signal Converter 641) (from 10 to 2.5 VDC with galvanic isolation), an adjustable voltage reference, a data logger (QLI50 Sensor Collector, manufactured by Vaisala Oyj), and a control PC. The mentioned electronics are located in the control room, and only the calibration lamp, the dark room temperature sensor, and the UV-B/A sensors are located in the dark room.

For absolute irradiance calibrations, the most important thing is controlling the current passing through the lamp. The lamp current is acquired by the voltage measurement over the reference shunt and controlled with the control room PC. The electrical circuit diagram of a calibration is shown in Fig. 1, where the main circuit is in bold. The system allows the current accuracy to be $\pm 0.001$ A. To ensure the accuracy, the multimeter and the shunt are sent every 1–2 years to SGS FIMKO Testing and Certification Services, Finland, for checking and calibration. During the last calibration (6–7 May 2014) the uncertainty of the shunt was $\pm 0.007\,\%$, which makes $\pm 0.007\,\Omega$m. The reported uncertainty was based on a standard uncertainty multiplied by a coverage factor $k = 2$, providing a level of confidence of approximately 95 %. The last calibration confirmed the specifications of the multimeter provided by the manufacturer, in which the accuracy of the voltage measurement over 24 h was 0.06 ppm.

LabVIEW System Design Software (LabVIEW), National Instruments, has been tailored to read the voltage drop of the shunt measured by the multimeter and to regulate the lamp current to the defined value. The regulation can be done in steps of 12 mA using the power supply (HP6675A). As there is a need for higher precision, an adjustable voltage reference with control voltage of 0–14 mV is connected to the external analogue input of the power supply to finely tune the current regulation (Fig. 1).

In order to safely monitor the voltage over the lamp, which results from the current regulation, a voltage limiter and a voltage converter are needed to give the right input signal to the data logger. Galvanic isolation is made at the same time with the conversion. The voltage readings can be used to monitor the long-term stability of the lamp and for quality control of the measurements. The readings from the voltage converter, temperature, and UV-B/A sensors are transferred via the data logger to the control PC.

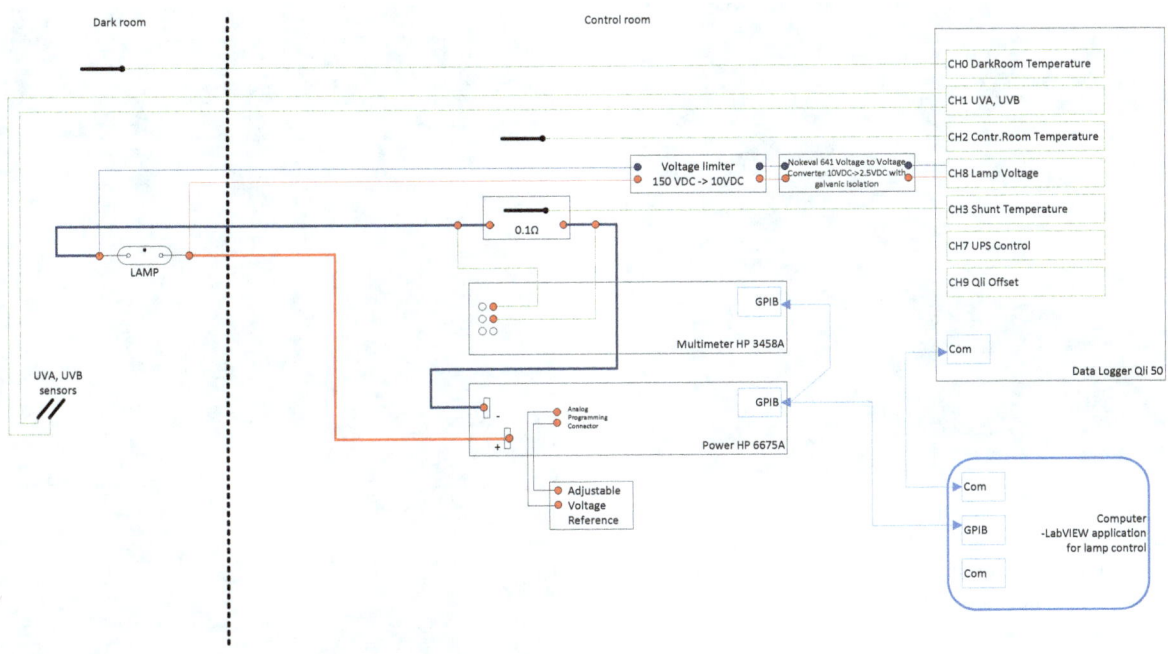

**Figure 1.** The electrical circuit of the optical laboratory in FMI-ARC. The main circuit is in bold. The vertical black dashed line denotes the separation between the dark room and the control room. Instruments on the left side are in the dark room, while the instruments on the right side are in the control room.

## 2.2 Laboratory for biological studies

The laboratory for biological studies is located in the next room of the optical laboratory. The room is equipped with machines needed, e.g., for snow and vegetation studies: a temperature chamber, a cold chamber, an ice cube maker, a fume hood, and a liquid nitrogen chamber.

The temperature in the temperature chamber (UT12, Thermo) can be regulated between ambient temperature $T$ ($+20\,°C$) and $+250\,°C$. The regulation range of the cold chamber (SRC 1812/3.1 B (L), Porkka) is between $+2$ and $+12\,°C$, and its dimensions are $1800\,\text{mm} \times 1200\,\text{mm} \times 2000\,\text{mm}$. The ice cube maker makes $21\,\text{kg}$ of ice per $24\,\text{h}$, and it can store $4\,\text{kg}$ of ice. The volume of the liquid nitrogen chamber is $35\,\text{L}$, the static working time is $130$ days, the working time is $80$ days, and the evaporation rate is $0.27\,\text{L}\,\text{day}^{-1}$. The fume hood is manufactured by IS VET.

## 3 Measurement procedures

### 3.1 UV spectroradiometer calibrations

The response of the Brewer spectroradiometer is determined by performing $1000\,\text{W}$ lamp measurements in the laboratory. The lamps are $1000\,\text{W}$ tungsten-filament incandescent halogen lamps of type DXW operated in vertical orientation. The bulbs have been installed in their sockets by Gigahertz Optik. A primary standard is used to transfer the calibration from the National Standard Laboratory MIKES-Aalto. Using the measurements of the Brewer, the irradiance scale is transfered to working standards, which are used for the calibration of the Brewer every 6 weeks. The procedure is described in more detail in Mäkelä et al. (2016) and Lakkala et al. (2008).

Before a calibration the Brewer is moved inside the dark room usually the day before. The multimeter is switched on then, which allows both the Brewer and the multimeter to stabilize around $15\,\text{h}$ before a calibration. The Brewer is placed on a trolley, which can be fixed and levelled in exactly the same place each time. The Brewer and the lamp were aligned under the same vertical optical axis by using an alignment jig and laser. The distance between the diffuser of the Brewer

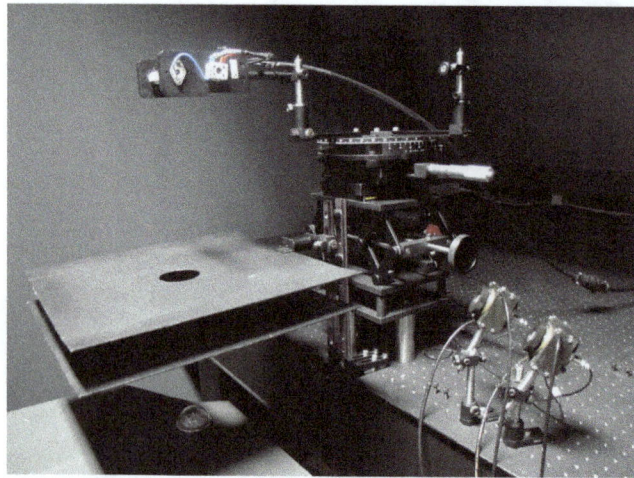

**Figure 2.** Lamp measurement with the Brewer spectroradiometer. The diffuser of the Brewer is in the bottom left corner under the baffles. The lamp is placed in the lamp holder over the baffles. The goniometer is seen in the middle, and the UV-A and UV-B sensors in the bottom right corner.

and the lamp is set to 50 cm. Two baffles are used between the lamp and the Brewer in order to reduce the effect of stray light. The measurement arrangements are shown in Fig. 2.

During a lamp measurement, the current is set to 8 A, with an accuracy of ±0.001 A. The current is controlled by the LabVIEW program, which allows the current to increase slowly for 2–3 min before reaching the final level. After measurements the current goes down slowly. The lamps typically need 15–20 min to stabilize before a measurement can start. The measurement itself takes around 17 min when scanning up to 365 nm. After the measurement, the lamp is left untouched until it has cooled down to near room temperature. The temperature of the control and dark rooms is set to 23 °C. The ventilation is on during the warming of the lamps but turned off during the measurements in order to avoid airflows around the lamp. The current, voltage, and room temperatures are recorded in a separate metadata file for each measurement. The intensity of the lamp is recorded using the UV-B and UV-A sensors of the laboratory, so that sudden changes can be noticed.

### 3.2 Stability of the multifilter UV instruments

The fixed set-up of the optical laboratory is also used for performing the stability checks of the NILU-UV multichannel radiometers of the FMI-ARC. As routine procedure, the stability of the channels of the NILU-UV is checked twice a year: in spring and in autumn. One-hundred-watt (100 W) OSRAM Radium lamps are mounted in lamp units produced by the manufacturer of the NILU-UV instrument. The lamp unit is connected to the circuit in the place of the lamp (Fig. 1). At least five lamps are used in order to detect the drift of the lamp from the drift of the instrument.

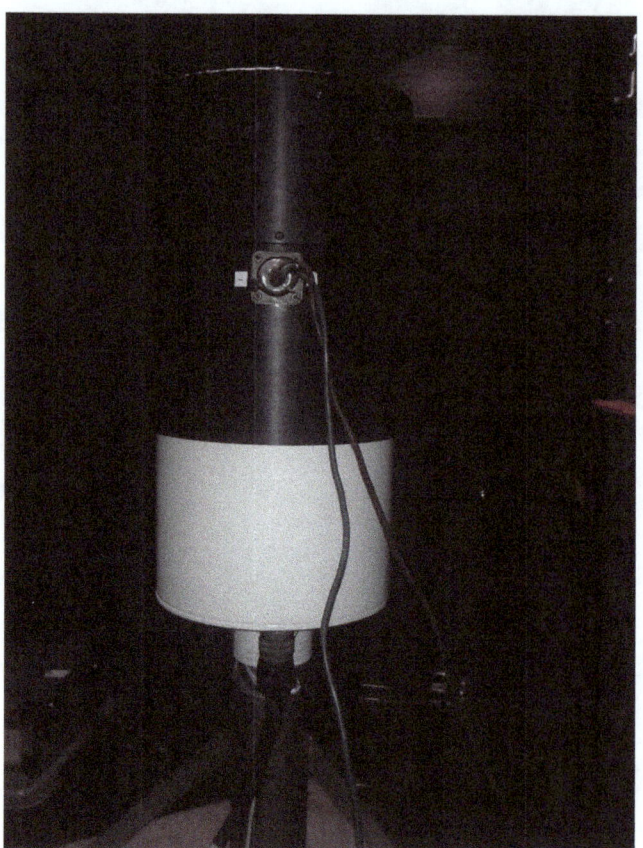

**Figure 3.** Lamp measurement of the NILU-UV radiometer. The NILU-UV is the white cylinder. The lamp is placed inside the black cylinder on the top of the NILU-UV.

The lamp is warmed up during 5 min in the dark box of the lamp unit at the side of the radiometer. If the lamp were warmed up at its measurement position on the top of the diffuser, the warming of the radiometer would affect the measurements (Lakkala et al., 2005). The current is increased slowly and set to 6 A by the operator. After the warming, the dark box including the lamp is placed above the diffuser, giving a vertical beam exactly to the same point of the diffuser during each measurement (Fig. 3). The data are recorded with a time step of 1 s for around 20 s, after which the dark box is removed from the diffuser and the lamp is left to cool down back to the room temperature.

### 3.3 Stability of the UV-VIS spectrometer

The fixed set-up of the optical laboratory is also used for monitoring the stability of the ASD field spectroradiometer and a Spectralon reference plate. The Spectralon reference plate is used as a reference for reflection measurements. The lamp measurements are performed once a year with similar 1000 W tungsten-filament incandescent halogen lamps of type DXW operated in vertical orientation as used for the calibration of the Brewer spectroradiometers (see Sect. 3.1).

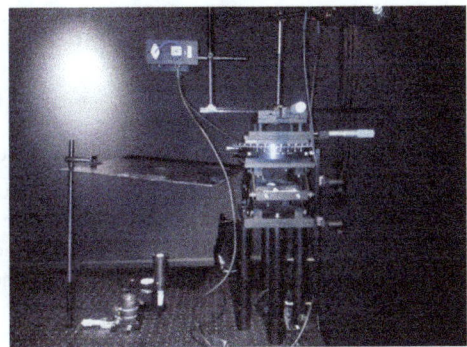

**Figure 4.** The remote cosine receptor measurement with the ASD FieldSpec Pro JR spectroradiometer.The entrance of the cosine receptor is the small white spot on the top of the black cylinder straight under the lamp. The light spot on the wall is reflection from the laboratory light, which is off during the measurements.

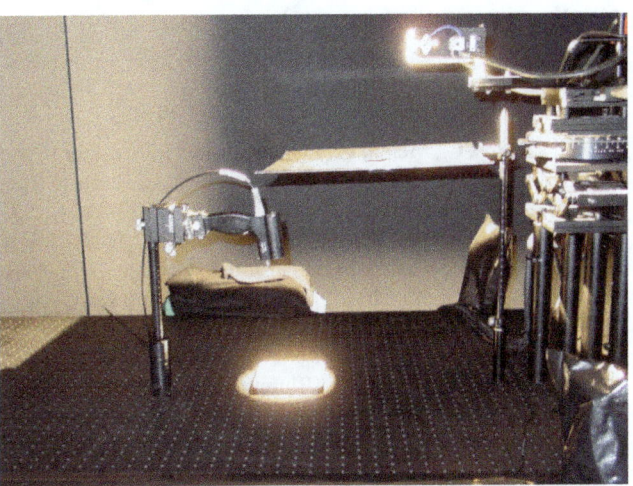

**Figure 5.** The reflectance measurement of the Spectralon reference plate with the ASD FieldSpec Pro Jr spectroradiometer. The reference plate in placed on the table, and the fibre optic cable is attached to the pistol grip fixed to the table in order to measure the reflected lamp radiation. The lamp is attached to the holder in the top right corner.

The calibration of the lamp is traceable to the National Standard Laboratory MIKES-Aalto.

Before the measurements, the ASD spectroradiometer is left to stabilize for at least half an hour in the room temperature. To monitor the stability of the ASD spectroradiometer, and to separate changes in the Spectralon plate from the changes in the spectroradiometer in the reflectance measurement, the spectrum of the lamp is measured with the remote cosine receptor (RCR). The RCR is fixed on the optical table at a vertical distance of 49 cm from the lamp (Fig. 4). One baffle is used to avoid stray light. Reflectance measurements of the Spectralon plates are used for monitoring of the changes in the field Spectralon plate and in the ASD spectroradiometer itself. Since the field Spectralon is subject to, e.g., dust, dirt, snow, rain, freezing, and mechanical stress from continuous use, its properties change. Both the field Spectralon plate and a reference Spectralon plate stored in laboratory conditions are measured. In the reflectance measurements the fibre optic cable is attached to a pistol grip fixed to the table, and the Spectralon plate is set in a fixed position. This set-up allows the fibre optic cable to be pointed to exactly the same spot on the Spectralon plates each time. The alignment is adjusted so that the radiation from the lamp is reflected from a Spectralon plate to the fibre optic cable (Fig. 5). The distance between the plate and the lamp is set to 65 cm.

## 4 Conclusions

The optical laboratory facilities at the FMI-ARC comprise a control room, a dark room, a facility for biological studies, and an office. They are ideal for calibration and characterization of optical instruments such as spectroradiometers, broadband and multichannel radiometers, and aurora cameras. The facilities promote the possibilities for multidisciplinary research. Several international groups have performed studies at FMI-ARC and used the facilities for stratospheric, snow, vegetation, and ionospheric studies during the Lapland Atmosphere – Biosphere Facility (LAPBIAT) project under the Improving Human Research Potential – Access to Research Infrastructures of the European Union (contract no. 025969-TA, http://www.sgo.fi/lapbiat). The facilities have served as a central research infrastructure of the Finnish Ultraviolet International Research Center (FUVIRC), where biologists could properly handle and store their samples.

In this work, we presented the set-up for calibration of the Brewer spectroradiometers and measuring the stability of the ASD spectroradiometer and the NILU-UV multichannel radiometers of the FMI-ARC. The facilities have also been used for characterizations of the instruments; e.g. the temperature and cosine response characterization of the Brewer spectroradiometer have been done in the dark room. Also the aurora cameras of the FMI's network are calibrated in the dark room. The stability of the travelling reference instrument of the NILU-UV Antarctic network was measured once a year in the optical laboratory (Lakkala et al., 2005).

In the dark room, a fixed set-up is made for vertical optical axis 1000 W DXW lamp measurements, and the electronics allow precise regulation of the current. The optical table is large enough for customized set-up for different optical instruments, and the rooms have enough space for temporary instruments in order to welcome research groups with different needs.

*Acknowledgements.* Tapani Koskela is acknowledged for the original design of the optical laboratory.

Edited by: N. Partamies

# References

Bais, A., Zerefos, C., Meleti, C., Ziomas, I., and Tourpali, K.: Spectral measurements of Solar Radiation and its Relation to Total Ozone, $SO_2$ and Clouds, J. Geophys. Res., 98, 5199–5204, 1993.

Heinilä, K., Salminen, M., Pulliainen, J., Cohen, J., Metsämäki, S., and Pellikka, P.: he effect of boreal forest canopy to reflectance of snow covered terrain based on airborne imaging spectrometer observations, Int. J. Appl. Earth Obs. Geoinf., 27, 31–41, 2014.

Høiskar, B., Haugen, R., Danielsen, T., Kylling, A., Edvardsen, K., Dahlback, A., Johnsen, B., Blumthaler, M., and Schreder, J.: Multichannel moderate-bandwidth filter instrument for measurement of the ozone-column anount, cloud transmittance, and ultraviolet dose rates, Appl. Optics, 42, 3472–3479, 2003.

Lakkala, K., Kyrö, E., and Turunen, T.: Spectral UV Measurements at Sodankylä during 1990–2001, J. Geophys. Res., 108, 4621, doi:10.1029/2002JD003300, 2003.

Lakkala, K., Redondas, A., Meinander, O., Torres, C., Koskela, T., Cuevas, E., Taalas, P., Dahlback, A., Deferrari, G., Edvardsen, K., and Ochoa, H.: Quality assurance of the solar UV network in the Antarctic, J. Geophys. Res., 110, D15101, doi:10.1029/2004JD005584, 2005.

Lakkala, K., Arola, A., Heikkilä, A., Kaurola, J., Koskela, T., Kyrö, E., Lindfors, A., Meinander, O., Tanskanen, A., Gröbner, J., and Hülsen, G.: Quality assurance of the Brewer spectral UV measurements in Finland, Atmos. Chem. Phys., 8, 3369–3383, doi:10.5194/acp-8-3369-2008, 2008.

Lakkala, K., Jaros, A., Aurela, M., Tuovinen, J.-P., Kivi, R., Suokanerva, H., Karhu, J., and Laurila, T.: Radiation measurements at the Pallas-Sodankylä Global Atmosphere Watch station – diurnal and seasonal cycles of ultraviolet, global and photosynthetically-active radiation, Boreal Environ. Res., 21, 427–444, 2016.

Mäkelä, J. S., Lakkala, K., Meinander, O., Kaurola, J., Koskela, T., Karhu, J. M., Karppinen, T., Kyrö, E., de Leeuw, G., and Heikkilä, A.: In search of traceability: two decades of calibrated Brewer UV measurements in Sodankylä and Jokioinen, Geosci. Instrum. Method. Data Syst. Discuss., doi:10.5194/gi-2015-40, in review, 2016.

McKinlay, A. F. and Diffey, B. L.: A reference action spectrum for ultraviolet induced erythema in human skin, CIE Research Note, CIE J., 6, 17–22, 1987.

Niemi, K., Metsämäki, S., Pulliainen, J., Suokanerva, H., Böttcher, K., Leppäranta, M., and Pellikka, P.: The behaviour of mastborne spectra in a snow-covered boreal forest, Remote Sens. Environ., 124, 551–563, doi:10.1016/j.rse.2012.06.008, 2012.

Pulliainen, J., Salminen, M., Heinilä, K., Cohen, J., and Hannula, H.-R.: Semi-empirical modeling of the scene reflectance of snow-covered boreal forest: validation with airborne spectrometer and lidar observations, Remote Sens. Environ., 155, 303–311, 2014.

Seckmeyer, G., Bais, A., Bernhard, G., Blumthaler, M., Booth, C., Disterhoft, P., Eriksen, P., McKenzie, R., Miyauchi, M., and Roy, C.: Instruments to Measure Solar Ultraviolet Radiation, Part 1: Spectral Instruments, Global Atmosphere Watch Report No. 125, World Meteorological Organization (WMO), Geneva, 30 pp., 2001.

Seckmeyer, S., Bais, A., Bernhard, G., Blumthaler, M., Johnsen, B., Lantz, K., and McKenzie, R.: Instruments to Measure Solar Ultraviolet Radiation, Part 3: Multi-channel filter instruments, Global Atmosphere Watch Report No. 190, World Meteorological Organization (WMO), Geneva, 51 pp., 2010.

Sukuvaara, T., Pulliainen, J., Kyrö, E., Suokanerva, H., Heikkinen, P., and Suomalainen, J.: Reflectance spectroradiometer measurement system in 30 meter mast for validating satellite images, IGARSS: 2007 IEEE International Geoscience and Remote Sensing Symposium, 23–28 July 2007, Barcelona, 2885–2889, 2007.

Webb, A., Gardiner, B., Martin, T., Leszcynski, K., Metzdorf, J., and Mohnen, V.: Guidelines for Site Quality Control of UV Monitoring, Global Atmosphere Watch Report No. 126, World Meteorological Organization (WMO), Geneva, 39 pp., 1998.

Webb, A., Gardiner, B., Leszcynski, K., Mohnen, V., Johnston, P., Harrison, N., and Bigelow, D.: Quality Assurance in Monitoring Solar Ultraviolet Radiation: the State of the Art, Global Atmosphere Watch Report No. 146, World Meteorological Organization (WMO), Geneva, 45 pp., 2003.

# The magnetic observatory on Tatuoca, Belém, Brazil: history and recent developments

**Achim Morschhauser**[1], **Gabriel Brando Soares**[2], **Jürgen Haseloff**[1], **Oliver Bronkalla**[1], **José Protásio**[2], **Katia Pinheiro**[2], and **Jürgen Matzka**[1]

[1]GFZ German Research Centre for Geosciences, Geomagnetism, Telegrafenberg, 14473 Potsdam, Germany
[2]Geophysics Department, Observatório Nacional, Rio de Janeiro, CEP, 20921-400, Brazil

*Correspondence to:* Achim Morschhauser (mors@gfz-potsdam.de)

**Abstract.** The Tatuoca magnetic observatory (IAGA code: TTB) is located on a small island in the Amazonian delta in the state of Pará, Brazil. Its location close to the geomagnetic equator and within the South Atlantic Anomaly offers a high scientific return of the observatory's data. A joint effort by the National Observatory of Brazil (ON) and the GFZ German Research Centre for Geosciences (GFZ) was undertaken, starting from 2015 in order to modernise the observatory with the goal of joining the INTERMAGNET network and to provide real-time data access. In this paper, we will describe the history of the observatory, recent improvements, and plans for the near future. In addition, we will give some comments on absolute observations of the geomagnetic field near the geomagnetic equator.

## 1 Introduction

The Tatuoca magnetic observatory (IAGA code: TTB) has a long history, and its roots go back to as early as 1933 when a temporal magnetic observatory was set up on the island of Tatuoca (Gama, 1955). Already at that time, the site of the observatory was chosen to fall within low magnetic latitudes (Gama, 1955), and an inclination of 18.18° was measured when a permanent magnetic observatory was opened on Tatuoca in 1954 (AGU, 1955). Eventually, the northward-moving equator passed the observatory in March 2013 (Fig. 1). The closest neighbouring observatory of Tatuoca is located in Kourou (IAGA code: KOU, French Guyana) at a distance of about 700 km north and 400 km west of Tatuoca (Fig. 1), and was installed in 1995 by the Institut de Physique

du Globe de Paris (IPGP). While the TTB observatory is currently under the full influence of the equatorial electrojet (EEJ), the KOU observatory is far enough from the magnetic equator to record this signal. Thus, subtracting the magnetic data recorded at the KOU observatory from those recorded at the TTB observatory will isolate the magnetic signal of the EEJ from the signal of the solar quiet (SQ) currents and the magnetospheric ring currents (cf. Manoj et al., 2006).

Moreover, the Tatuoca observatory is located within the South Atlantic Anomaly (SAMA) (Hulot et al., 2015), and shows a strong secular variation of almost $200\,\mathrm{nT\,yr^{-1}}$ in the radial component as predicted, for example, by IGRF-12 (Thébault et al., 2015).

The National Observatory of Brazil (ON) and the German Research Centre for Geosciences (GFZ) are currently preparing the Tatuoca observatory to join the INTERMAGNET network. This will add a third observatory to the INTERMAGNET equatorial observatories, besides Huancayo and Addis Ababa. In 2015 and 2016, two trips were organised to Tatuoca in order to equip the observatory with modern instrumentation, to train the local observers in a different type of absolute measurement, and to update data processing routines. Also, real-time data become more important with respect to applications for space weather monitoring and directional drilling (Buchanan et al., 2013), and a long-term goal of this project is to provide real-time data of the TTB observatory.

In this paper, we will first give an overview of the observatory location and its infrastructure. Then, we will summarise the history of the observatory, with a focus on the technical and operational state before the initiation of this project.

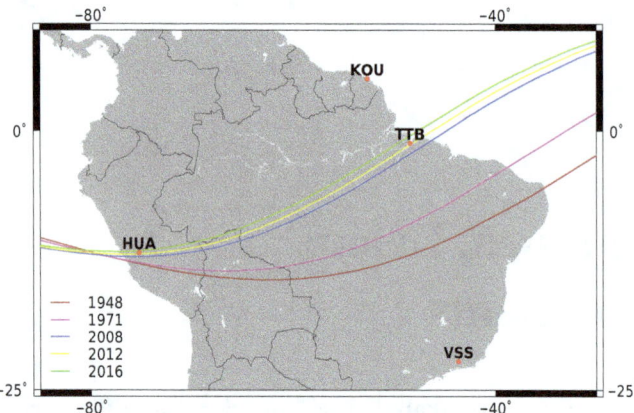

**Figure 1.** The location of the Tatuoca (TTB) observatory is shown along with INTERMAGNET observatories in the region (Kourou (KOU), Huancayo (HUA), and Vassouras (VSS)). In addition, the location of the geomagnetic equator according to IGRF is shown for different years between 1948 and 2016, as indicated by the colours.

This summary is followed by a description of the recent improvements and the current state of the observatory before we shortly comment on the data and their availability. The paper concludes with a summary and outlook.

## 2 History of the observatory

The history of the Tatuoca magnetic observatory started in 1925 when the ON considered installing a permanent station within the equatorial region. This goal was reinforced by a recommendation of the International Union of Geodesy and Geophysics (IUGG) in 1933, after an Assembly in Lisboa. At this date, two La Cour type magnetographs were provided by the International Polar Year Commission to the National Observatory of Brazil (Gama, 1955). Subsequently, Mr Marquez (ON) was in charge of finding a location free from artificial magnetic disturbances, and he chose a small island owned by the Brazilian government and close to the city of Belém. The magnetic station on Tatuoca operated only from September 1933 to January 1934 due to lack of funding. Important results from these recordings were published in 1951 by the Temporary Commission on the Liquidation of the Polar Year 1933–1934 (Olsen, 1951; IUGG, IPY, 1950). In the subsequent years, the Tatuoca project had been halted due to budget limitations, especially during World War II (Gama, 1955).

In 1951, UNESCO offered a Ruska field theodolite magnetometer inductor for absolute measurements to the ON, with the condition that the Brazilian government will finance the necessary buildings. The construction of the variometer and absolute houses on Tatuoca was completed in 1953, and the office and other buildings were ready in 1954. In this year, a magnetograph was supplied by the Inter American Geodetic Survey which was installed by W. C. Parkinson (Depart-

ment of Terrestrial Magnetism, Carnegie Institution of Washington) and by L. I. Gama, the director of ON (AGU, 1955).

Many tests on the absolute and variometer measurements were performed to check the feasibility of a magnetic observatory on Tatuoca. However, due to logistic problems, this was achieved only in 1957 when J. Kozlosky (Inter American Geodetic Survey) visited the island for a few days. On 19 August 1957, the Tatuoca magnetic observatory started its regular operation, and is also listed as a station of the International Geophysical Year 1957–1958 (Nicolet and Doyen, 1959). The observatory has been providing continuous data except for data gaps between 1979 and 1980 due to technical problems and renovation work (Ferreira, 1990). Hourly mean data from Tatuoca were published each year in internal reports of ON. In addition, most hourly mean values from 1957 to 1959, 1964 to 1965, and from 1990 to 1999 are published at the World Data Centre Edinburgh (WDC). During most of this time (1970s to 1990s), José Teotônio Ferreira was responsible for data processing at the observatory. As well, Luiz Muniz Barreto, who was responsible for the Tatuoca and Vassouras observatories during six decades, and director of ON in the 1970s and 1980s, needs to be mentioned here.

In May 1996, a digital automatic station was installed in Tatuoca and the classical variometers were disabled. However, the digital station presented problems in February 1997 and the classical variometers were reactivated (Ferreira, 1998). Finally, the last magnetogram from the classical variometers on Tatuoca was obtained on 13 May 2007. The classical variometers stopped working due to the lack of photograph paper, which was out of production. As a substitute, a LEMI 417 variometer was installed for continuous vector measurements. Also, a POS-1 scalar magnetometer was installed in 2007, which stopped working in 2013 due to insufficient power supply.

## 3 Observatory location and set-up

The Tatuoca magnetic observatory is located at 1.205° S and 311.487° E (geodetic coordinates) on a small island within the Amazon Delta in the state of Pará in Brazil, and the island is located a 1 h boat trip away from the port of Icoaraci close to the city of Belém (Figs. 1 and 2). Further, the island of Tatuoca has an approximate size of 460 m by 300 m, and is largely covered by dense vegetation (Fig. 3). A big advantage is that the island is exclusively used by the observatory and owned by the Brazilian government. Therefore, it is well protected from any artificial disturbances, and had never been relocated during its 65 years of existence.

On the island of Tatuoca, several buildings are located which are related to the observatory (Fig. 3). From south to north, there is a residential house for the observatory staff, an electronics house with the batteries and solar regulators as well as a small office and a storage house with diesel gener-

**Figure 2.** Satellite image of Tatuoca island and its surroundings (Copyright 2017 Landsat/Copernicus, TerraMetrics and Google). The city of Belém is located to the south of Tatuoca island and the observatory.

**Figure 4.** The variometer and absolute houses of Tatuoca observatory are shown. These are located at the north-eastern corner of the island (Fig. 3).

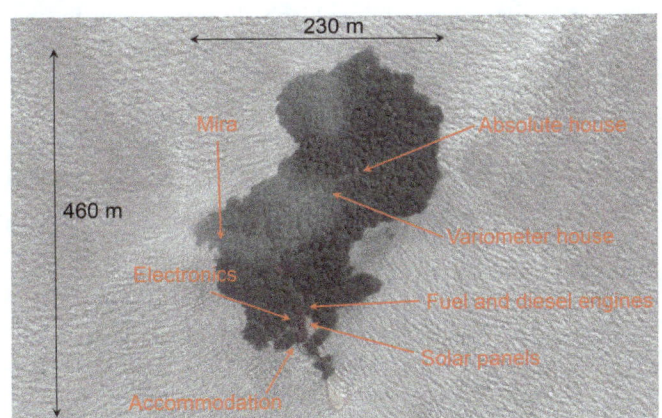

**Figure 3.** Satellite image of Tatuoca island (copyright 2016 Google). The main buildings and infrastructure of the observatory are marked and annotated.

ators. In the north-eastern corner of the island, the absolute and variometer houses are located (Fig. 4).

A schematic drawing of the variometer house is shown in Fig. 5. The variometer house consists of an outer corridor (light grey) and an inner insulated room (dark grey). The insulation dampens the temperature variation to between 30 and 35 °C (in 2016), with a maximum daily variation of 3 K. The inner room is equipped with two large solid pillars and one smaller pillar in the south-eastern corner (Fig. 5, all shown in black). In addition, two wooden shelves are located in the southern part of the inner room (black-grey checked). On the easternmost of the solid pillars, a LEMI-417 vector fluxgate magnetometer is located (yellow circle), and the electronics of this instrument is located near the entrance of the variometer house (yellow rectangle). The LEMI system is powered by a 45 Ah lead-acid battery which is charged by a dedicated 30 W solar panel on the roof of the variometer

house. Further, a POS-2 proton gradiometer was located in the south-eastern corner of the variometer house (shown in white, Fig. 5). This instrument was never in operation and was removed in October 2016 (see below).

The absolute house has an approximate size of 4.8 by 8.0 m, and is roughly oriented in N–S direction. It houses 10 pillars, 4 of which are located at the northern end, including the main pillar, and 6 of which are located at the southern end. The latter pillars carry several historic instruments, including the Ruska theodolite donated by UNESCO. The main pillar is equipped with a ZEISS 020B theodolite in degree scale to which a Canadian EDA fluxgate magnetometer had been attached. The EDA fluxgate had an analogue current reading, and therefore the absolute measurements had to be performed with the zero residual method (Newitt et al., 1996, p. 43ff). As described in Sect. 4, the fluxgate has been replaced with a digital instrument during our first trip in November 2015. For absolute measurements, an azimuth mark is located at a distance of 150 m to the southwest. Further, a GEM System GSM-19 proton Overhauser magnetometer is available for measuring the magnetic field intensity. Until recently, the time of the absolute measurements was taken from an analogue wall clock which is regularly set according to the GPS time of the LEMI electronics in the variometer house.

In total, there are three observers and one cook who swap shifts in teams of two each week. Therefore, the observatory is usually occupied by two persons who do two consecutive absolute measurements on 3 days each week. In addition, the head of the observatory and one technician are both located in Belém, and frequently visit the island. For this purpose, and for transporting goods and fuel to the island, the observatory owns a small motorboat.

Concerning power supply, the observatory is equipped with recently upgraded solar panels of nominal 324 W to-

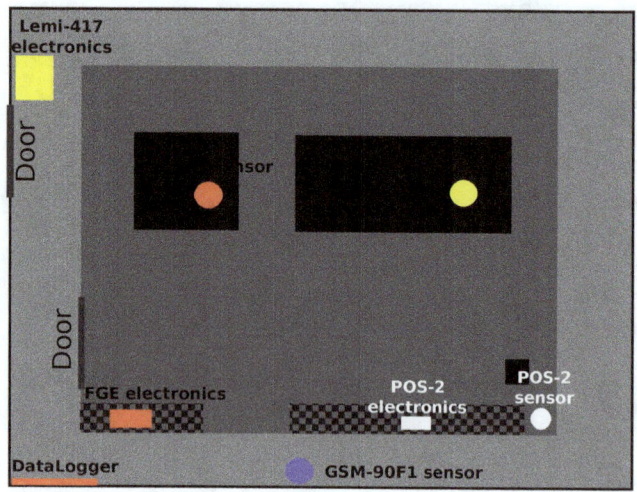

**Figure 5.** Schematics of the variometer house. The insulated inner room is shown in dark grey. Further, the equipment that has been installed prior to 2015 is shown in white (removed 2015) and yellow (in operation after 2015), the equipment installed in 2015 is shown in red, and the equipment installed in 2016 is shown in blue. Black areas refer to solid stone pillars in the variometer house, black checked areas refer to shelves, and doors are shown in black as well.

**Figure 6.** The thermally insulated inner room of the variometer house: the new FGE sensor is placed on the pillar in the front (left), and the LEMI 417 sensor is placed on the pillar in the back. To the right, a part of the shelf with the FGE electronics is visible.

tal, charging eight 165 Ah lead-acid batteries, i.e. 1320 Ah. In addition, there exist two diesel generators of 5 and 6 kW at 120 V, which can also be used to charge the batteries. The diesel generator directly powers the lights in the variometer and absolute houses via a dedicated electric cable system. The batteries provide energy mainly for the accommodation building via a 127 V inverter. In parallel, the batteries power the recently installed equipment (Sect. 4).

## 4   Recent improvements

With the intention to prepare the Tatuoca Observatory to join the INTERMAGNET network, a team of ON and GFZ visited the observatory for two weeks from 17 to 27 November 2015. During this time, new instruments were installed and new methods for absolute measurements were introduced. During a follow-up visit from 24 to 28 October 2016, some further improvements to the instrumentation and absolute measurements were made, as described below.

### 4.1   Variometer house

A Technical University of Denmark (DTU) FGE fluxgate variometer was installed in the variometer house on 21 November 2015 (Pedersen and Merenyi, 2016; Rasmussen and Lauridsen, 1990), and baselines have been available for this variometer since 22 November 2015. As shown in Figs. 5 and 6, the FGE was installed on the existing western socket, at a distance of about 2.2 m from the LEMI-417 sensor. For testing purposes and as a backup system, the

LEMI was kept in operation. The FGE was oriented to magnetic north (HDZ) by minimising the output of its unbiased $Y$-sensor while an appropriate bias field was chosen for the $X$ (horizontal north) and $Z$ (vertical down) channels in order to extend the dynamic range of the readings to the available range of $\pm 10\,\mu$T. The FGE electronics was first placed on the south-eastern shelf and moved to the south-western shelf in October 2016, at a distance of 2.4 m from the FGE sensor (Fig. 5). At the time of installation, the FGE electronics box was also modified to house a MinGEO ObsDAQ 24 bit analogue to digital converter. Any additional electronic equipment was placed in the south-western corner of the outer corridor (Fig. 5). This equipment consists of a RaspberryPi datalogger system and transformers for powering the FGE and the datalogger. The RaspberryPi has the advantage of low power consumption and easy availability. For more details on the datalogger system, please refer to Morschhauser et al. (2017) in this issue.

Absolute scalar measurements in the variometer house are useful for checking the calibration and resolution of the variometer data. In Tatuoca, a POS-1 and POS-2 were previously installed. However, no consistent readings could be obtained when testing these instruments. Therefore, we have removed the non-operational POS-1 and POS-2 electronics and sensors in November 2015. As a replacement, a GEMSystem GSM-90F1 Overhauser magnetometer was installed in October 2016 (blue symbols in Fig. 5). We used this magnetometer to check that there is no indication of a temperature effect on the variometer by comparing the field intensity readings of both instruments.

### 4.2   Power supply

The newly installed electronics in the variometer house (the FGE fluxgate magnetometer, the GSM-90F1 Overhauser magnetometer, and the datalogging system) are powered by the existing solar cells and batteries which are located in the southern part of the island (Fig. 3). In order to trans-

**Figure 7.** The plastic tube is shown that is used to protect the power supply and fibre optics cable running from the electronics house (visible at the far end in the image) to the variometer house (not visible).

mit power over a distance of 150 m, the 12 V direct current (DC) of the lead-acid batteries is converted to 220 V alternate current (AC) using a commercial 300 W inverter. The power supply line consists of a 3-wire 1.5 mm$^2$ power cable (H05RN-F 3X1.5) which was installed in a protective plastic tube (Fig. 7). This plastic tube was shallowly buried and can later be used as ductwork for future installations. As lightning occurs frequently near the equator, currents may be induced in the power line by nearby lightning strikes. The currents may easily destroy the sensitive electronics in the variometer house. Therefore, the installation was protected by FURSE (ESP240-16A/BX) overcurrent protectors at each end of the power line. The grounding of these protectors was improved in October 2016 by installing three 2–2.5 m long copper rods with a diameter of 12 mm which were connected to the FURSE via 16 mm$^2$ copper cables.

### 4.3 Data transmission

In the same building in which the batteries are located (labelled "electronics" in Fig. 3), a netbook and a 3G router were installed. The netbook can connect to remote servers using a reverse SSH tunnel via the 3G network. Indeed, increasing coverage by mobile telecommunication network makes data transmission easy and cheap even in more remote places where expensive solutions (satellites, direct link, dedicated landlines) would have been the only alternative before. However, the SIM card that was used to transfer the variometer data stopped working from 4 February to 20 July 2016. Since then, data transfer has been reliable thanks to a new SIM card. The laptop is also used as a backup for the variometer data and displays a daily magnetogram for the local staff to check the correct operation of the system. Since October 2016, the absolute measurements are also manually stored in the netbook and transmitted to a remote server. In this way, quasi-definitive data can be produced with reduced latency. Due to initial problems with a fibre-optical link between the variometer house and the electronics house, the

data were manually downloaded from the RaspberryPi datalogger in the variometer house on a daily basis via an ethernet link. Since February 2017, the fibre optics link is fully operational and data are synchronised to a remote server every 15 min. In the future, we plan to implement near-real-time data transfer using a message protocol such as MQTT (Message Queue Telemetry Transport) (Bracke et al., 2017).

### 4.4 Absolute measurements

In the absolute house, changes were kept at a minimum level while making some significant improvements: first, the EDA fluxgate (E.D.A. electronics Ltd., Ottawa, CA) was replaced by a DTU model G fluxgate and electronics (serial number 0151, sensor PIL 7451) on 24 November 2015 after eighteen absolute measurements to determine baselines for the FGE variometer were made.

Second, the absolute house was cleared from a number of magnetic and non-magnetic objects on 26 November 2015. As a result, potential future movement of magnetic objects and associated changes in the level of the observatory (showing up as apparent changes in the baselines) can be avoided. Also, a clean absolute house makes it easier to identify new and potentially magnetic objects that have accidentally been forgotten. Before and after removing these objects, five absolute measurements were taken. These 10 absolute measurements revealed a difference in the absolute level of the observatory of $+1.5$ nT in the horizontal component ($H$) and $+0.4'$ ($\approx 3$ nT) in the declination ($D$) after the removal of these objects while no difference was found for the vertical component. We note that any change in the absolute level should not exceed one nT in order to preserve the accuracy of the secular variation data from the TTB observatory (Matzka et al., 2010). This could have been achieved by correcting all future or past data with an appropriate constant offset. However, there are strong indications that the absolute level of the observatory was not stable to better than 3 nT in the previous periods, and therefore the previous data have not been corrected for this relatively low change in the observatory's absolute level. Instead, the baseline was adopted by introducing a baseline jump corresponding to the jump in the measured absolute values.

As a consequence of installing the model G fluxgate, the residual method of absolute measurements was introduced (Jankowsky and Sucksdorff, 1996, p. 89; Worthington and Matzka, 2017, this issue). In this way, the accuracy of the available ZEISS theodolite 020B can be fully used by exactly positioning the horizontal (vertical) circle to full arcminutes during the declination (inclination) measurement. Otherwise, the resolution of the angular readings would have to be estimated to 0.1 arcmin for the 020B theodolite. In particular, three pairs of absolute measurements are done per week, and the time is taken from a wall clock set according to the LEMI GPS in the variometer house. However, this clock is magnetic and had to be located far enough from the

**Sensor up, telescope north**

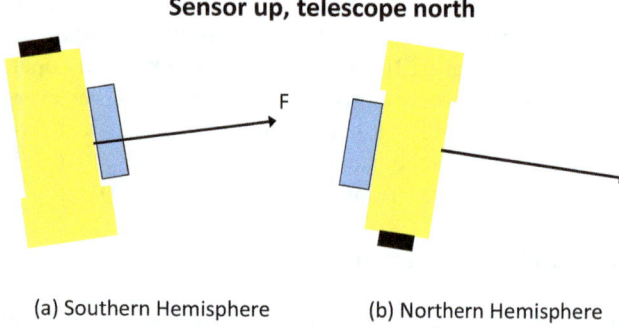

(a) Southern Hemisphere          (b) Northern Hemisphere

**Figure 8.** The position "sensor up, telescope north" is shown for a theodolite near the magnetic equator on the Southern Hemisphere **(a)** and the Northern Hemisphere **(b)**. Depending on the magnetic hemisphere, these positions differ by 180°.

observer, making it hard to read. Therefore, it was replaced by an almost non-magnetic stopwatch in October 2016. This stopwatch allows the time to be easily read with 1 s accuracy and is set according to the system time of the netbook in the electronics house. In turn, the netbook's system time is synchronised via NTP with its GPS and several remote NTP servers.

## 5    Special considerations for absolute measurements near the magnetic equator

The standard concepts and observation routines of absolute measurements are challenged at the equator for a number of reasons. Mainly, these challenges result from the trivial fact that inclination is close to zero near the magnetic equator.

A first problem arises as the telescope is nearly vertical during inclination measurements and a zenith ocular is needed to read the vertical circle for positions where the telescope points upwards (for an alternative method, see Brunke and Matzka, 2017). This situation is made even more complicated by the fact that the widely used Zeiss Theo 020B has no degree numbers on the vertical circle from 162 to 179° and from 181 to 198°. Thus, only the minute marks can be read from the vertical circle if the telescope is pointing down. A slow and cumbersome remedy is to count the number of degree marks between the closest numbered mark and the desired telescope position. Another method is to assume a feasible degree number (e.g. the same one as with the last absolute measurement) and to compare the results of the absolute measurement (baselines, sensor offset, collimation angles) with the previous absolute measurements. In this way, a wrong reading will lead to inconsistent absolute measurements and can easily be identified. Then, the corresponding erroneous reading of the vertical circle must be corrected by a full degree or even multiples of it, and the correct absolute value can be calculated.

Another problem near the magnetic equator arises as formulas to calculate inclination from DI-flux measurements differ in sign for the northern and southern hemispheres (note that Eq. (5.4) of Jankowsky and Sucksdorff, 1996, p. 95 has the wrong sign for the Southern Hemisphere, as well as other sign errors (Matzka and Hansen, 2007)). When the geomagnetic equator is passing the observatory location due to secular variation, it may even happen that an observatory changes its magnetic hemisphere during a single absolute measurement due to the additional daily variation.

Further, telescope positions during inclination measurements are typically denoted "sensor up, telescope north" and so on. If the inclination is very shallow, however, it is not easy to identify whether the telescope actually points south or north, and whether the sensor is positioned up or down relative to the telescope. Here, a simple rule can help to find the correct position: in the Northern Hemisphere, the north-pointing telescope will always point upwards, and in the Southern Hemisphere, the north-pointing telescope will always point downwards (Fig. 8). This still may lead to some confusion if an observatory is changing magnetic hemispheres due to the movement of the magnetic equator. Then, certain positions, e.g. "Sensor up, telescope North" will instantaneously be rotated by 180 degrees (see Fig. 8). However, observers might not realise this situation immediately due to the slow change in inclination and they might report readings in mixed up positions. In this situation, we recommend from our experience that the observers should follow their normal procedure of measurement and any corrections for mixed up positions can be applied during the calculation of baselines.

Still, absolute observations near the magnetic equator do not only make the measurement process more complicated. Since the vertical component is close to zero, the levelling of the telescope is not very critical for declination measurements at the magnetic equator. On the other hand, levelling errors can cause significant problems for observatories at middle to high latitudes, and usually happen due to inexperienced or careless observers.

Although sun observations are not routinely carried out at geomagnetic observatories, they are sometimes necessary for performing accurate absolute measurements as they allow geographical north to be accurately and independently determined. In this case, the standard methods that involve the leading and trailing limb of the sun are not practicable near the geographic equator, where the sun is moving almost vertically. Special considerations on sun observations are detailed in Wienert (1970, p. 136).

## 6    Data

All available digital variometer data of Tatuoca have been processed along with the available absolute measurements. These data include the recordings of the LEMI variometer

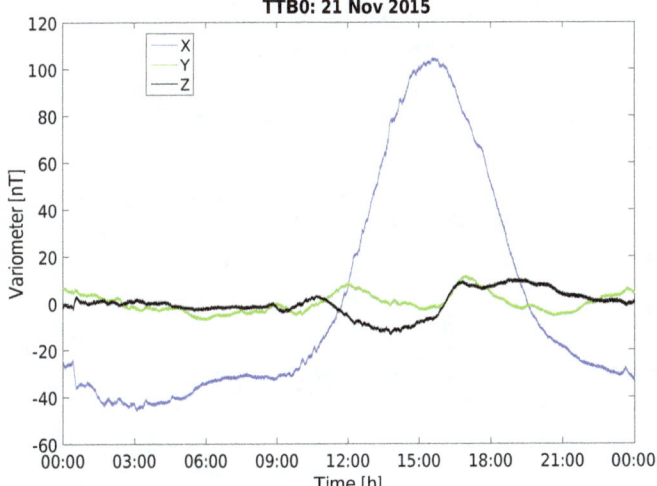

**Figure 9.** First full day of variometer data of the DTU FGE variometer raw data with 1 Hz resolution. Time is in UTC. The $X$-sensor is roughly oriented to magnetic north, and the increased amplitude during daytime due to the equatorial electrojet is visible.

**Table 1.** Variances of the base values for the LEMI sensor and the FGE sensor. The variances have been calculated for the horizontal field component ($H$), declination ($D$), and the vertical field component ($Z$) for four different periods.

| Instrument: | LEMI | | FGE | | |
|---|---|---|---|---|---|
| Period: | I | II | II | III | IV |
| $H_0$ [nT] | 2.74 | 2.40 | 2.29 | 9.44 | 0.76 |
| $D_0$ [arcsec] | 36.60 | 28.36 | 27.55 | 48.62 | 43.43 |
| $Z_0$ [nT] | 3.32 | 3.22 | 1.40 | 3.06 | 2.78 |

Period I is the time until the first visit (2 June 2008 to 17 November 2015).
Period II is the time from after the first visit until the internet connection was lost (28 November 2015 to 3 February 2016).
Period III is the time when internet connection was lost to before the second visit (5 February 2016 to 23 October 2016).
Period IV is the time after the second visit until a lightning strike occurred (29 October 2016 to 30 December 2016).

from June 2008 to December 2016 and the recordings of the FGE variometer from November 2015 to January 2016. These data will soon be made available at the German Research Centre for Geosciences (GFZ) and the World Data Centre (WDC). Here, we will give a short example of the observed daily variations and present the preliminary base values of the observatory.

On 21 November 2015, the first full day of data was recorded by the DTU FGE variometer. The recorded variations of the $X$ (roughly geomagnetic north), $Y$ (roughly geomagnetic east), and $Z$ (vertical down) components are shown in Fig. 9. On this single day, the signal of the equatorial electrojet (EEJ) is visible as an increase in the $X$-sensor readings during daytime (time in UTC), underlining the importance of the observatory for studying the EEJ.

The preliminary base values of the observatory are shown in Fig. 10 for the horizontal ($H_0$) field, the declination ($D_0$), and the vertical field ($Z_0$). The base values presented here have not been checked for outliers caused by transposed digits or other mistakes in the absolute measurements after January 2016. On the left (Fig. 10a), the base values for the LEMI-417 are shown. Two abrupt changes in the base values (especially $D_0$) can most likely be attributed to a realignment of the LEMI sensor to geomagnetic north. Further, the base values of the FGE sensor are shown in Fig. 10b, but with a significantly different scaling than for the LEMI sensor. Here, the vertical red lines indicate the period for which direct data transmission from Tatuoca was not possible, and the vertical black lines indicate the beginning and end of the visits to Tatuoca.

The variances of the preliminary base values were estimated by first linearly detrending the data. This detrend-

ing was done separately for periods when the base values changed abruptly. Then, the standard deviation was calculated for different periods and for the horizontal field, the declination, and the vertical field. In Table 1, the resulting standard deviations are summarised. Overall, the base values are stable to within 3 nT, but very large outliers occur frequently. For the period before our first visit (period I: 2 June 2008–17 November 2015), only LEMI data are available, and standard deviations are a bit higher compared to period II, which spans the time from after the first visit to before the internet connection was lost (28 November 2015–3 February 2016). Also, the variances of the base values as derived from the FGE sensor are slightly smaller than those of the LEMI sensor, confirming the quality of the FGE instrument. As the internet connection was lost until the second visit (period III: 5 February 2015–23 October 2016), the variances of the base values significantly increased in all three components to a level that would be problematic for an INTERMAGNET observatory. Mainly, the reason is that no immediate feedback could be given to the observatory staff carrying out the absolute measurements, underlining the importance of regular data transmission for ensuring data quality. After our most recent visit in October 2016 (period IV: 29 October 2016 to 30 December 2016), the quality of absolute measurements has improved, although significant scatter still occurs in $D_0$ and $Z_0$. However, it is expected that a more detailed investigation and correction or removal of misreadings in the absolute measurements in the course of preparing the definitive data of 2016 will lead to a significantly better standard deviation.

## 7   Summary and outlook

Since 2015, the National Observatory (ON) of Brazil and the German Research Centre for Geosciences (GFZ) have collaborated in preparing the Tatuoca magnetic observatory to become a member of the INTERMAGNET network of mag-

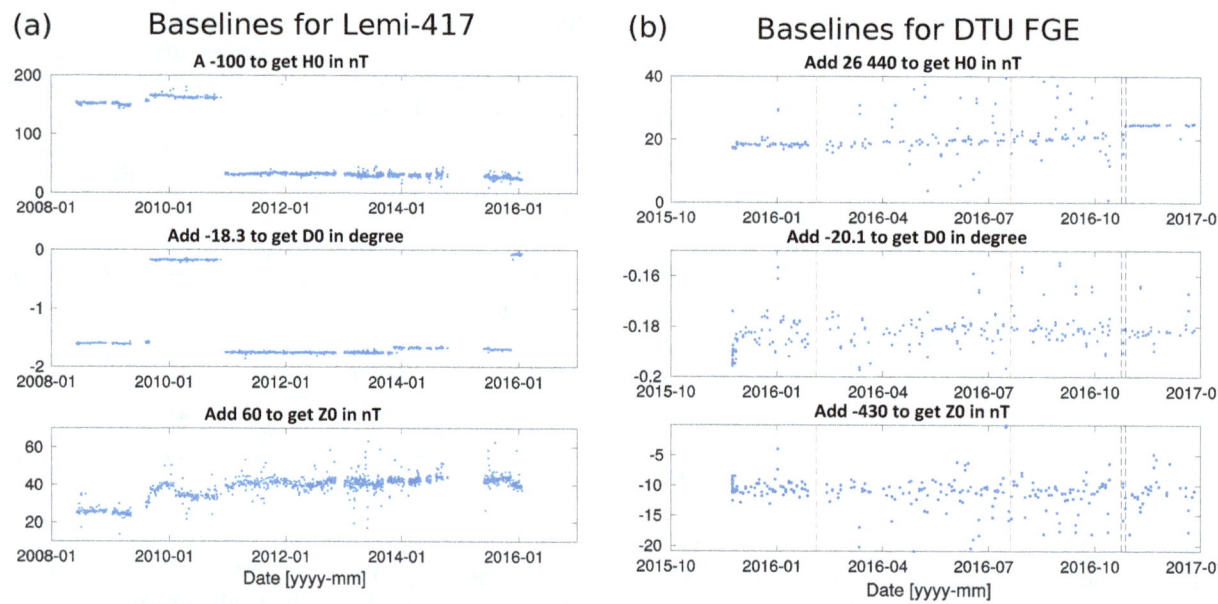

**Figure 10.** Base values for the Tatuoca observatory are shown for **(a)** the LEMI-417 variometer, and **(b)** the DTU FGE variometer. For the period between the vertical red lines, regular data download was not available. Further, vertical black lines indicate the visit to Tatuoca in October 2016.

**Figure 11.** The power cable between the variometer house and the electronics house was damaged a few metres from the variometer house due to currents induced by a lightning strike close to Tatuoca observatory on 31 December 2016.

netic observatories. INTERMAGNET has defined criteria for quality control and data checking, and provides centralised infrastructure for data distribution (Love and Chulliat, 2013). Thus, our efforts will add an observatory adhering to high data quality standards at an interesting location within the magnetic equator and the South Atlantic magnetic anomaly.

As of the end of 2016, a new DTU suspended variometer is installed on Tatuoca along with a modern datalogging system and a GemSystems GSM-90F1 scalar magnetometer. Further, a 3G modem is used to transmit the data to central servers on a daily basis. As well, the EDA fluxgate magnetometer on the ZEISS 020B theodolite in the absolute house was replaced with a DTU fluxgate model G. This lat-

ter change allowed the residual method of absolute measurements to be introduced, increasing the accuracy of absolute measurements. Definitive base values have been calculated for the period from 2008 to 2015, and preliminary baselines are available for 2016. Most of the time, the base values are stable to within 3 nt, but very large outliers exist. Also, we experienced how important it is to provide immediate feedback to the observatory staff in order to assure high-quality absolute measurements. This is particularly important, as a variety of peculiarities complicate absolute measurements near the equator. For example, missing degree marks at some vertical telescope positions make the readings prone to errors.

On the 31 December 2016, a lightning strike hit the island of Tatuoca. As a consequence, severe currents were induced in the power cable between the variometer house and the electronics house (Fig. 11). Although the installed lightning protection was preventing these currents from destroying the equipment, additional currents were induced in the 10 m ethernet cable that was attached to the datalogging system. As a consequence, the RaspberryPi was destroyed and the FGE electronics was damaged. Similarly, the inverter, the netbook, and solar charge controllers were destroyed, probably due to induced currents in the cables leading to the batteries and solar panels. This event underlines the importance of lightning protection at magnetic observatories, and we fully repaired the damage in February and March 2017.

Although the observatory is in a promising state, further improvements are required for it to become a reliable member of the INTERMAGNET network. First, the stability of the baseline can still be improved. Second, the power supply chain for data recording should become independent from the power supply chain that is available for the housing of the observatory staff.

In addition to these major tasks, there exists various smaller improvements that we may consider in the future. For example, we may include a direct illumination of the theodolite for better readability by using LEDs, and attaching the LEMI variometer to the variometer power supply. The latter will eliminate any potential signal of the DC current that powers the LEMI via solar cells. Also, the temperature stability of the sensor and electronics in the variometer house could be improved by installing an electrical heating and additional insulation around the sensor and electronics.

*Author contributions.* AM, GBS, JH, JP, KP, and JM were actively participating in the described work and visits to the observatory, OB was programming the datalogger and is responsible for data transmission, AM wrote the article, and KP and JM coordinated the activities.

*Competing interests.* The authors declare that they have no conflict of interest.

*Special issue statement.* This article is part of the special issue "The Earth's magnetic field: measurements, data, and applications from ground observations (ANGEO/GI inter-journal SI)". It is a result of the XVIIth IAGA Workshop on Geomagnetic Observatory Instruments, Data Acquisition and Processing, Dourbes, Belgium, 4–10 September 2016.

*Acknowledgements.* We thank Hernández Quintero and one anonymous reviewer for their detailed reviews and suggestions that helped to improve the manuscript. Also, we thank Jean Rasson and the team at the Dourbes Observatory for organising the INTERMAGNET workshop in 2016. Katia Pinheiro acknowledges the support of FAPERJ (Bolsa Jovem Cientista do Nosso Estado-E_06/2015) and Gabriel Brando acknowledges CNPq.

Edited by: Jean Rasson

# References

AGU: New geophysical observatory, Tatuoca, Belem, Brazil, J. Geophys. Res., 60, 117–117, https://doi.org/10.1029/jz060i001p00117-01, 1955.

Bracke, S., Gonsette, A., Rasson, J., Poncelet, A., and Hendrickx, O.: Automated observatory in Antarctica: real-time data transfer on constrained networks in practice, Geosci. Instrum. Method. Data Syst. Discuss., https://doi.org/10.5194/gi-2017-17, in review, 2017.

Brunke, H.-P. and Matzka, J.: Numerical Evaluation of magnetic absolute Measurements with arbitrary distributed DI-Fluxgate Theodolite Positions, Geosci. Instrum. Method. Data Syst. Discuss., https://doi.org/10.5194/gi-2017-3, in review, 2017.

Buchanan, A., Finn, C. A., Love, J. J., Worthington, E. W., Lawson, F., Maus, S., Okewunmi, S., and Poedjono, B.: Geomagnetic Referencing – The Real-Time Compass for Directional Drillers, Oilfield Review, 25, 32–47, 2013.

Ferreira, J. T.: Resultados Magnéticos de Tatuoca – 1989, Publicaçõ Especial do Observatório Nacional, 1990.

Ferreira, J. T.: Resultados Magnéticos de Tatuoca – 1997, Publicaçõ Especial do Observatório Nacional, 1998.

Gama, L. I.: Installation of the Tatuoca Magnetic Observatory, vol. 6, Publicações do Serviço Magnético, 1955.

Hulot, G., Sabaka, T., Olsen, N., and Fournier, A.: The Present and Future Geomagnetic Field, in: Treatise on Geophysics, Elsevier BV, 5, 33–78, https://doi.org/10.1016/b978-0-444-53802-4.00096-8, 2015.

IUGG, IPY: Geomagnetic K-indices, Assoc. Terr. Mag. Electr., Bull. No. 12d, 1950.

Jankowsky, J. and Sucksdorff, C.: Guide for magnetic measurements and observatory practice, IAGA, International Association of Geomagnetism and Aeronomy (IAGA), ISBN: 10-9650686-2-5, 1996.

Love, J. J. and Chulliat, A.: An International Network of Magnetic Observatories, EOS, 94, 373–374, https://doi.org/10.1002/2013eo420001, 2013.

Manoj, C., Lühr, H., Maus, S., and Nagarajan, N.: Evidence for short spatial correlation lengths of the noontime equatorial electrojet inferred from a comparison of satellite and ground magnetic data, J. Geophys. Res., 111, A11312, https://doi.org/10.1029/2006JA011855, 2006.

Matzka, J. and Hansen, T. L.: On the Various Published Formulas to Determine Sensor Offset and Sensor Misalignment for the DI-flux, Publ. Inst. Geophys. Pol. Acad. Sci., C-99, 152–157, 2007.

Matzka, J., Chulliat, A., Mandea, M., Finlay, C. C., and Qamili, E.: Geomagnetic Observations for Main Field Studies: From Ground to Space, Space Sci. Rev., 155, 29–64, https://doi.org/10.1007/s11214-010-9693-4, 2010.

Morschhauser, A., Haseloff, J., Bronkalla, O., Müller-Brettschneider, C., and Matzka, J.: A low-power data acquisition system for geomagnetic observatories and variometer stations, Geosci. Instrum. Method. Data Syst. Discuss., https://doi.org/10.5194/gi-2017-23, in review, 2017.

Newitt, L. R., Barton, C. E., and Bitterly, J.: Guide for Magnetic Repeat Station Surveys, IAGA, International Association of Geomagnetism and Aeronomy (IAGA), ISBN 10-9650686-1-7, 1996.

Nicolet, M. and Doyen, P.: Geopgraphical Distribution of the International Geophysical Year Stations, in: Annals of the International Geophysical Year, vol. VIII, Paergamon Press Ltd., 1959.

Olsen, J.: Results of magnetic observations made at Tatuoca (Brazil), September 1933–January 1934, Temporary Commission on the Liquidation of the Polar Year (TCLPY) 1932–33, International Meteorological Organization, 1951.

Pedersen, L. W. and Merenyi, L.: The FGE Magnetometer and the INTERMAGNET 1 Second Standard, Journal of the Indian Geophysical Union, 2, 30–36, 2016.

Rasmussen, O. and Lauridsen, E.: Improving baseline drift in flux-gate magnetometers caused by foundation movements, using band suspended fluxgate sensors, Phys. Earth Planet. In., 59, 78–81, https://doi.org/10.1016/0031-9201(90)90211-f, 1990.

Thébault, E., Finlay, C. C., Beggan, C. D., Alken, P., Aubert, J., Barrois, O., Bertrand, F., Bondar, T., Boness, A., Brocco, L., Canet, E., Chambodut, A., Chulliat, A., Coïsson, P., Civet, F., Du, A., Fournier, A., Fratter, I., Gillet, N., Hamilton, B., Hamoudi, M., Hulot, G., Jager, T., Korte, M., Kuang, W., Lalanne, X., Langlais, B.; Léger, J.-M., Lesur, V., Lowes, F. J., Macmillan, S., Mandea, M., Manoj, C., Maus, S., Olsen, N., Petrov, V., Ridley, V., Rother, M., Sabaka, T. J., Saturnino, D., Schachtschneider, R.,

Sirol, O., Tangborn, A., and Thomson, A., Tøffner-Clausen, L., Vigneron, P., Wardinski, I., and Zvereva, T.: International Geomagnetic Reference Field: the 12th generation, Earth Planet. Space, 67, 217 pp., https://doi.org/10.1186/s40623-015-0228-9, 2015.

Wienert, K. A.: Notes on geomagnetic observatory and survey practice, Earth Sciences, UNESCO, 5, p. 79, 1970.

Worthington, E. W. and Matzka, J.: USGS Experience with the Residual Absolutes Method, Geosci. Instrum. Method. Data Syst. Discuss., https://doi.org/10.5194/gi-2017-24, in review, 2017.

# 3

# Vehicular-networking- and road-weather-related research in Sodankylä

**Timo Sukuvaara, Kari Mäenpää, and Riika Ylitalo**

Finnish Meteorological Institute, P.O. Box 503, 00101 Helsinki, Finland

*Correspondence to:* Timo Sukuvaara (timo.sukuvaara@fmi.fi)

**Abstract.** Vehicular-networking- and especially safety-related wireless vehicular services have been under intensive research for almost a decade now. Only in recent years has road weather information also been acknowledged to play an important role when aiming to reduce traffic accidents and fatalities via intelligent transport systems (ITSs). Part of the progress can be seen as a result of the Finnish Meteorological Institute's (FMI) long-term research work in Sodankylä within the topic, originally started in 2006.

Within multiple research projects, the FMI Arctic Research Centre has been developing wireless vehicular networking and road weather services, in co-operation with the FMI meteorological services team in Helsinki. At the beginning the wireless communication was conducted with traditional Wi-Fi type local area networking, but during the development the system has evolved into a hybrid communication system of a combined vehicular ad hoc networking (VANET) system with special IEEE 802.11p protocol and supporting cellular networking based on a commercial 3G network, not forgetting support for Wi-Fi-based devices also. For piloting purposes and further research, we have established a special combined road weather station (RWS) and roadside unit (RSU), to interact with vehicles as a service hotspot. In the RWS–RSU we have chosen to build support to all major approaches, IEEE 802.11, traditional Wi-Fi and cellular 3G. We employ road weather systems of FMI, along with RWS and vehicle data gathered from vehicles, in the up-to-date localized weather data delivered in real time. IEEE 802.11p vehicular networking is supported with Wi-Fi and 3G communications.

This paper briefly introduces the research work related to vehicular networking and road weather services conducted in Sodankylä, as well as the research project involved in this work. The current status of instrumentation, available services and capabilities are presented in order to formulate a clear general view of the research field.

## 1 Introduction

The vehicular-networking-related research work in Sodankylä started within the EUREKA Celtic CARLINK (Wireless Traffic Service Platform for Linking Cars) project (Sukuvaara and Nurmi, 2009), established in 2006. The architecture development basis combined both vehicular ad hoc network (VANET) and infrastructure-based networking with roadside fixed network stations. The conceptual idea of multiprotocol access networking was used for combining Wi-Fi and GPRS networking. As a result, the CARLINK project designed and piloted one of the first operating vehicle-to-vehicle (V2V) and vehicle-to-infrastructure (V2I) communication architectures.

The concept of hybrid vehicular access network architecture were successfully studied, developed and evaluated in the CARLINK project. The general idea of the continuation project EUREKA Celtic Plus WiSafeCar (Wireless traffic Safety network between Cars) (Sukuvaara et al., 2013) was to overcome the limitations of communications by upgrading communication methodology. Wi-Fi was upgraded with the special vehicular WAVE (Wireless Access in Vehicular Environments) system based on the IEEE 802.11p standard amendment (IEEE Std. 802.11p, 2009) and GPRS with 3G communication. The architecture was employed with a set of more sophisticated services, tailored for traffic safety and convenience. The set of example services was also adjusted to be compliant with services proposed by the Car-

**Figure 1.** Combined RWS–RSU system.

2-Car Communication Consortium (C2C-CC) (Baldessari et al., 2007) and ETSI standardization for the "day one set of services" (ETSI, 2010). Especially the newly founded IEEE 802.11p-based vehicular access network system underwent an extensive set of test measurements, both with V2V and V2I communications. The measurements demonstrated that the IEEE 802.11p clearly has better general performance and behavior in the vehicular networking environment compared to the traditional Wi-Fi solutions used for this purpose. The pilot platform deployment proved that the new system also operates in practice and that the pilot services defined can be provided properly. In the deployment, the overlay cellular network (3G) played an important role, and this hybrid method would be an attractive solution for the ultimate commercial architecture. The WiSafeCar project drew an outline for the commercially operating intelligent vehicular access network architecture, with a general deployment proposal.

Even though the commercial deployment did not take place, the developed system served as the basis for a more advanced project, the EUREKA Celtic Plus CoMoSeF (Cooperative Mobility Services of the Future) project (Sukuvaara et al., 2015), along with other intelligent traffic-related research. The focus in the CoMoSeF project was on near-the-market services and multi-standard communication. The aim was not only to serve vehicles but also to exploit vehicle-originating data to ultimately enhance the same services. Similarly, roadside units (RSUs) do not just serve the vehicles as connectivity points; they also host road weather station (RWS) capabilities to provide additional data for the services. Both of these properties are combined in the Finnish Meteorological Institute approach to employing vehicular networking architecture to provide route weather information

for vehicles passing our combined RWS–RSU. The enhanced RWS–RSU was also studied in the Northern Periphery Programme (NPP)-funded SNAPS (Snow, Ice and Avalanche applications) project, where it represented the winter traffic data and enhanced service source for bypassing vehicles as well as online customers of local stakeholders. The Sodankylä RWS is equipped with up-to-date road weather measurement instrumentation, compatible with (but not limited to) the equipment of operational RWS. The procedure was to design, develop and test both the local road weather service generation and the service data delivery between RWS and vehicles. The vehicle passing the combined RWS–RSU is supplemented wirelessly and automatically with up-to-date road-weather-related data and services, and at the same time possible vehicle-oriented measurement data are delivered upwards to the database to be used as a component of weather information. IEEE 802.11p is the primary communication protocol, but traditional Wi-Fi communication is also supported, together with cellular 3G access as a backbone. Furthermore, the winter traffic data gathered from the vehicles were studied in the Interreg IV A Nord Intelligent Road project. More advanced road weather services to be delivered directly to vehicles were intensively studied in the EU Seventh Framework Programme for Research and Technological Development (FP7) project FOTsis (European Field Operational Test on Safe, Intelligent and Sustainable Road Operation). As the result of all these projects and research work, the interactive RWS station, together with research vehicles, forms the pilot system in Sodankylä, acting as a real-life test bed for the present and yet-to-come demonstration systems.

| Parameter | Sensor | Measurement height/depth | Measurement period |
|---|---|---|---|
| Temperature | PT100 | 2 m | Oct. 2011 → |
| Humidity | HMP45D | 2 m | Oct. 2011 → |
| Wind speed and direction | Thies Clima 2D Ultrasonic Anemometer | 6.5 m | Oct. 2011 → |
| Soil moisture profile | Stevens Hydra Probe II | -1,-5,-10,-20,-30,-50,-100,-200,-300 cm | Oct. 2011 → |
| Soil temperature profile | Stevens Hydra Probe II | -1,-5,-10,-20,-30,-50,-100,-200,-300 cm | Oct. 2011 → |
| Present weather and visibility | Vaisala PWD22 | 2.4 m (Oct. 2012 - 7.9.2012: 2.6 m) | Oct. 2011 → |
| Road weather camera | Axis 221 camera | | Oct. 2011 – Oct. 2013 |
| Road surface state (remote) | Vaisala DSC111 | | Nov. 2012 → |
| Road surface temperature (remote) | Vaisala DST111 | | Nov. 2012 → |
| Road surface state and temperature | Vaisala DRS511 | 0 cm | Sept. 2012 → |
| Wind speed and direction | Vaisala WA15 | 6.3 m | Nov. 2012 → |
| Air humidity | Vaisala HMP155 | 4.5 m | Sept. 2012 → |
| Air temperature | Vaisala HMP155 | 4.5 m | Sept. 2012 → |
| Soil temperature | Vaisala DTS12 | -40 cm | Sept. 2012 → |
| Present weather and visibility | Vaisala PWD22 | 6 m | Sept. 2012 → |
| Infrared camera | Zavio B7210 Full HD | | Nov. 2012 → |
| Soil temperature profile | LISTEC SEC 15 d-LIST sensor | 0–3m | Nov. 2013 → |

**Figure 2.** Collection of RWS measurements.

## 2 Research road weather station

FMI has constructed a special combined RWS–RSU in northern Finland, near its facilities in Sodankylä. The station, viewed in Fig. 1, is equipped with up-to-date road weather measurement instrumentation. The general objective is to design, develop and test both the local road weather service generation and the service data delivery between RWS and vehicles. The collection of RWS measurements is listed in Fig. 2.

The IEEE 802.11p VANET standard is used as the primary communication entity. Traditional Wi-Fi (IEEE 802.11g/n) and cellular networking (3G) are used as reference methods for the existing operative solution and as the alternative communication methods if the VANET network is not available.

The interaction between vehicle and RWS represents the typical V2I communication. The vehicle passing the RWS–RSU is supplemented wirelessly and automatically with up-to-date road-weather-related data and services, and at the same time possible vehicle-oriented measurement data are delivered upwards. As seen in Fig. 3, the local server in RWS–RSU hosts the station operations. It is linked with a NEC LinkBird-MX modem for IEEE 802.11p communication attempting, but it also has an internal Wi-Fi modem, and both of these communication channels actively seek the passing vehicle communication systems. The local server also gathers measurement data from two different measurement entities, the Vaisala ROSA road weather measurement system and FMI weather station measurements. The data from these sources, together with vehicle-oriented data, are sorted and further delivered to FMI local facilities through the 3G communication link. The advanced services are developed in FMI facilities and delivered back to the RWS–RSU to be further delivered to vehicles. The messaging system and operational procedure are presented in a simplified format in Fig. 4. The same software entity maintains the data delivery between RWS and vehicles, and RWS and FMI site, while gathering and updating the local weather data of RWS–RSU.

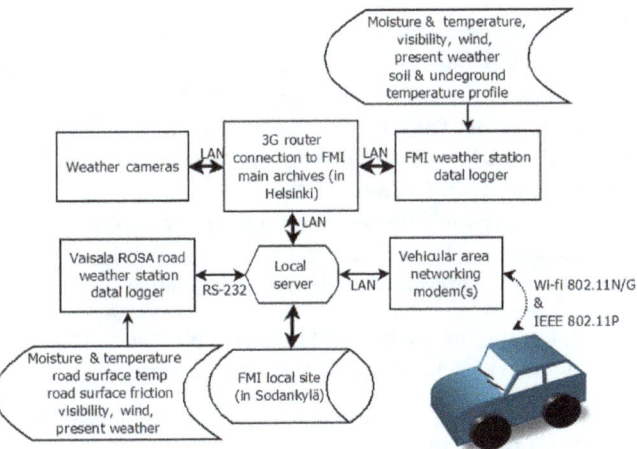

**Figure 3.** Communication entity of RWS–RSU.

The communication system, originally presented by Mäenpää et al. (2013), supports the operations in IEEE 802.11p, traditional Wi-Fi and 3G environments. The communication software has been generated with Python general-purpose, high-level programming language. Python version 2.7.3 has been used throughout our development process. All the operations run in parallel Python .py-modules. Basically all the communication elements use the same operation module, presented in Fig. 4. Depending on the usage profile (RWS, vehicle in V2I, vehicle in V2V) a different kind of initiation process is required. The RWS–RSU has an infinite-loop Python operation procedure, which has been initiated before starting any other elements. Therefore it is expected to perform a specialized eternal loop of its network operations already before any vehicle is about to initiate communication. One module generally supports only one communication protocol, so in order to enable parallel operation of 802.11p and 802.11n one must initiate parallel modules for this. Finally, the 3G communication cannot be initiated in this manner, as it is not practical to broadcast data in a

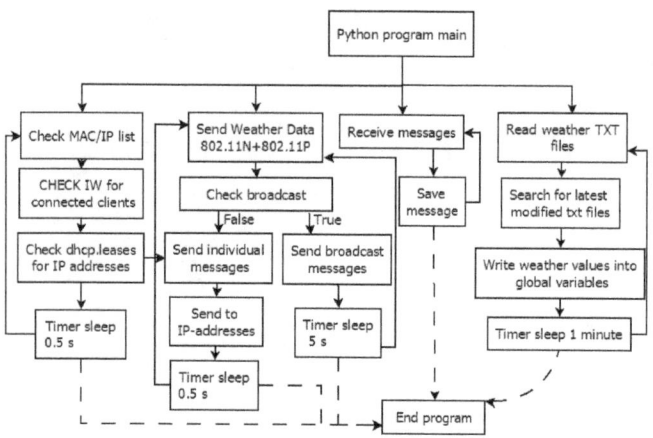

**Figure 4.** Operational process in RWS–RSU.

cellular network and ultimately not allowed by the commercial network operators. The 3G operation is arranged simply by forcing end users to fetch up-to-date RWS data in pre-defined intervals from the RWS stations nearby. Therefore, RWS only needs to ensure that the up-to-date data are always stored in the RWS download folder.

Figure 5 presents the devices and their connections in V2I communication. The operational procedure for the communication can be presented in the following steps:

0. All programs are initiated both in RWS and in the vehicle. Devices are connected according to Fig. 5.

1. Vehicle radios are constantly searching for nearby IEEE 802.11p/Wi-Fi networks.

2. When one is found, the vehicle radio forms a connection and the data exchange between the computers in RWS and vehicle can begin.

2a. Neither RWS nor the vehicle knows anything about the IEEE 802.11p/Wi-Fi network status. They can only see if the IP address is "real" and active or not.

3. When the connection between vehicle and RWS devices has been established and the IP of the vehicle PC is visible for the RWS host computer, the latter starts pushing messages to the vehicle PC's IP at a constant rate.

4. When the connection is lost, the IP address disappears and messages will not be sent anymore.

5. Up-to-date RWS data are stored and updated regularly to the download folder, in order to support 3G-based data fetch by the vehicles out of range.

After this procedure the cycle begins again and the vehicle radio starts searching for the nearby IEEE 802.11p/Wi-Fi networks.

The server software is the same for both Wi-Fi (IEEE 802.11n) and IEEE 802.11p communication. In the software

only a minor difference exists between the protocol procedures, in terms of different IP and message delivery rate. The complete server side code is presented in Fig. 4. As stated before, different protocols are launched in the parallel Python software modules. During the communication tests only User Datagram Protocol (UDP) messages have been used, but the Transmission Control Protocol (TCP) messages are supported as well. Third-generation (3G) communication is purely based on TCP messages.

There are two threads that run at all times inside the RWS server: a weather-condition-monitoring script and a message-sending script. The weather monitor just reads the data and saves them into a table that the messaging script can read. This is done in order to speed up the sending of messages.

The vehicle computer uses the same Python communication modules as RWS, presented in Fig. 4. When starting the vehicle application program, the user chooses the transmission protocol (UDP/TCP), the communication protocol (Wi-Fi/802.11p), the delay between messages and the delay for the program startup. Mac list is only checked if the server's internal Wi-Fi is chosen as the messaging platform. The messages received from the vehicles passing the RWS are currently only being printed to the screen.

On the client side we have two to three threads running at the same time:

1. The Wi-Fi connection is only used during IEEE 802.11n communication.

2. For system evaluation purposes, the GPS values are constantly being monitored and saved into a GPS table. This table is used when a message is received in order to pinpoint the location where the message was received. We can also monitor the speed and direction from the GPS data and see how many messages are lost during transit from the numbers that are included in each message.

3. The 3G communication is conducted by the vehicle. The vehicle PC has a simple Python module running in parallel with other modules, which fetches the nearest RWS data in pre-defined intervals. Time stamps of the different data contents are compared to select the most recent one.

## 3 The vehicular measurements

In order to fulfil the concept of serving vehicles and exploiting their data, the measurements are also conducted in vehicles and the data collected from there. Our vehicle data consist of mainly pilot-type service data like accident warning information, with more systematic measurement data of friction measurements, external temperature sensors and vehicle telematics data collected from a controller area network (CAN) bus. The accident warnings are simply initiated by

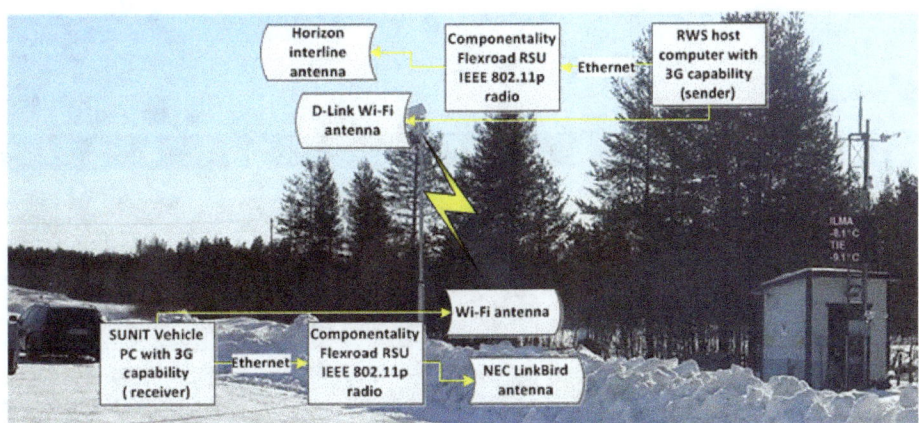

**Figure 5.** Devices and their connections in IEEE 802.11p communication.

**Figure 6.** Teconer friction measurement instrument mounted onto the vehicle.

pushing an emergency button in the vehicle computer unit, to be later on integrated into the vehicle's internal systems. The friction measurements and telematics data represent more sophisticated vehicle observations.

FMI uses two different optical friction monitoring sensors in its road weather services. The Vaisala DSC111 friction monitoring instrument is tailored for fixed friction measurements. It has been deployed permanently in the Sodankylä special RWS, introduced in the previous chapter. From the mobile friction monitoring perspective, it serves as a reference measurement.

The mobile friction monitoring is conducted with Teconer RCM411 instrumentation (viewed in Fig. 6). RCM411 has been designed for quality control and optimization of winter maintenance. RCM411 is also suitable for runway condition reporting. The sensor can be installed on a vehicle in order to monitor the surface conditions in real time. RCM411 detects all typical surface states, like dry (green line color on the map), moist (light blue), wet (dark blue), slushy (violet), snowy (white) and icy (red). RCM411 also measures water and ice layer thicknesses in fractions of millimeters up to

3 mm. A model based on the surface type and amount is used to estimate the coefficient of friction. An acceleration-based $\mu$ TEC friction meter can be integrated into the same user interface installed in a cell phone.

Friction monitoring occurs on the measuring vehicle continuously. The friction measurement data are collected from the measuring vehicle in pre-defined intervals through 3G communication or through IEEE 802.11p or Wi-Fi communication whenever entering in the range of Sodankylä RWS. Friction data of other vehicles or from the RWS can be delivered back to the vehicle as reference data. Currently there is no application deployed for this purpose, and this is not in the scope of the project.

Telematics data collected from the vehicle CAN bus has been recently employed for our vehicle data contents. At the moment only the temperature data are exploited, but the possibility of using vehicular telematics data as a source or at least an indicator of meteorological services is actively being sought.

In addition to existing vehicular data sources, also the Taipale Telematics Sensor system is processed, which can

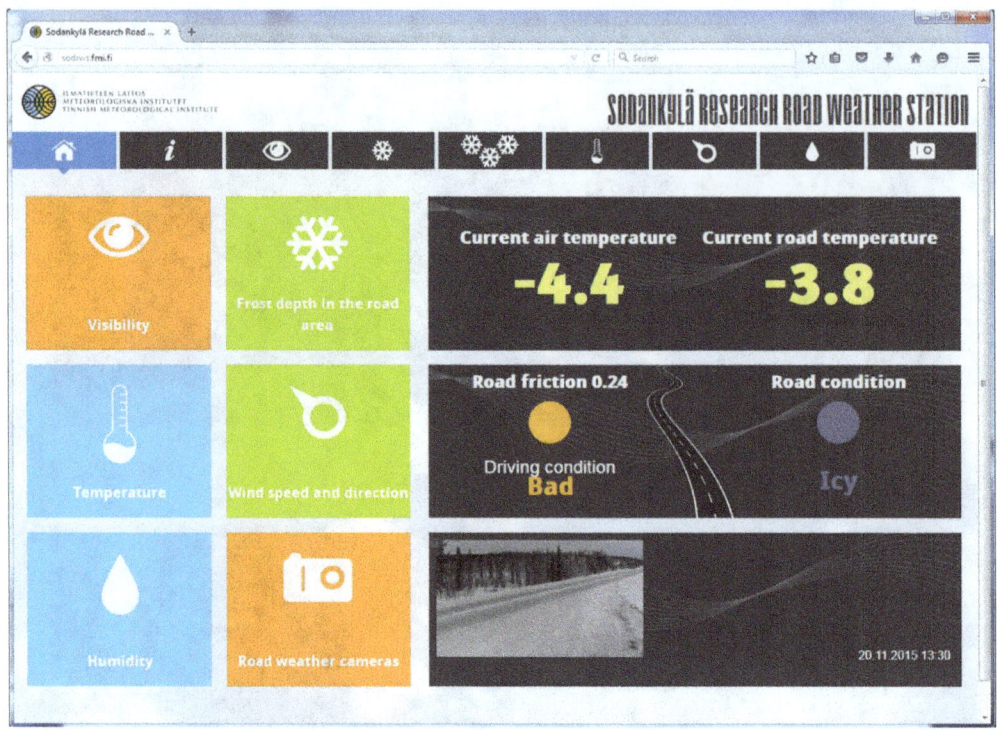

**Figure 7.** Road weather station website at http://sodrws.fmi.fi

be used to fuse the data of different external data sources. At the moment only the navigation and temperature data are gathered from the Sensior, but the additional sensor instrumentation is under consideration.

## 4   Measurement data

Vehicular-networking- and road-weather-related measurements generated in Sodankylä RWS and supporting infrastructure consist of operative example RWS services as well as specially tailored pilot measurements.

The operative RWS services are gathered on our public RWS website, found at http://sodrws.fmi.fi and viewed in Fig. 7. The historical data series captured from the RWS are presented in our public local database, at http://litdb.fmi.fi/rws.php. The website contents are tailored also to mobile devices with Android-based operating systems as well as the iPhone and Jolla, aiming to present our vision of a road weather service user interface scalable for different environments. In addition to this, the measurement data are gathered into historical time series, to be exploited in future research. An example of such a data set, road frost data from the winter 2014–2015, is presented in Fig. 8. The frost measurement is conducted with multiple temperature sensors buried at different depths, indicating frost when temperature is below 0°C. In the warm periods and at the end of the winter season, frost first melts from the ground level, which can clearly be seen in Fig. 8.

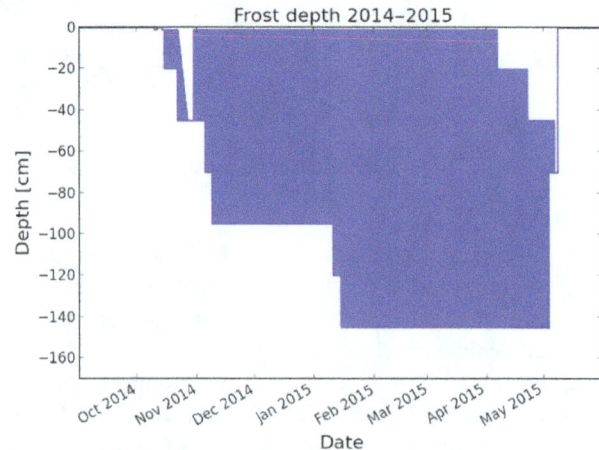

**Figure 8.** Frost depth data from the RWS measurements.

As an example of the pilot measurements in Sodankylä, the data throughput estimation measurements conducted between combined RWS–RSU and the passing vehicle are presented in Figs. 9 and 10. In this measurement the focus was on the IEEE 802.11p-based VANET communication, comparing it to the traditional Wi-Fi-based communication in the same environment and conditions (based on IEEE 802.11g standard). On the RWS–RSU side the host computer located at the station was employed to broadcast data for the passing vehicles in pre-defined packet size and interval. Many

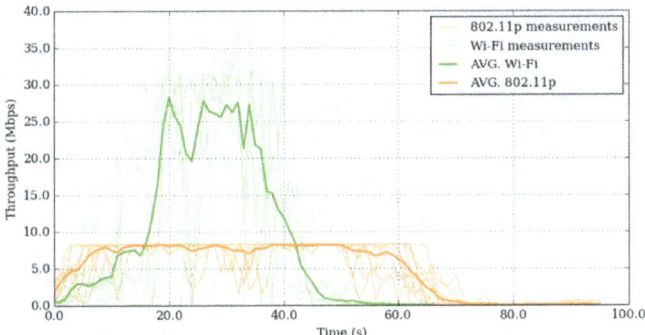

**Figure 9.** Data throughput from combined RWS–RSU to vehicle with 80 km h$^{-1}$ speed.

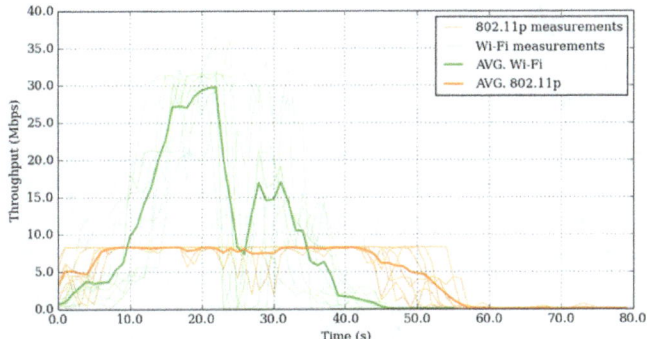

**Figure 10.** Data throughput from combined RWS–RSU to vehicle with 100 km h$^{-1}$ speed.

different combinations were briefly tested, until the optimal rate (1500 byte packets in 1 ms interval) was found and further used in the measurements. Figure 9 presents the results with 80 km h$^{-1}$ vehicle speed, and Fig. 10 results with 100 km h$^{-1}$. The green colored line is the Wi-Fi measurement average, and the lighter green lines are the Wi-Fi measurements. Similarly the solid orange is the IEEE 802.11p average measurement, and the lighter orange is the measurements.

It can be seen that at both speeds the communication window is rather harmonized with IEEE 802.11p, obviously with faster 100 km h$^{-1}$ speed resulting in a shorter communication window. The cumulative average throughput during the communication window for 802.11p was 467 Mb in tests with 80 km h$^{-1}$ vehicle speed and 382 Mb with 100 km h$^{-1}$ speed. In an additional singular test with larger antennas better performance was clearly achieved in terms of range and cumulative throughput. The cumulative average throughput for Wi-Fi communications had a larger fluctuation than the IEEE 802.11p measurements, but the window for 80 km h$^{-1}$ Wi-Fi was 602 and 488 Mb for 100 km h$^{-1}$. The predictable performance of 802.11p is more a important advantage compared to smaller absolute capacity. Nevertheless, the size of the communication window in all the measurements is clearly large enough for supporting our combined RWS–RSU scenario. The details of the measurements, analysis and architecture deployment strategies based on the results are presented in Sukuvaara (2009), Sukuvaara et al. (2013), Sukuvaara et al. (2015) and Sukuvaara (2015).

## 5 Conclusions

This paper has introduced the research work related to vehicular networking and road weather services conducted in Sodankylä, bound to our concept of an interactive road weather station as a service hotspot for road weather services and data collection. FMI's combined road weather station and roadside unit acts as a central infrastructural element of such a V2V and V2I communication platform, supported with areal

infrastructure and observing vehicles. The aim is to utilize road weather systems of FMI, along with RWS data and the data gathered from vehicles, in the up-to-date localized weather data delivered to the vehicles in real time. IEEE 802.11p-based vehicular networking is the primary channel, supported with parallel traditional Wi-Fi and 3G communications. In the future, 4G and 5G communication will be employed and tested as well.

Our research shows that our approach of hybrid communication offers a considerable approach for serving vehicles with real-time weather and traffic information. An extensive set of road weather measurements has also been conducted, to be exploited as part of road weather services of FMI as well as part of vehicular networking research. Detailed and more specific data contents with local area weather data can be delivered to vehicles in service hotspots located beside the road. Whenever outside the range of any RWS, 3G cellular data ensure that the most critical information related to weather and traffic is always up to date. As a summary, our approach of combined RWS–RSU represents our imagination of merging modern road weather services and vehicular intelligence, and it is a respectable test bed for future road weather and networking services as well.

FMI's combined road weather station and roadside unit in Sodankylä is the unique research platform combining very advanced road weather measurements with a versatile collection of the most common wireless communication methodologies used in vehicular environment. Together with harsh Arctic road weather conditions it represents an incomparable development environment and pilot RWS station within the field of ITSs (intelligent transport systems) and vehicular networking.

*Acknowledgements.* This work has been supported by a number of different research projects funded by the Technology Advancement Agency of Finland (TEKES), the European Union EUREKA cluster program Celtic Plus, the European Union FP7, Interreg IV A Nord and the Northern Periphery Programme. The authors wish to thank all the financiers and project partners in this work.

Edited by: M. Syrjäsuo

## References

Baldessari, R., Bödekker, B., Brakemeier, A., Deegener, M., Festag, A., Franz, W., Hiller, A., Kellum, C., Kosch, T., Kovacs, A., Lenardi, M., Lübke, A., Menig, C., Peichl, T., Roeckl, M., Seeberger, D., Strassberger, M., Stratil, H., Vögel, H.-J., Weyl, B., and Zhang, W.: "Car 2 Car Communication Consortium Manifesto, Version 1.1.", available at: http://www.car-to-car.org/ (last access: 16 January 2016), 2007.

ETSI Standard: ETSI ES 202 663 V1.1.0 (2010-01) Intelligent Transport Systems (ITS), European profile standard for the physical and medium access control layer of Intelligent Transport Systems operating in the 5 GHz frequency band, European Telecommunications Standards Institute, Sophia Antipolis, France, 2010.

Finnish Meteorological Institute: Observations at the Arctic Research Centre, available at: http://litdb.fmi.fi (last access: 15 June 2015), 2011.

IEEE Std. 802.11p/D9.0: Draft Standard for Information Technology – Telecommunications and information exchange between systems – Local and metropolitan area networks – Specific requirements, Part 11: Wireless LAN Medium Access Control (MAC) and Physical Layer (PHY) specifications, Amendment 7: Wireless Access in Vehicular Environments (2009) Institute of Electrical and Electronics Engineers Inc., New York, 2009.

Mäenpää, K., Sukuvaara, T., Ylitalo, R., Nurmi, P., and Atlaskin, E.: Road Weather Station acting as a wireless service hotspot for vehicles, 9th International conference on Intelligent computer Communication (ICCP 2013), 5–7 September 2013, Cluj-Napoca, Romania, 2013.

Sukuvaara, T.: "Development, implementation and evaluation of an architecture for vehicle-to-vehicle and vehicle-to-infrastructure networking", Doctoral Thesis, University of Oulu, Department of Communications Engineering, Oulu, Finland, 118 pp., 2015.

Sukuvaara, T. and Nurmi, P.: "Wireless traffic service platform for combined vehicle-to-vehicle and vehicle-to-infrastructure communications", IEEE Wireless Commun., 16, 54–61, December 2009.

Sukuvaara, T., Ylitalo, R., and Katz, M.: " IEEE 802.11p based vehicular networking operational pilot field measurement", IEEE J. Select. Areas Commun., 31, 409–418, September 2013.

Sukuvaara, T., Mäenpää, K., and Ylitalo, R.: "Service Hotspot for delivering local road weather services for vehicles", in: proceedings of 21th World Congress on ITS, 5–9 October 2015, Bordeaux, France, 2015.

# 4

# Radiometric flight results from the HyperSpectral Imager for Climate Science (HySICS)

**Greg Kopp**[1], **Paul Smith**[1], **Chris Belting**[1], **Zach Castleman**[1], **Ginger Drake**[1], **Joey Espejo**[1], **Karl Heuerman**[1], **James Lanzi**[2], **and David Stuchlik**[2]

[1]Laboratory for Atmospheric and Space Physics, University of Colorado, Boulder, CO 80303, USA
[2]NASA Wallops Flight Facility, Wallops Island, VA 23337, USA

*Correspondence to:* Greg Kopp (greg.kopp@lasp.colorado.edu)

**Abstract.** Long-term monitoring of the Earth-reflected solar spectrum is necessary for discerning and attributing changes in climate. High radiometric accuracy enables such monitoring over decadal timescales with non-overlapping instruments, and high precision enables trend detection on shorter timescales. The HyperSpectral Imager for Climate Science (HySICS) is a visible and near-infrared spatial/spectral imaging spectrometer intended to ultimately achieve $\sim 0.2\,\%$ radiometric accuracies of Earth scenes from space, providing an order-of-magnitude improvement over existing space-based imagers. On-orbit calibrations from measurements of spectral solar irradiances acquired by direct views of the Sun enable radiometric calibrations with superior long-term stability than is currently possible with any manmade spaceflight light source or detector. Solar and lunar observations enable in-flight focal-plane array (FPA) flat-fielding and other instrument calibrations. The HySICS has demonstrated this solar cross-calibration technique for future spaceflight instrumentation via two high-altitude balloon flights. The second of these two flights acquired high-radiometric-accuracy measurements of the ground, clouds, the Earth's limb, and the Moon. Those results and the details of the uncertainty analyses of those flight data are described.

## 1 Introduction

The 2007 NRC Decadal Survey for Earth Science (NRC, 2007) calls for shortwave spatial/spectral Earth-scene measurements with radiometric accuracy and SI-traceability of better than 0.2 % for Earth-climate studies on decadal timescales. These accuracies, being nearly ten times better than current on-orbit capabilities, will establish benchmark measurements of solar radiation scattered by the Earth, provide reference calibrations for other on-orbit instruments, and initiate a climate-data record to be used for future climate-policy decisions.

Current space-based imaging systems have radiometric uncertainties of $\sim 2\,\%$ or greater and are limited by the accuracies and stabilities of spaceflight calibration lamps, atmospheric-correction uncertainties needed for vicarious ground-scene calibrations, and degradation of solar diffusers used for on-orbit instrument-sensitivity tracking. Three prominent and long-duration Earth-imaging NASA instruments, the Moderate Resolution Imaging Spectroradiometer (MODIS), the Sea-viewing Wide Field-of-View Sensor (SeaWiFS), and the Advanced Very High Resolution Radiometer (AVHRR), have radiometric accuracies for their reflective solar bands of $\sim 2\,\%$ (see Guenther et al., 1996; Xiong et al., 2005a, b, c, on MODIS and Barnes and Holmes, 1993; Barnes and Zalewski, 2003, on SeaWiFS) and only cover discrete spectral bands. The National Polar-orbiting Operational Environmental Satellite System (NPOESS) National Polar-orbiting Partnership's Visible Infrared Imaging Radiometer Suite (VIIRS) has similar discrete-band coverage as MODIS, with slightly better radiometric accuracies of 1.2 to 1.6 % (Xiong et al., 2014). Hyperion (Pearlman et al., 2000), with continuous spectral coverage from 400 to 2500 nm and 10 nm spectral resolution, has a 3.5 % radiometric uncertainty (Beiso, 2002). With similar spectral coverage and resolution, the Airborne Visible/Infrared Imaging Spectrometer (AVIRIS) has an uncertainty on the order of

4 % (Green et al., 1998). The $M^3$ and the hyperspectral visible to shortwave infrared (VSWIR) imaging spectrometer for the Hyperspectral Infrared Imager (HyspIRI) Decadal Survey mission have radiometric uncertainties of 5 % (HyspIRI Mission Concept Team, 2015).

The HyperSpectral Imager for Climate Science (HySICS) is a prototype instrument to demonstrate a new means of achieving $\sim 0.2$ % ($1\sigma$) on-orbit radiometric accuracies. This hyperspectral imager utilizes a solar cross-calibration technique whereby outgoing Earth radiances of solar-reflected light are ratio-ed to the incoming spectral solar irradiance (SSI) with $< 0.2$ % relative uncertainty. Unlike other solar-calibrated instruments that rely on indirect-sunlight measurements from attenuating diffusers, the HySICS acquires direct solar-radiance measurements to achieve reduced uncertainties. This solar cross-calibration approach relies on precisely known attenuation of the incident solar radiance by $10^{-4.7}$. Attenuations of this magnitude are achieved using a combination of different-sized apertures, electronically adjustable detector integration times, and spectral filters having known transmissions from in-flight calibrations, as described by Smith et al. (2011). This spatial/spectral instrument spans the shortwave spectral-range with a single focal-plane array (FPA) for reduced mass, volume, power, and cost of potential future spaceflight instrumentation. Two high-altitude balloon flights from above most of the Earth's atmosphere demonstrated the ability to acquire spatial/spectral ground-scene images that were cross-calibrated using SSI measurements to provide radiometrically calibrated SI-traceable data cubes.

In this article, we provide an overview of the HySICS instrument and describe the solar cross-calibration approach relying on precisely characterized attenuation methods (Sect. 2), summarize the two completed high-altitude balloon flights (Sect. 3), detail the data-analysis methods and estimated uncertainties (Sect. 4), and present resulting data cubes of Earth ground scenes and the Moon acquired during Flight 2 (Sect. 5).

## 2  HySICS instrument

An eventual spaceflight instrument to achieve the 2007 Decadal Survey's solar-reflected Earth-radiance measurement requirements, needed for climate studies, would likely be designed to achieve desired ground-scene characterizations having a 0.5 km spatial resolution and 100 km cross-track field of view (FOV) while spanning the 350 to 2300 nm spectral range with 6 nm spectral resolution. Acquiring such measurements from low Earth orbit formed the driving requirements for the HySICS spatial/spectral imager, mandating a 10° FOV and a 0.02° instantaneous FOV (IFOV). The HySICS is based on an Offner imaging spectrometer incorporating a precision $\sim 10^{-5}$ attenuation system to enable direct measurements of both the Earth and Sun despite their greatly disparate radiances (Kopp et al., 2013, 2014).

**Table 1.** HySICS performance specifications.

| Parameter | Value |
|---|---|
| Effective focal length (EFL) | 82.2 mm |
| Field of view (FOV) | 10° |
| Instantaneous FOV (IFOV) | 0.02° |
| Point spread function (PSF) | 90 % energy in 30 µm pixel |
| Average slit width | 28.297 µm |
| Offner magnification | 1 : 1.006 (object : image) |
| Spectral range | 350–2300 nm |
| Spectral resolution | 6 nm, constant, Nyquist-sampled |
| Aperture diameters | 20, 10, and 0.5 mm |
| Nominal frame rate | 14 Hz |

### 2.1  Optical system

The optical design of the pushbroom HySICS imaging-spectrometer is representative of state-of-the-art hyperspectral imagers, featuring a four-mirror anastigmat (4MA) telescope followed by an Offner spectrometer. The instrument-performance parameters are shown in Table 1, and a schematic of the optical layout, which is an evolution of that described by Espejo et al. (2011), is shown in Fig. 1.

A precision NIST-calibrated aperture is the first element in the optical train, precisely determining the collecting area for the light entering the instrument. This front-most aperture location allows the most accurate radiometry by reducing uncertainties in estimates of scatter and diffraction effects, which must be corrected to provide low radiometric uncertainties. Diffraction from the precision-aperture's knife edge is well understood theoretically, but scatter is surface dependent and must be measured for the actual optics. Both have been characterized to reduce uncertainties and correct for light losses at the detector. There are no view-limiting baffles in front of the aperture as these can cause additional diffractive, scattering, and glint effects that are difficult to model and correct. A six-element rotatable aperture wheel allows selection of any of the HySICS's six circular apertures. A 20 mm diameter aperture is used to acquire sufficient signal for Earth-scene radiances, while a 0.5 mm diameter solar-calibration aperture provides a relative attenuation of $10^{-3.2}$ due to the two apertures' geometric areas. Two of each in addition to a 10 mm and a blank-off aperture provide redundancy and a dark mode. Each of the six aperture locations in this wheel has a separate thermistor to allow corrections for thermal expansion of the aperture area.

Immediately following the aperture wheel is a similar wheel containing attenuation filters. These share a common filter-wheel thermistor. A Hg / Ar pen-ray lamp mounted in one of the filter-wheel positions provides occasional spectral calibrations of the downstream spectrometer. Independent control of the aperture and filter wheels allows any aperture-and-filter combination from the six of each installed in each of the two wheels.

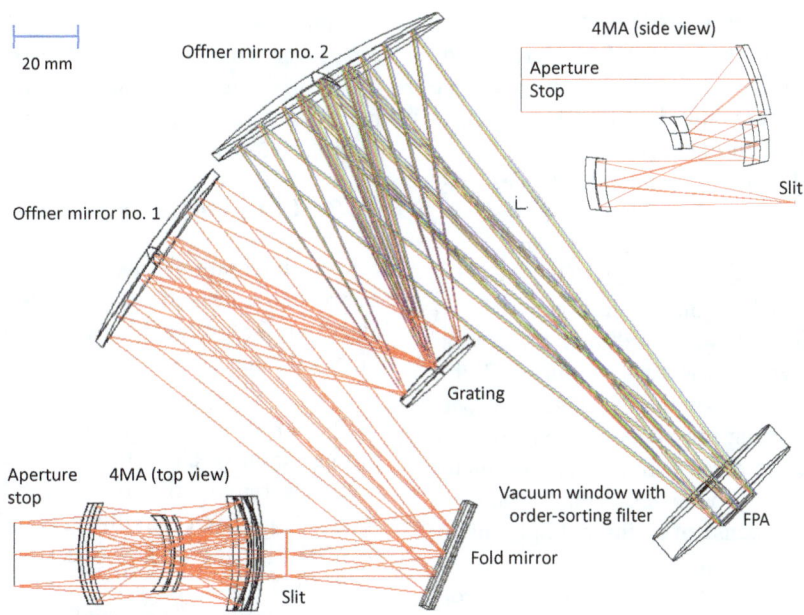

**Figure 1.** Optical layout of the HySICS shows the 4MA telescope followed by a grating-based Offner spectrometer that images onto a full-spectral-range HgCdTe focal-plane array with a three-region order-sorting filter on the back surface of its vacuum entrance window. The Offner and 4MA have nearly orthogonal optical-axis planes to reduce polarization sensitivity. The main picture shows a top view of the entire optical path, while a side view of the 4MA itself is shown in the upper right inset. The physical entrance aperture is positioned at the system's aperture stop. The spectrometer entrance slit is shown in its correct (albeit unconventional) orientation.

The compact 4MA telescope following the aperture and filter wheels uses aspherical diamond-turned aluminum mirrors with electroless-nickel coatings. A protected-aluminum topcoat is magneto-rheological finish (MRF) post-polished for reduced scatter from each element. The fully reflective system eliminates the need for chromatic corrections over the HySICS's broad spectral range. The 4MA mirrors and housing incorporate precision-machined mounting tabs and alignment pins for mechanical robustness and low sensitivity to thermal distortions. This telescope is designed to produce a distortion-free image of a spatial scene onto a 0.028 mm wide slit, providing a slit-width-limited spatial resolution of 0.02° from its 82.2 mm effective focal length.

The precision 0.028 mm × 14.40 mm rectangular spectrometer slit was micro-machined by NIST/Boulder. The Offner-facing surface is coated with carbon nanotubes to reduce back reflections. The absolute slit width was calibrated by NIST/Gaithersburg, as this parameter and its uncertainties are important when reconstructing disk-integrated solar irradiances from cross-slit scans of the spatially resolved Sun.

The Offner spectrometer uses independent primary and tertiary mirrors. The secondary element, a convex reflective 100-ln mm$^{-1}$ ruled grating, provides spectral dispersion, low scatter, and broadband efficiency. This efficiency is obtained via a "sawtooth" pattern with four "teeth" of repeated grating regions, each of which contains blaze angles that smoothly and monotonically vary across the region from being optimized for the shortest to the longest wavelengths of the HySICS's spectrum. A baffle enclosure machined from black plastic and a zero-order trap limit stray light inside the spectrometer's enclosing housing. Optical testing demonstrates a spectral line full width at half maximum (FWHM) of < 36 μm at 633 nm, corresponding to a spectral resolution of 3.7 nm with a spectral scale of 103.31 nm mm$^{-1}$ at the Offner's focal-plane array detector. Smile and keystone distortions are below measurable limits across the FPA.

A three-region order-sorting filter prevents overlap of different orders of diffraction. Region 1, for wavelengths less than 634 nm, is clear; Region 2 passes wavelengths ≥ 634 nm with a 10 nm FWHM transition-region for blocking second-order diffraction; and Region 3 passes wavelengths ≥ 1188 nm with a 39 nm FWHM transition region for blocking third-order diffraction. This filter is coated on the side of the substrate facing the FPA and is mounted 2.7 mm from the FPA's front surface to improve cutoff sharpness in the converging Offner beam. The order-sorting filter also serves as the entrance window to the FPA's vacuum enclosure that is needed to allow cryogenic-temperature operations of the detector.

The 480 × 640 pixel substrate-removed 16-bit HgCdTe Teledyne FPA spans the desired spectral range and meets the majority of needed specifications. The 30 μm pixels closely match the spectrometer slit width. FPA quantum-efficiency ranges from 0.38 e$^-$ ph$^{-1}$ at 350 nm to 0.76 e$^-$ ph$^{-1}$ at 2213 nm, with a cutoff wavelength of 2500 nm. Teledyne reported 16 data numbers (DNs) of read noise, a 12 e$^-$ DN$^{-1}$

gain, and a $692\,000\,e^-$ full well for the delivered device; actual results differed slightly, as described in Sect. 4.1. An operating temperature of $150\,K$ is achieved via a cryo-cooler and a vacuum-enclosure surrounding the FPA. A small ion-pump helps maintain vacuum during flight.

All optics, as well as the aperture and filter wheels, are mounted to a thick aluminum baseplate. Three independently controlled thermoelectric coolers (TECs) reduce thermal gradients of the near-ambient-temperature optics. The entire instrument is encased in a thick aluminum housing for contamination control and thermal stability during integration and test with the balloon gondola as well as during flight. A small door opens for flight observations, which are performed at flight altitude ambient pressures of $\sim 3\,m\,Torr$. A depolarizing door-mounted entrance window can optionally reduce instrument sensitivity to polarized scenes when the door is closed (albeit at the expense of additional light losses due to the window surfaces).

A separate electronics box contains all the controlling components for the HySICS optical module. This 1-atmosphere nitrogen-pressurized enclosure is maintained during flight since not all off-the-shelf electronic components are intended for near-vacuum operations. The FPA electronics are mounted in this box in close proximity to the FPA for reduced noise. Five-hundred gigabytes of solid-state memory arranged as a redundant array of independent disks store all data redundantly during flight, allowing up to 8 h of continual, uncompressed, 14 Hz imagery from the FPA.

## 2.2 Solar attenuation system

Three methods collectively provide the required $10^{-4.7}$ attenuation for directly viewing the Sun: reducing optical-entrance aperture size, decreasing detector integration times, and inserting attenuating filters. The specifics of these three attenuation methods are detailed below. The attenuations collectively provided by the aperture ratio and the integration-time methods proved sufficient for the needed solar-attenuation range, making the filter-based attenuation method unnecessary; nevertheless, that system was incorporated in the HySICS and flight validated as well.

### 2.2.1 Aperture attenuation method

Changing from an entrance-aperture diameter of 20 mm for viewing Earth scenes to 0.5 mm for viewing the Sun provides a geometric attenuation level of $10^{-3.2}$. Optical-system complexities disfavor the use of larger apertures, while diffraction-loss uncertainties start to preclude the use of significantly smaller ones to obtain greater attenuation ratios via this method.

The HySICS apertures are diamond-turned nickel-coated aluminum, providing a very sharp aperture edge with nearly negligible scatter. The six installed apertures have entrance diameters of 20, 10, and 0.5 mm, with two each of the largest

and smallest. All are calibrated by NIST/Gaithersburg for geometric area using a non-contact optical technique to achieve the desired attenuation uncertainties, with the limiting factor being the 0.06 to 0.08 % ($1\sigma$) relative area uncertainties of the solar-viewing 0.5 mm diameter apertures.

### 2.2.2 Integration-time attenuation method

Shorter integration times are used for solar viewing than Earth-scene measurements. These are enabled by the FPA electronics, reproducible detector linearity, and an electronic global shutter to avoid spatial smear during image integration.

The FPA's controlling electronics demonstrate $< 14\,ns$ timing stability and linearity to $< 10^{-6}$ over the integration time range from $16.8\,\mu s$ to $34.4\,ms$, providing $10^{-3.3}$ solar-attenuation capability. The FPA response itself, unsurprisingly, has higher non-linearities but nevertheless demonstrates sufficient linearity stability to allow corrections for operation over the large applied intensity range.

### 2.2.3 Filter attenuation method

Spectral filters capable of roughly $10^{-1}$ attenuations can be calibrated on-orbit via lunar observations. Greater filter-based attenuations are precluded by low lunar-radiance levels that would limit the accuracies of these on-orbit calibrations. On-orbit spectral filter calibrations using the Sun are also possible because of the large integration-time range achievable with the FPA. These solar-based calibrations benefit from the use of the same small aperture (and thus the same optical path) used in operations when acquiring solar observations with the filters.

The three ionically colored Schott glass filters in the HySICS were polished to 0.1 nm RMS surface roughness and $\lambda/4$ flatness to reduce induced scatter and distortion. The balloon-flight filter selection includes

1. NG4 (a neutral-density filter with $\sim 0.1$ transmission),

2. NG5 (a neutral-density filter with $\sim 0.3$ transmission), and

3. BG25 (a high-transmittance filter in the UV and IR).

## 3 High-altitude balloon flights

The HySICS was flown on two high-altitude balloon flights to demonstrate its ability to cross-calibrate Earth-scene radiances to the spectral solar irradiance. Each of the $\sim 9\,h$ flights maintained a float altitude of $39\,000\,m$ ($120\,000\,ft$) to acquire SSI measurements in the near-absence of attenuations or scatter by the Earth's atmosphere.

## 3.1 Balloon-system design

The HySICS was mounted on a two-axis gimballed pointing system able to track the Sun and Moon for calibrations and able to maintain a fixed-angle nadir view for scanning along the ground as the balloon drifted. The pointing system was mounted near the center of a large rectangular-frame gondola that was suspended from the balloon itself. A rotator mechanism between the balloon and gondola provided coarse azimuthal pointing ($\pm 3°$) of the latter, while the gondola-based WASP provided fine-pointing of the instrument itself.

### 3.1.1 WASP system

The Wallops Arc Second Pointer (WASP) is a two-axis altitude-azimuth gimbal-based pointing system designed to achieve nearly arc-second accuracy levels for balloon payloads (Stuchlik, 2015a, b). This system was provided courtesy of HySICS co-investigators D. Stuchlik's and J. Lanzi's team at NASA's Wallops Flight Facility (WFF). With the HySICS center-of-mass aligned within the WASP gimbal-axes to $\pm 25\,\mu m$, the system is able to point the instrument accurately at the ground, Sun, and Moon and track each of these objects while acquiring the needed measurements and calibrations.

The WASP generally provided $< 10$ arcsec of pointing accuracy during Flight 2, meeting the HySICS pointing requirements. The most critical pointing requirements are driven by scanning the Sun or the Moon lengthwise along the HySICS slit to obtain FPA flat fields by positioning the same portion of the Sun or Moon on each pixel in the FPA's spatial direction. Pointing knowledge and after-the-fact corrections are not sufficient for this flat-fielding calibration method; real-time pointing accuracy is needed. The WASP system achieved approximately 8 arcsec ($1\sigma$) pointing deviations across the $\pm 6°$ range about disk center for along-slit solar scans, acquiring the needed flat-field calibrations. Accuracies of 2 arcsec ($1\sigma$) across the $\pm 1.5°$ range for cross-slit solar-scans were achieved, with these scans intended to acquire solar-irradiance measurements by spatially integrating sequential images across the solar disk via post-flight ground-based data processing. The WASP provided 0.7 arcsec ($1\sigma$) along-slit stability and 2.0 arcsec cross-slit stability when staring at the Sun. While the WASP is generally capable of yet more accurate pointing, that provided during flight was sufficient for the HySICS's purposes.

The WASP was also able to inertially track the Moon using an on-board ephemeris. This new pointing-system capability enabled flat-fielding calibrations using the Moon while operating with the same 20 mm aperture (and thus optical paths) and integration-time parameters as used for Earth-scene observations.

### 3.1.2 Gondola

The balloon gondola is a rectangular-frame structure that houses the entire payload, consisting of the HySICS instrument, the WASP, 27 lead-acid batteries to supply power, all telemetry and tracking equipment, thermal enclosures, several crush pads for landing, and ballast. The net mass of the payload and gondola is 2300 kg (5000 lb), including 540 kg (1200 lb) of ballast.

The gondola was designed and built at the University of Colorado's Laboratory for Atmospheric and Space Physics (LASP), using a combination of 80–20 aluminum and square aluminum tubing. The structure is 3 m in height and contained within a 4.3 m diameter region when the crush pads are installed on all but the top of the gondola's six rectangular sides. During flight, the entire structure is suspended by the azimuthal rotator that provides coarse pointing.

The WASP and HySICS are centrally located in the gondola such that the HySICS can view nadir for observing the Earth and greater elevation angles for solar and lunar measurements. Once expanded at altitude, the overhead Helium-filled balloon restricts viewing to elevation angles $< 60°$. A remotely controlled caging mechanism locks the WASP to the gondola frame for launch and parachute-descent landing.

## 3.2 Flight summaries

Both high-altitude balloon flights were performed out of Fort Sumner, NM, and supported by the Columbia Scientific Balloon Facility (CSBF). Upper-atmosphere winds limit Fort Sumner balloon flights to a few weeks in the springtime and fall, while CSBF schedules limit support at Fort Sumner to only the fall launch season. Ground winds generally limit launches to early mornings. Upper-atmosphere wind speeds determine flight duration and allow only a narrow timeframe of a couple of weeks for lengthy flights needed for many other programs' nighttime viewing. HySICS observations allow a more extended launch window, since the Sun and Earth are the primary targets and both can be viewed shortly after the morning launches; nighttime observations are not needed.

Lunar observations, however, are needed, as they allow flat-fielding using the same optics as for ground viewing. While low lunar phases are beneficial for the higher radiances provided near full moon, such nighttime-acquired flat fields would be separated temporally from the Earth-ground scenes and would also require longer flight durations. Instead, higher lunar-phase angles were chosen to acquire the flat-field calibrations at similar instrument temperatures and times to the acquired ground scenes. Launch windows at less than 90° lunar phase were desired so that likely flight durations would include daytime solar observations along with early-morning or late-evening lunar observations. Unfortunately, these were precluded by high ground winds preventing launch attempts during the HySICS flight campaigns. Instead, both flights occurred with a higher-than-desired lunar phase. The low lunar-signals due to these high phases limited achieving the desired low uncertainties for flat-fielding and filter calibrations with the Moon. Nevertheless, all intended observations were acquired to demonstrate all aspects of and the achievable capabilities of the HySICS solar cross-calibration methods. (Such lunar-phase restrictions would be alleviated from space-borne platforms having more extended lunar-observing times.)

Flight 1 occurred on 29 September 2013, with launch at 13:30 UT and landing at 22:13 UT. A float altitude of 37 100 m (121 800 ft) was reached for this engineering flight, during which the HySICS and WASP attempted all needed measurements. The gondola was recovered and returned to LASP for refurbishment. No damage to the instrument occurred during this flight or landing. Flight 2 launched at 15:36 UT on 18 August 2014, reached a float altitude of 37 200 m (122 000 ft) at 17:52 UT, was powered off at 23:52 UT, and landed early on the following day. Despite a rough landing, post-recovery checkout revealed that the instrument was unharmed and all optical alignments were maintained, validating the HySICS's robust design.

## 3.3 Flight observations

The HySICS has three primary observation targets, each containing various observation-modes as well as several internal-instrument calibrations.

### 3.3.1 Ground scans

These cross-track scans, with the ground track and speed determined by the balloon velocity from the aloft winds, provide samples of the desired data from an eventual flight instrument. During Flight 2, four ground scans were acquired. The two in the morning included a mix of the New Mexico high desert with broken clouds, while the two in the afternoon were predominantly of high, thin clouds. Three-dimensional data cubes of these scans were created in ground processing after all radiometric calibrations were applied.

Several scans of the Earth limb were also obtained on this flight. These scans provide spatial–spectral information through the vertical extent of the Earth's atmosphere. The Earth limb itself was largely occulted by the tops of bright cumulus clouds at the near-horizontal look-angle for these scans. Some such scans also included the Moon as it was setting.

### 3.3.2 Solar scans

Along-slit scans enable flat-fielding of the FPA by placing the same portion of the Sun on each spatial element of the array. Cross-slit scans build up an entire data cube of the Sun, enabling the spatially integrated solar irradiance to be determined and allowing SI-traceability to SSI (provided on an absolute scale by other measurements or models), as detailed in Sect. 4.4. Since demonstrating the solar cross-calibration

method was the primary purpose of these flights, solar scans dominated the flight observation time. Near local noon the Sun's elevation was greater than 60°, so solar observations could not be acquired due to glint or occultation by the large overhead balloon. At these times, either lunar or ground scenes were acquired instead.

### 3.3.3 Lunar observations

Similar to those done with the Sun, along-slit scans enable flat-fielding of the FPA by placing the same portion of the Moon on each spatial element of the array. The lunar scans can be done with the larger Earth-viewing aperture, potentially providing a more appropriate flat field to be applied to ground scans than those obtained from solar scans. Additionally, spectral-filter transmission is calibrated during flight by quick successive measurements with each filter in and out of the optical path while tracking a fixed position of the Moon.

### 3.3.4 Internal-instrument calibrations

Internal-instrument calibrations and diagnostics helped track instrument functionality, stability, and performance in flight. Spectral calibrations were made intermittently throughout the flights by briefly illuminating the instrument's Hg / Ar pen-ray lamp. Pointing stability was quantified by attempting to maintain the instrument slit at a fixed position on the edge of the lunar limb for an extended period. At this position, lunar intensity is very sensitive to cross-slit variations in pointing, providing a diagnostic of pointing stability. An along-slit scan at this lunar position quantified the instrument's alignment relative to the WASP's elevation (altitude) direction.

## 4 Flight 2 data analysis and uncertainties

The intent of Flight 2 was to quantify the radiometric uncertainties to which HySICS-acquired Earth scenes could be related to known spectral solar irradiances. The HySICS spatial/spectral ground images, $S_{meas\_obj}(\lambda)$, which are measured in units of instrument DNs, are converted to physical units of spectral solar irradiance (such as $W\,m^{-2}\,nm^{-1}$) by applying a scale factor for an on-orbit-determined unit-conversion factor, $C(\lambda)$ (in units of spectral solar irradiance per DN), and the instrument's unit-less, ground-calibrated radiance-attenuation factor, $A(\lambda)$, which corrects for the optical throughput and integration times used for solar vs. Earth viewing according to the following:

$$S_{SI}(\lambda) = S_{meas\_obj}(\lambda) A(\lambda) C(\lambda), \tag{1}$$

where $S_{SI}(\lambda)$ represents the radiance of the observed scene in SI-traceable, physical units. The unit-conversion factor $C(\lambda)$ has the form

$$C(\lambda) = SSI(\lambda) / S_{meas\_Sun}(\lambda), \tag{2}$$

where SSI($\lambda$) is the spectral solar irradiance (provided by an independent spaceflight instrument or a solar model), and $S_{\text{meas\_Sun}}(\lambda)$ is the HySICS's in-flight measurement of the SSI in DNs acquired by spatially integrated cross-slit scans of the solar disk. Equation (1) is thus effectively a ratio of two in-flight HySICS measurements, $S_{\text{meas\_obj}}(\lambda)/S_{\text{meas\_Sun}}(\lambda)$, and calibration factors, $A(\lambda)$, to account for solar- and Earth-scene attenuations. Being a ratio, accurate on-orbit knowledge of common-mode instrument efficiencies are not critical for acquiring radiometrically accurate ground measurements. Since the needed solar and Earth measurements can be acquired in close temporal sequence, by using this on-orbit solar cross-calibration method, HySICS's SI-traceable measurements of ground scenes are not susceptible to potential long-term in-flight degradation of the instrument optics. This method ties the long-term accuracy of the HySICS to the accuracy to which the SSI is known and the long-term stability of the instrument's attenuation systems. The latter is based on physical components, such as geometric aperture-sizes and electronic timing, such as that controlling detector integration times; both are inherently very stable.

Since the factors in Eq. (1) are independent, their individual uncertainties are evaluated separately and root-sum-squared for each final scene-dependent uncertainty. These correction factors and their uncertainties are derived from component- and instrument-level characterizations from both pre-flight laboratory-based calibrations and in-flight calibrations of the instrument, which are described in this section.

## 4.1 Focal-plane array corrections and uncertainties

The initial data-analysis step is to apply corrections to the raw data images. Applying all such corrections gives $S_{\text{meas\_obj}}(\lambda)$ in Eq. (1). The initial corrections are detector-specific and are typical of any FPA-based instrument so are only cursorily mentioned in this sub-section for completeness.

### 4.1.1 Bad-pixel removal

Non-responsive pixels and badly fluctuating pixels, defined as those with a measurement-to-measurement standard deviation of more than $5\sigma$ greater than the sensor-wide average standard deviation, are filled using an average of all properly operating neighboring pixels. The HySICS FPA had 732 pixels needing such corrections. These are sufficiently few that they do not greatly influence subsequent statistics based on full-FPA data using their corrected values.

### 4.1.2 Read noise

Read noise for the Teledyne sensor is determined using a traditional photon-transfer measurement (Janesick, 2001) of a constant radiant-power source provided by blackbody radiation from a uniform, warm, temperature-stabilized target. This target is measured at various exposure levels by varying

the integration time from 33.6 $\mu$s to 34.4 ms. A corresponding dark image, acquired during a prior measurement of a 77 K target to eliminate blackbody radiation, is subtracted from each exposure in the photon-transfer measurement. The measured noise on each pixel, given by the standard deviation of 50 repeated measurements, is dominated by read noise at the shortest integration times and by shot noise at the longest integration times. Although a true zero integration time cannot be achieved, the noise versus signal level for each pixel is curve-fit to an expected photon-transfer curve to extrapolate to its true read noise. The sensor-wide average read noise is 8.3 DN.

With a gain of $\sim 12\,e^-\,\text{DN}^{-1}$ (see Sect. 4.1.5), read-noise uncertainties are thus based on random fluctuations around $100\,e^-$. Since these are of similar amplitude across the array and are independent of incident signal, read noise causes a higher relative uncertainty at low signal levels, such as the extreme portions of the spectral range where the solar signal and the detector response are both low. Higher signal levels, such as can be achieved from brighter scenes or longer integration times, reduce the effects of read-noise uncertainties. By acquiring all solar calibrations at both short and long integration times, read-noise in select wavelength ranges is greatly improved. Similarly, since read noise is a random statistical fluctuation, acquiring repeated images of the same scene reduces the effects of read noise as the reciprocal square root of the number of images. Such integration-time variations and multiple-image acquisitions are not possible when viewing the ground during flight, since balloon-track motion between frames causes either a different ground scene (for static nadir-viewing) or a different look-angle of the same ground scene (if actively tracking) to be measured by non-simultaneous successive frames; however, multiple-image acquisitions are implemented for HySICS calibrations using static sources such as the Sun and Moon. Thus, read noise mainly contributes to the ground-measurement uncertainties at shorter wavelengths.

### 4.1.3 Dark and thermal-background corrections

Both dark-noise and thermal-background signals from the surrounding instrument scale with integration time and are dependent on instrument or FPA temperature. Thermistors monitor the FPA and several of the nearby instrument-components. Laboratory characterizations of the dark signal enable corrections for both internal-FPA and background-thermal effects.

The HySICS FPA's inherent dark signal is sufficiently low that it is difficult to detect in the presence of any background light. A cold target placed in front of the imager while keeping the sensor housing at $-25\,°C$ reduced such background signals but did not completely eliminate them sufficiently. Dark current, which increases with FPA operating-temperature, was therefore measured at elevated operating temperatures of 165 K and warmer. These measurements were extrapolated to the FPA's nominal 150 K operating-temperature, yielding a dark current of $350\,e^-\,s^{-1}$ and resulting uncertainties of 0.29 DN for the longest HySICS integration times (34.4 ms) used. These inherent dark-signal uncertainties are well below the quantization limit of the device.

Background signals were also corrected during flight. Following all data acquisitions, the blanked aperture wheel position blocked incoming light for 100 exposures. These consist only of dark current, instrument-thermal-background contributions, and imager-fixed-pattern noise. They are acquired at the same integration time and nearly the same temperatures as the data frames themselves. The average of these dark exposures is subtracted from the data frames, thereby removing background offsets with the exception of possible thermal offsets caused by temperature differences between when the data and the dark measurements where acquired. These temperature dependencies are in turn corrected via in-flight thermal-background measurements using portions of the array viewing dark space during solar and lunar scans. From multiple such scans, FPA sensitivities to instrument thermal effects are determined as a function of surrounding instrument-component temperatures. All raw HySICS data images are thus corrected for thermal background based on the instrument temperatures at the actual time of data acquisition, using the instrument-temperature dependencies determined from these dark-space observations.

Although these thermal-background signals are largest at the longer-wavelength portion of the FPA's sensitivity, they influence the entire array uniformly, since the FPA has no long-wave rejection filter over the portions used only for shorter-wavelength readout, making the above corrections necessary for all portions of the spectrum. While the dark current is very small and contributes nearly insignificantly to the net HySICS uncertainties, the thermal-background signal contributes to shot noise (described in Sect. 4.2.1).

### 4.1.4  Linearity corrections

Deviations from linearity are determined individually for each FPA pixel in laboratory testing using varying levels of incident-light intensity and integration times. If temporally stable, non-linearities can be corrected once characterized. These corrections are applied to the images after the bad-pixel, dark, and thermal-background corrections.

Sensor linearity is measured in two steps: (1) The electronically determined integration time is measured directly using timing pulses from the sensor's field-programmable gate array's digital output signal and (2) the response of the FPA itself is measured using a stable light source while varying the now-known electronically controlled integration time. The former verifies the timing of the controlling electronics, which are, as expected for oscillator-based signals, very linear and stable. The latter step includes the effects of FPA pixel-well or amplifier-signal saturation and is a function of the net signal on each pixel. To characterize these non-

linearities, the sensor is illuminated by a stable FEL lamp while the electronically controlled integration time is varied and the resulting signal levels are measured. A linear curve-fit is used to determine the expected signal level on each pixel, and deviations from that fit with signal level are considered non-linearities in that pixel's response. The curve fit uses only the most linear portion of the data at less than 50 % of the FPA's full well. Repetition of this measurement using various FEL-lamp intensities ensures that the deviation from linearity has an FPA signal-level dependence rather than an integration-time dependence.

The resulting non-linearities and uncertainties are detailed in Sect. 4.3.2, where the non-linearity corrections, uncertainties, and intensity range and the resulting dominant determinants of the overall instrument attenuation uncertainty based on the integration-time method are discussed.

### 4.1.5  Pixel-dependent gain determinations

Sensor gain, or the conversion $[e^-\,DN^{-1}]$ from FPA DNs to electrons $[e^-]$ and thus photons, is determined from a photon-transfer measurement in laboratory testing on a pixel-by-pixel basis using statistics of each pixel's variations at different intensity-exposure levels. This was done using the same experimental setup as the read-noise measurement described in Sect. 4.1.2. In the larger-signal regime, where pixel noise is dominated by shot noise, sensor gain is defined as the ratio of signal level to pixel-noise variance. The previously determined read-noise variance is subtracted from the measured pixel-noise variance so that the residual noise is that due solely to shot noise. For each pixel, the signal level and pixel noise are determined using 50 exposures repeated at ten different signal levels. The experiment is repeated 100 times to determine the average gain and to reduce the shot-noise measurement uncertainty. The sensor-wide average pixel gain is $12.01\,e^-\,DN^{-1}$, with an average uncertainty per pixel of $0.12\,e^-\,DN^{-1}$, or < 0.003 % uncertainty in the shot-noise calculation at a signal level of 15 % (10 000 DN) of full scale. This pixel-dependent correction is applied to each pixel in the array but is an insignificant contributor to the net uncertainties.

### 4.1.6  Flat-field corrections

Flat-fielding the HySICS sensor requires a full-system calibration, since it is affected by the collective efficiencies of all upstream optics. This calibration therefore needs to be performed separately for the smaller solar-viewing aperture and the larger Earth-viewing aperture, as light passing through the two apertures interacts with different portions of the downstream optical elements in the instrument. These differences are accounted for via the flat-field calibrations and are corrected in post-processing of the data.

Although different apertures are used for the two scans, the flat-fielding procedure for both is to use a stable light

**Table 2.** Flat-fielding uncertainties.

| Parameter | Measurement uncertainty (%) | | | Measurement uncertainty (%) | | |
|---|---|---|---|---|---|---|
| | 550 nm | 1000 nm | 2000 nm | 550 nm | 1000 nm | 2000 nm |
| | Solar flat field | | | Lunar flat field | | |
| Peak variation uncertainty | 0.41 | 0.15 | 0.18 | 4 | 1.1 | 1 |
| Pointing accuracy | 0.065 | 0.059 | 0.063 | 0.32 | 0.28 | 0.28 |
| Blackbody radiation correction | 0 | 0 | 0 | 0 | 0 | 0 |
| Background level correction | 0.053 | 0.007 | 0.012 | 0.007 | 0.001 | 0.001 |
| Total | 0.418 | 0.161 | 0.191 | 4.013 | 1.135 | 1.038 |

source that can be swept across every pixel on the sensor, enabling a measurement of the relative response, or gain, of each. In space, the only available sufficiently stable light sources are the Sun and the Moon, which are used for the small- and large-aperture flat-field calibrations respectively. In both cases, a slice near the center of the solar or lunar disk is scanned in the along-slit direction from one edge of the imager to the other, while images are continuously captured at the instrument's nominal 14 Hz cadence used for ground-scene measurements. The flat-field calibration scans $\pm 6.5°$ from the boresight, going $1.5°$ outside the HySICS's FOV in both directions to ensure full spatial coverage. At a scan rate of $5.88 \, \text{arcmin s}^{-1}$, the along-slit flat-field scan requires 102 s to complete, during which time both sources are considered stable even during high rates of change in lunar phase. (Diffraction from the small solar-viewing aperture spatially blurs the Sun so that fine detail due to solar oscillations or granulation, which vary on 5 to 10 min timescales, are not observable and sensitivity to pointing errors is small. The lunar flat-field scan, however, is more sensitive to pointing and slit-alignment errors because of large intensity-variations across the lunar crescent and – unlike viewing the Sun through the small aperture – there is very little diffraction to spatially blur the image.) The resulting data are a series of images containing the solar or lunar spectrum stretched across the full sensor in the spectral direction and gradually moving through the sensor's entire spatial direction with each successive image. This technique essentially sweeps an identical spectrum across the spatial direction of the array, by which the spatial-direction flat-fielding is accomplished. (The spectral-direction flat-fielding is done at a later data-processing stage when the HySICS measurements are calibrated to the independently known SSI.) Since both sources extend over multiple spatial pixels, the pixel-to-pixel signal comparison can be repeated multiple times and utilized in measurement averaging as well as providing a basis for uncertainty estimations. A total of 31 spatial positions across the solar disk are applied to the flat-field correction, while only 9 positions across the narrow lunar crescent during Flight 2 are used. Uncertainties for select wavelengths are summarized in Table 2.

As with read noise, since the flat-field calibrations are acquired using static sources, they can benefit from multi-acquisition scans to reduce random uncertainties and at different integration times to improve signal in spectral regions having lower sensitivity such as the visible. These approaches, described in more detail in Sect. 4.2.6, were not performed for the flat-field calibrations of either the Sun or the Moon during Flight 2 and, as a result, the flight-acquired flat-field uncertainties dominate all others at the shorter wavelengths where instrument sensitivity is low. For flat-field calibrations using the Sun, cross-slit scans of which did benefit from multi-acquisition scans at different integration times and thus have low uncertainties for most other parameters, the flat-field uncertainties dominate at the shorter wavelengths and are comparable to diffraction at the longer wavelengths, so would greatly benefit from multi-acquisition scans. Lunar flat-field calibrations were marginal because of the high lunar phase during the time of the flight, giving low lunar signal and small spatial extent. These along-slit lunar scans are not only low in signal, but very sensitive to pointing, particularly since the large aperture used for lunar flat-field calibrations does not spatially blur the lunar image due to diffraction as the smaller aperture does to the solar image. Where the acquired flat-field uncertainties exceed the array's intrinsic 3.3 % pixel-to-pixel variations, such as in the shorter-wavelength portion of the visible, they were clipped at this intrinsic value.

Multi-acquisition flat-field calibrations at different integration times for both the Sun and the Moon and a lower lunar phase-angle would greatly improve the uncertainties demonstrated by Flight 2. Nevertheless, in spectral regions having high signal, the flat-field uncertainties acquired during this flight are $< 0.2 \%$, demonstrating the flat-field calibration method capabilities and showing promise of achieving desired lower uncertainties across a broader spectral range with the suggested multi-acquisition approach.

Flat-field corrections are applied to ground scenes and cross-slit solar scans, and thus these uncertainties directly affect those data. Measurements that rely purely on relative measurements, such as calibrations of aperture ratio

(Sect. 4.3.1) and filter transmission (Sect. 4.3.3), are not affected by these flat-field uncertainties.

## 4.2 Instrument uncertainties

Section 4.1 discussed corrections from the FPA and their associated uncertainties. Further contributions to $S_{\text{meas\_obj}}(\lambda)$ in Eq. (1) account for higher-level aspects of the HySICS's optical performance, which was evaluated at both component and integrated levels. The component-level tests validated or refined modeled performance at each stage during assembly. The tests indicated expected performance for most components, including the 4MA, apertures, filters, slits, and Offner mirrors. High-level uncertainties, such as those from photon counting, diffraction, optical scatter, varying optical paths, opto-mechanical or thermal effects on spectral scale, and polarization sensitivity are described in the following sub-sections.

### 4.2.1 Shot noise

Shot noise arises from photon-counting statistics and varies as the reciprocal square root of the signal, including that from any thermal background. As with read noise, it is reduced via multiple-image acquisitions for calibrations of static sources, namely the Sun and the Moon, but cannot be similarly reduced for single-acquisition images of the ground. Shot noise is the dominant source of uncertainty for ground scenes across the majority of the spectrum.

### 4.2.2 Diffraction

HySICS directly measures outgoing Earth-reflected shortwave radiances and incoming solar radiances. By spatially integrating radiances from the entire solar disk, which are acquired from cross-slit scans of the Sun, the HySICS measurements are calibrated to the independently known incoming SSI. To provide an accurate spatial integration, HySICS data analysis needs to correct for radiative losses, such as due to stray light and diffraction, that may cause differences in the amount of light reaching the FPA when viewing the Sun as opposed to ground scenes. Losses from diffraction are higher for the solar-viewing configuration than for ground viewing because of the smaller aperture used for solar observations. (Figure 2 illustrates the noticeably larger diffraction that must be accounted for when using the 0.5 mm solar aperture compared to the 20 mm Earth-viewing aperture. At 1000 nm, the diffraction limit from each is $\sim 8$ and 0.2 arcmin, respectively.) Spatial scans sweeping across and then well away from the Sun provide scatter and diffraction characterizations. These, in addition to lab measurements of the same, enable corrections to facilitate accurate determinations of the net SSI based on cross-slit scans of the Sun.

Diffraction can be modeled well with a NIST-quoted uncertainty of $\sim 1.8\%$ for simple circular-aperture geometries (Shirley et al., 2002). The effects of scatter, however, are very

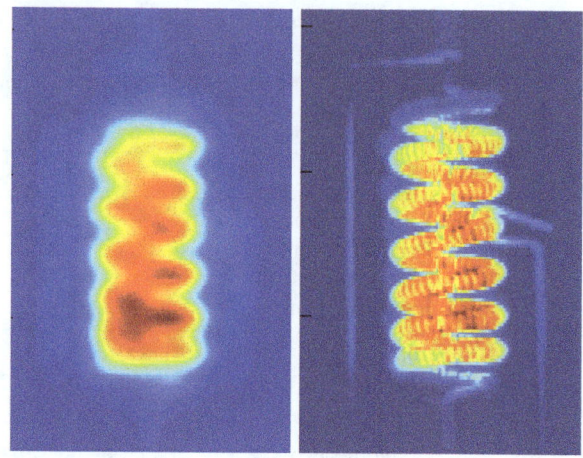

**Figure 2.** These scans of a lab FEL-lamp filament show the blurring caused by diffraction when using the 0.5 mm aperture (left panel) vs. the 20 mm aperture (right panel). These effects must be accounted for in spatially integrated spectral solar irradiance determinations.

instrument-specific and can be more difficult to model in advance to sufficient levels of accuracy. Lab measurements using the setup shown in Fig. 3 helped determine these contributions by characterizing the HySICS system's diffraction and scatter properties. This experiment occults the light coming directly through the aperture, but captures most of the light scattered or diffracted from it, and then re-images that light onto a separate FPA. Sample results are shown in Fig. 4 and match the expected angular dependence due to diffraction alone, indicating that diffraction, as opposed to scatter, is the dominant source of this indirect light for the as-built HySICS.

By modeling the diffracted light and verifying the model with lab measurements, the expected light losses are accounted for when spatially integrating the solar disk to obtain a value that can be correctly calibrated to the independently known SSI. A 1.8% uncertainty on this correction is allocated as per NIST diffraction-estimate uncertainties. Because diffraction scales with wavelength, these corrections begin to dominate the solar-calibration uncertainties at the longer wavelengths but never greatly exceed the contributions from read and shot noise. The HySICS's small solar-viewing aperture was chosen such that uncertainties due to diffraction may be the limiting uncertainty at the longest wavelengths but would not dominate across the spectrum, effectively balancing desirable greater-attenuation capabilities afforded by smaller apertures with the increased uncertainties expected from them. This balance established the HySICS attenuation levels achievable via aperture ratios to $\sim 10^{-3}$.

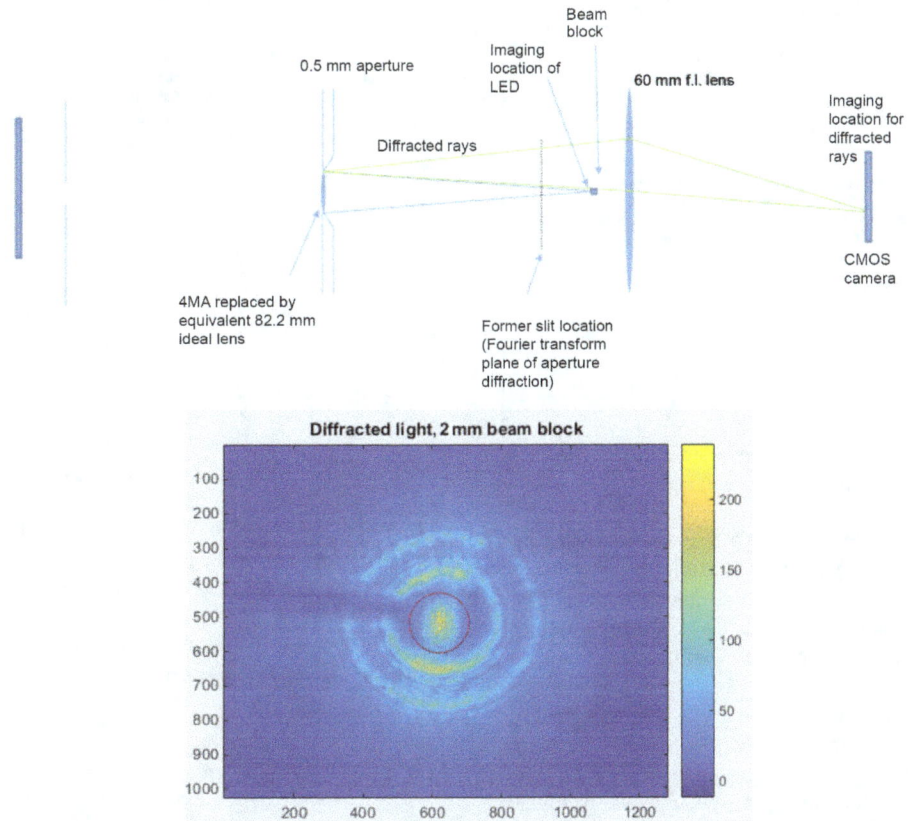

**Figure 3.** Lab scatter- and diffraction-characterization setup (upper schematic) gives the 2-D pattern shown in the lower image when using a 2 mm beam block behind the 0.5 mm solar-viewing aperture. This beam block occults the un-diffracted and un-scattered light incident on the aperture from a distant, nearly collimated light source (left side of upper schematic). Most light that is diffracted or scattered by the aperture edges passes around the beam block to be reimaged onto the camera, helping to quantify the intensity and spatial pattern of that light. The innermost Airy rings from typical aperture-edge diffraction are visible in the lower image. Since incident sunlight will diffract and scatter similarly, thus spreading some light beyond the edges of the solar-disk image, these effects must be corrected when determining net solar-irradiance values via spatial integrations from HySICS's cross-slit scans of the solar disk. Lab measurements such as these, combined with diffraction models to account for wavelength sensitivity and extend the spatial extent, help reduce uncertainties for those corrections. (The nearly horizontal radially extending dark region to the left of image center in the lower image is due to the support for the beam block.)

### 4.2.3  Spectral-scale corrections

Since radiometric uncertainties are dependent on the product of the instrument's spectral accuracy and the derivative of the measured spectrum, knowledge of, and corrections for, the spectral scale (or wavelength position) are characterized and applied.

Intermittent measurements using the HySICS's internal pen-ray lamp throughout the flight allow spectral calibrations based on this narrow-band source to verify spectral-scale accuracy or correct for possible wavelength-position fluctuations due to thermal or mechanical changes. Independent control of the three TECs regulating optical-bench temperature reduced thermal gradients during Flight 2 and thus reduced variations in the spectral scale. The spectral scale when at altitude shifted by only 3 nm, with variations across all wavelengths being maintained to < 1 nm across the spectral range.

The spectral corrections were interpolated to the times of observations. Of particular importance are the corrections at the times of solar calibrations, as the Sun has more abrupt spectral variations than ground scenes, and uncertainties in the spectral scale near the edge of a large spectral variation can give a correspondingly large radiometric uncertainty at wavelengths near spectral lines. Since the HySICS uses an FPA that spectrally bins the incident light from the spectrometer into 3 nm regions defined by the size of the FPA pixels, small potential spectral shifts in the incident light coupled with large spectrally dependent changes in signal near the sharp edges of these pixel-defined spectral bins can affect radiometric uncertainties.

Both the effects of this pixel-delineated spectral binning and those from thermal or mechanical instrument distortions are included in estimates of the HySICS's wavelength-position uncertainties, which are plotted as a function of wavelength in Fig. 6 for solar observations and in Fig. 7

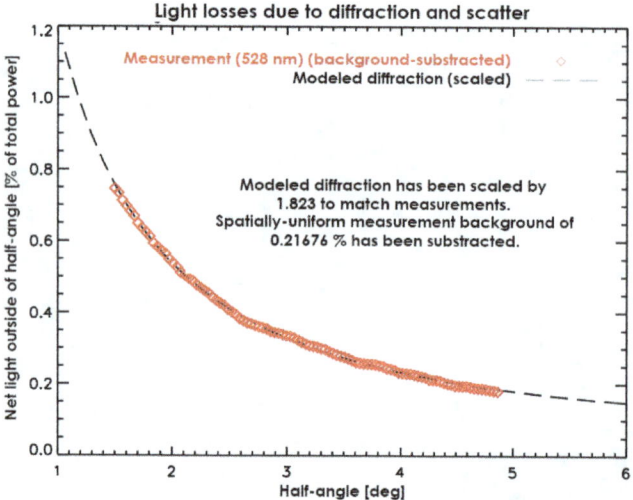

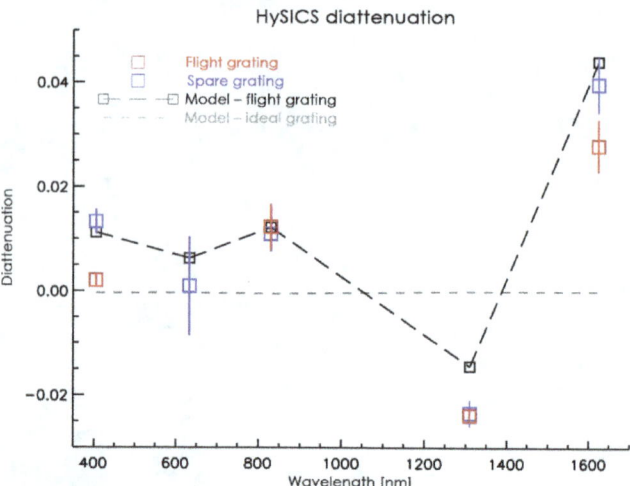

**Figure 4.** An idealized model estimating the net amount of light falling outside of various angles due to diffraction alone (black dashed curve) is scaled to match lab measurements including both scatter and diffraction (red diamonds) at 528 nm. The measurements match the angular dependence expected from diffraction, indicating that the majority of this measured light loss is mainly due to diffraction rather than scatter, which would have a less-well-modeled relation to angle. Correcting for light losses via this validated diffraction model reduces the uncertainties in solar-radiance measurements, improving the HySICS's determination of spectral solar irradiance.

**Figure 5.** Diattenuation of the HySICS at the integrated-instrument level is limited by high grating-induced polarization, which is as large as 4 % at wavelengths above 1000 nm. A Zemax model based on the flight-grating measurements (black, long dashes) shows the lower instrument diattenuations expected using a less-polarization-sensitive grating (gray, short dashes).

for ground-scene measurements. Because of the in-flight spectral-calibration corrections via the internal pen-ray lamp, wavelength-position uncertainties are rarely the dominant contributor to the net radiometric uncertainties, although they do increase at the shortest wavelengths, where the Sun has more spectral-absorption lines, as well as near 820 nm, where the Sun has several absorption lines and the HySICS has low sensitivity.

For a spaceflight instrument regularly acquiring Earth observations, the spectral scale determined by the pen-ray calibrations could be validated by select Earth-atmospheric spectral lines under certain viewing conditions to help distinguish them from surrounding spatial or spectral features, such as by observing these lines from uniform bright background clouds or dark oceans or viewing them near the Earth limb by off-pointing from nadir. Oxygen molecules provide some such possible spectral lines, with one HySICS balloon flight even showing an $O_2$ line in emission against darker space in an Earth-limb scene. Although spectral-scale corrections are a small source of uncertainty and predominantly affect solar calibrations rather than Earth observations, such Earth-atmospheric spectral-line observations could help verify the instrument's in-flight spectral scale during normal observations.

### 4.2.4 Brightness offset

The FPA has a background-level offset that varies linearly with the measured signal. This offset is detectable by observing the extreme-most ultraviolet spectral column of the sensor, which, at 320 nm, is below the reflectivity cutoff for the instrument mirrors and, therefore, is effectively a dark column. All pixel values in this column should remain nearly constant, showing mainly dark-current and fixed-pattern noise. Instead, they consistently decrease by up to 120 DN when other portions of the array are observing extremely bright signals. This background-level decrease is also observed in all dark pixels during a solar scan, including columns neighboring the dark column as well as regions of the sensor viewing dark space up to 9.5° away from the Sun. Lab measurements of FEL and LED sources show similar effects.

This "brightness offset" of the background level, as measured on the dark column, is characterized using flight data. A matrix of background-level reduction versus sensor signal is generated from all large power-level transitions during the flight, such as when the solar disk moves out of the instrument FOV during a flat-field scan or when it comes into- or out-of-view during an irradiance scan. The amount of background-level reduction is linear with the amount of light detected by the sensor, regardless of its spectral distribution or spatial location, with the background level changing by $-2.3 \times 10^{-7}$ DNs per DN of signal. Because there is some dependence on the integration time of the sensor, this slope was determined for all integration times used during flight. The slope has an uncertainty of 14 % based on the

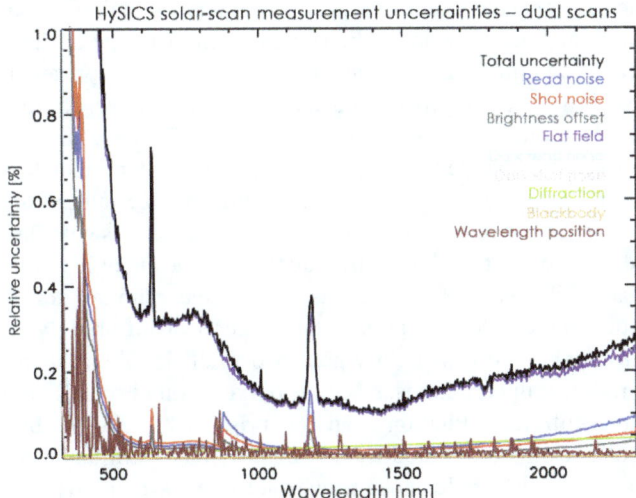

**Figure 6.** Contributions to net relative uncertainty (black) when observing the Sun during Flight 2 are shown as a function of wavelength. With the exception of the flat-field uncertainties, these plotted uncertainties are the result of two consecutive cross-slit solar scans acquired using specific integration times for the short- and long-wavelength spectral regions to reduce the uncertainties within each. Flat-field uncertainties due to low signal levels dominate across the spectrum but could be reduced with similar multi-image, dual-scan techniques applied to those calibrations.

standard deviation from multiple background-level characterization measurements. However, being as the brightness offset itself is a small correction, this does not directly translate into a similar-magnitude contribution to the overall radiometric measurement uncertainties.

The resulting brightness-offset corrections, which are dependent on the total signal on the sensor as well as the integration time, are applied to each image acquired. The relative uncertainties in this correction are greatest for measurements having low signals, so they predominantly affect Earth ground scenes, where they are generally the second-largest contributor to net uncertainties.

#### 4.2.5  Polarization Sensitivity

Accurate radiometric measurements of scenes having unknown polarization rely on the instrument having low polarization sensitivity (Lukashin et al., 2015). The HySICS was designed to reduce polarization sensitivity by orienting the optical plane of the 4MA perpendicularly to that of the spectrometer, such that reflection-induced diattenuation in the former is nearly offset by that in the latter. This was effective with the exception of the custom-ruled grating, the primary HySICS optical component that did not meet expected performance. Along with having low efficiency in the visible, polarization tests of this grating showed a much larger sensitivity than anticipated, with the net instrument-diattenuation results plotted in Fig. 5. Despite the orthogonal orientation of

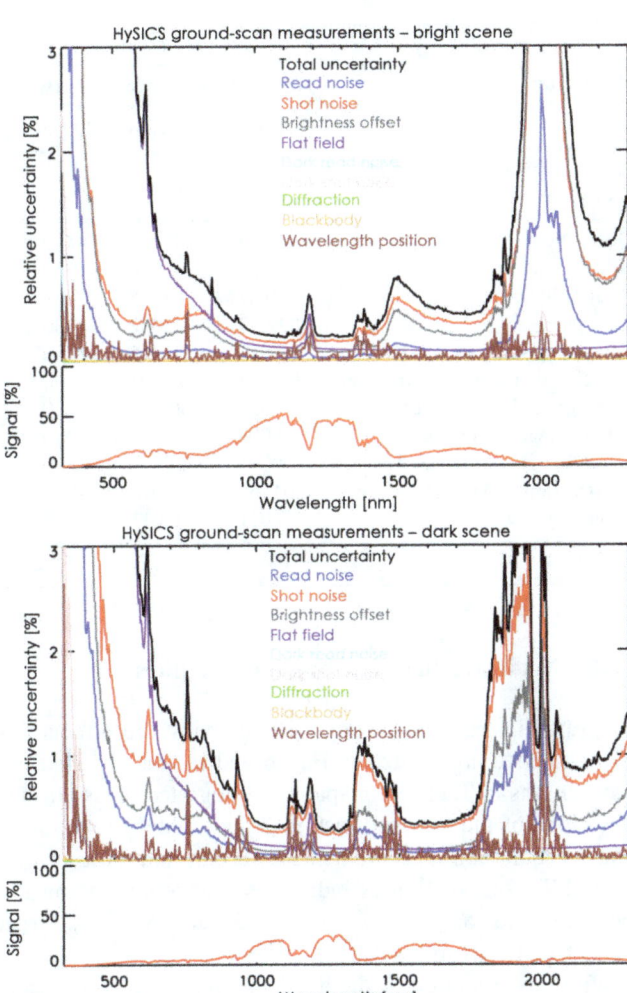

**Figure 7.** Contributions to net relative uncertainty (black) when observing a bright (top) and dark (bottom) Earth scene during Flight 2 are shown as a function of wavelength. The small, inset lower plots (red) indicate the signal strength from each scene relative to full scale of the instrument's FPA. Shot noise is generally the dominant uncertainty across the majority of the spectrum for ground scenes, which do not benefit from multiple-image or dual-scan acquisition techniques.

the 4MA to the Offner optics, this grating limits the instrument's desired low polarization sensitivity, particularly in the near infrared.

If measuring randomly polarized scenes, this internal-instrument polarization sensitivity has no effect on radiometric accuracy but, for scenes of unknown polarization amplitude and orientation, the radiometric uncertainties can potentially be as large as the instrument's diattenuation itself in the specific – albeit highly improbable – case of a 100 % polarized incident signal oriented along, or perpendicular to, the direction of the instrument's greatest polarization sensitivity.

**Table 3.** Solar-irradiance scan uncertainties.

| Parameter | Measurement Uncertainty (%) | | |
|---|---|---|---|
| | 550 nm | 1000 nm | 2000 nm |
| Image-dependent uncertainties (solar scans) | | | |
| Read Noise | 0.044 | 0.034 | 0.049 |
| Shot noise | 0.054 | 0.026 | 0.033 |
| Flat-field correction | 0.41 | 0.15 | 0.19 |
| Hot pixel | 0.0003 | 0.002 | 0.0005 |
| Wavelength bin location | 0.027 | 0.015 | 0.009 |
| Blackbody radiation correction | 0.0001 | 0.0001 | 0.0001 |
| Background level correction | 0.035 | 0.014 | 0.019 |
| Dark image read noise | 0.003 | 0.002 | 0.004 |
| Dark image shot noise | 0.0001 | 0.0001 | 0.0001 |
| Diffraction (0.5 mm Aperture) | 0.01062 | 0.018 | 0.0378 |
| Pointing accuracy | 0.011 | 0.011 | 0.011 |
| Total | 0.419 | 0.159 | 0.204 |

### 4.2.6   Net instrument-imaging uncertainties

Integrated-instrument uncertainties showing the effects described above are plotted in Fig. 6 and tabulated for select wavelengths in Table 3 for spatially integrated cross-slit observations of the Sun and in Fig. 7 and Table 4 for sample single-acquisition measurements of bright and dark ground scenes. These two figures indicate the uncertainties on the measurements $S_{\mathrm{meas\_SSI}}(\lambda)$ in Eq. (2) and $S_{\mathrm{meas\_obj}}(\lambda)$ in Eq. (1), respectively.

With the exception of the flat-field and diffraction uncertainties, the solar-scan uncertainties benefit from multiple-image acquisitions and a dual-scan approach. Multiple, repeated measurements of the same scene particularly reduce the effects of read and shot noise and from the brightness offset caused by the small vs. large apertures at the shorter wavelengths where the HySICS's response is lowest. Two back-to-back scans of the Sun, one using longer integration times to increase signals at wavelengths shorter than 850 nm and one with integration times better matched to the higher signals at longer wavelengths, followed by spectrally combining the scans in post-processing improves the signal in select portions of the spectrum. The improvements from multiple image acquisitions and the dual-scan approach are possible only because the Sun can be viewed repeatedly with the same instrument look-angles, so it provides a static in-flight calibration source. These techniques would also be applicable to reducing flat-field uncertainties but were not performed on Flight 2, so the solar-calibration results shown are dominated by the flat-field uncertainties.

Balloon-flight motions over the ground prevent applying these beneficial uncertainty-reduction techniques to ground scenes, so uncertainties must be based on single-image acquisitions. Despite the larger aperture and the longer integration times for ground scenes, the lower radiances of these sin-

gle images have larger relative uncertainties than those from the Sun, since they do not benefit from multi-image or dual-scan techniques. Typical net uncertainties from representative bright (cloud-filled) and dark (desert- and vegetation-filled) ground scenes are plotted in Fig. 7; these are the net scene-dependent uncertainties in the measurement factor $S_{\mathrm{meas\_obj}}(\lambda)$ from Eq. (1). The dominant uncertainties are from shot and read noise, with the brightness offsets with flat-field uncertainties dominating at shorter wavelengths. Note that while the dark scene has higher uncertainties across much of the spectral region, at the longer near-infrared (NIR) wavelengths, it has slightly greater overall signal and therefore lower uncertainties, since the darker ground scenes emit more infrared radiation than the brighter (in the visible) scenes from colder clouds, although these are still large due to the very low reflectance signals at these wavelengths.

### 4.3   Attenuation-system uncertainties

In addition to many of the instrument-level uncertainties for various observation scenes and modes described in Sect. 4.1 and 4.2, characterizing the radiometric uncertainties to which ground-scene radiances can be referenced to the spectral solar irradiance also involves quantifying the uncertainties from the three intensity-attenuation methods used to enable solar vs. Earth viewing. These attenuation methods, represented by the correction factor $A(\lambda)$ in Eq. (1), have additional uncertainties that are described in this section.

The total attenuations demonstrated during Flight 2 were capable of a net $10^{-7.1}$ reduction in incident radiance, much greater than the $\sim 10^{-4.7}$ needed for solar-radiance attenuations. Tables 5–7 give a numerical breakdown of the attenuation-system uncertainties for three select wavelengths across the instrument's spectral range. The dominant contributors to these uncertainties include low signal levels due to low FPA and grating efficiencies at certain wavelengths and high light-source variations for laboratory calibrations in the UV and visible, which would be straightforward to improve in future calibrations. Some instrument-specific uncertainties could be reduced by decreasing the large attenuation range demonstrated here. Forgoing the filter attenuation system, for example, would provide a net $10^{-6.2}$ attenuation, which is still larger than required for solar viewing. Using only the other two attenuation systems has demonstrated a $\sim 2\times$ improvement in radiometric accuracies over existing spaceflight instrumentation for an average across most of the visible and NIR spectral regions, with a $\sim 6\times$ improvement demonstrated in some regions. Further HySICS uncertainty reductions are expected from identified improvements in lab calibrations and spectrometer grating design.

Results and uncertainties from the individual attenuation-methods are detailed in the following subsections.

**Table 4.** Ground-scan uncertainties.

| Parameter | Measurement uncertainty (%) | | | Measurement uncertainty (%) | | |
|---|---|---|---|---|---|---|
| | 550 nm | 1000 nm | 2000 nm | 550 nm | 1000 nm | 2000 nm |
| Image-dependent uncertainties (ground scans) | Bright pixel | | | Dark pixel | | |
| Shot noise | 0.41 | 0.21 | 3.9 | 1.11 | 0.34 | 1.46 |
| Read noise | 0.097 | 0.032 | 1.5 | 0.38 | 0.075 | 0.57 |
| Flat-field correction | 3.1 | 0.155 | 0.098 | 3.1 | 0.155 | 0.098 |
| Diffraction (20 mm Aperture) | 0.000144 | 0.00036 | 0.00054 | 0.000144 | 0.00036 | 0.00054 |
| Wavelength bin location | 0.041 | 0.029 | 0.12 | 0.043 | 0.066 | 0.28 |
| Background level correction | 0.29 | 0.098 | 4.4 | 0.58 | 0.11 | 0.86 |
| Blackbody radiation correction | 0.0002 | 0.0001 | 0.004 | 0.0004 | 0.0001 | 0.0006 |
| Dark image read noise | 0.007 | 0.002 | 0.1 | 0.027 | 0.005 | 0.04 |
| Dark image shot noise | 0.019 | 0.007 | 0.266 | 0.06 | 0.014 | 0.089 |
| Total | 3.142 | 0.282 | 6.077 | 3.366 | 0.402 | 1.815 |

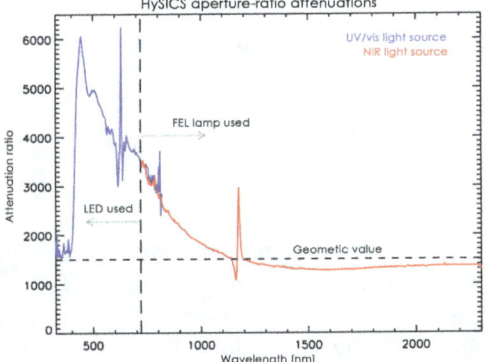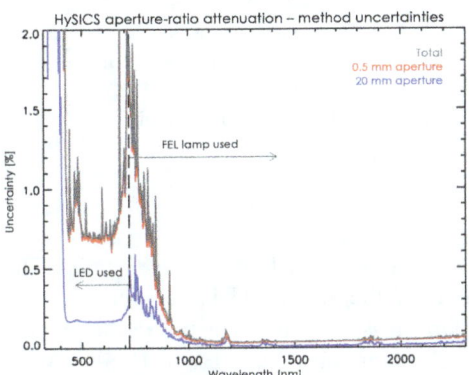

**Figure 8.** Different illuminations of the optical surfaces by the 0.5 and 20 mm apertures, mainly being affected by a boundary between grating blaze regions, cause the spectrally dependent attenuations that differ from the nominal geometric-ratio value due to the aperture-attenuation method, as shown in the left panel. The blue and the red curves are based on different lab light sources that provide peak power at shorter and longer wavelengths, respectively. The right panel gives the uncertainties in these attenuations. The large peak in uncertainties between the two light sources used is due to low signals from each and could be improved with additional calibration light sources.

### 4.3.1 Aperture-ratio uncertainties due to optic-surface-area illumination differences

The baselined $10^{-3.2}$ attenuation due to aperture-area ratios was demonstrated, with measured results plotted in Fig. 8. This figure also gives the corresponding uncertainties as a function of wavelength, while Table 5 details a breakdown of their contributing components. Although this aperture-ratio attenuation method relies on geometry and so should be nearly spectrally flat, at shorter wavelengths the method causes much more than the $10^{-3.2}$ attenuation expected from aperture-area ratios alone.

The small and large apertures respectively used for the Sun and Earth measurements illuminate different areal portions of the HySICS optical surfaces and thus have different throughput efficiencies that must be accounted for when transferring the solar-based radiometric scale to radiances from ground measurements. While most optical surfaces are sufficiently uniform or similarly illuminated to not be greatly affected by these different areal-illumination effects, the spectrometer grating is the dominant cause of current HySICS spectrally dependent efficiency variations between the two aperture-illumination regions.

To achieve the broad spectral range and high throughput efficiencies required with a single-spectrometer design, varying grating-blaze-angles are needed. The fabricated balloon-flight grating contains a sawtooth pattern of four discrete regions, with the blaze-angle varying monotonically across each. The small aperture used for solar measurements illuminates a boundary between two such regions to a much greater proportional degree than the larger aperture does, so it is more sensitive to symmetric alignment on this boundary. A slight misalignment on the edge of this "tooth" in the sawtooth grating pattern will preferentially favor the corresponding extreme-blaze-angle at the edge of the region in that misalignment direction, thus making the system more

**Table 5.** Aperture attenuation-method uncertainties.

| Uncertainty parameter | Measurement uncertainty (%) | | |
|---|---|---|---|
| | 550 nm | 1000 nm | 2000 nm |
| Read noise (0.5 mm aperture) | 0.054 | 0.005 | 0.0018 |
| Read noise (20 mm aperture) | 0.013 | 0.003 | 0.001 |
| Shot noise (0.5 mm aperture) | 0.12 | 0.012 | 0.004 |
| Shot noise (20 mm aperture) | 0.0095 | 0.003 | 0.002 |
| Diffraction (0.5 mm aperture) | 0.01062 | 0.018 | 0.0378 |
| Diffraction (20 mm aperture) | 0.000162 | 0.000288 | 0.000558 |
| Dark image read noise (0.5 mm aperture) | 0.0038 | 0.0004 | 0.0001 |
| Dark image read noise (20 mm aperture) | 0.0009 | 0.0002 | 0 |
| Dark image shot noise (0.5 mm aperture) | 0.0003 | 0.0001 | 0.0001 |
| Dark image shot noise (20 mm aperture) | 0.0001 | 0 | 0 |
| Background level correction (0.5 mm aperture) | 0.0006 | 0.0015 | 0.0003 |
| Background level correction (20 mm aperture) | 0.0007 | 0.0038 | 0.0016 |
| Light-source variation (0.5 mm aperture) | 0.68 | 0.022 | 0.007 |
| Light-source variation (20 mm aperture) | 0.17 | 0.012 | 0.004 |
| Measurement variation (0.5 mm aperture) | 0 | 0.058 | 0.031 |
| Measurement variation (20 mm aperture) | 0 | 0.037 | 0.016 |
| Dark image offset | 0 | 0.88 | 0.88 |
| Short exposure uncertainty | 0.78 | 0.78 | 0.78 |
| Total | 0.713 | 0.077 | 0.052 |

sensitive to either the shortest or longest wavelengths. The relative throughput for the small and large apertures was characterized in lab measurements with the results shown in Fig. 8. These effects are accounted for in HySICS's radiometric results as part of the aperture-ratio portion of the full attenuation-system correction, $A(\lambda)$. The spectral dependence shows much lower efficiency in the visible and higher efficiency in the NIR, suggesting the small region of the grating illuminated by the solar-viewing aperture is biased toward the NIR-blaze edge of one grating region rather than equally split between the two it straddles.

These laboratory calibrations were performed using two light sources, with one peaking in the visible and the other in the NIR, to span the full spectral region. The intermediate visible-to-NIR spectral region had low intensity from both lamps. Combined with the strong increase in attenuation and resulting lower intensities at shorter wavelengths when using the small aperture, these low light-source intensities limited the relative uncertainties in this visible-to-NIR spectral region, resulting in the large uncertainty peak shown in Fig. 8. Further calibrations with a broader range of bright lamp sources, particularly near the visible-to-NIR transition, could improve the uncertainties shown. More significantly, reducing the large spectral dependence of this aperture-ratio correction by using smoothly varying but non-monotonic blaze-angles via a more expensive custom-made grating rather than the four-region sawtooth one used here, should reduce much of these aperture-ratio uncertainty issues in the first place and is planned for a future HySICS instrument.

The aperture-ratio attenuation technique is inherently nearly independent of wavelength. That HySICS demonstrated the technique to well less than the needed uncertainties through most of the NIR spectral region shows promise that this attenuation method would be equally applicable over the entire spectral range with a more uniformly blazed grating and further laboratory characterizations.

### 4.3.2 Integration-time uncertainties

Correcting for non-linearities while varying the FPA's electronically controlled integration times was more successful than initially anticipated, achieving a demonstrated attenuation of $> 10^{-3}$ with a 0.05 % uncertainty for generally used exposure levels and a maximum of 0.12 % uncertainty that could accommodate extremely bright Earth-scenes. As described in Sect. 4.1.4, these corrections rely on characterizations of the electrically controlled integration-timing signals and the FPA's resulting response to various saturation levels.

The electronic timing signals show deviations from linearity that are $< 2 \times 10^{-7}$ for integration times from 16.8 μs to 8.62 ms, spanning a range of $10^{2.7}$ in integration times, and $< 1.6 \times 10^{-6}$ over the full range from 16.8 μs to 35.23 ms, spanning a $10^{3.3}$ range. These timing-signal deviations from linearity are relatively insignificant.

The results from the characterizations of the FPA's response described in Sect. 4.1.4, whereby the FPA's signal levels are determined from multiple, repeated measurements of an input FEL-lamp source at different exposure times, are shown in Fig. 9. Since the electronic shutter has nearly neg-

**Table 6.** Integration-time attenuation-method uncertainties.

| Uncertainty parameter | Bright scene (53 % FS) [%] | Max. int. (75 % FS) [%] |
|---|---|---|
| Electronic linearity | 0.00016 | 0.00016 |
| Gain non-linearity | 0.050 | 0.120 |
| Total | 0.050 | 0.120 |

ligible non-linearity across this range, these deviations from linearity that manifest mainly at greater exposure times (i.e. greater signal levels) are due to non-linearities in the detector response and/or readout-amplifier electronics. The average of the deviations plotted in Fig. 9 (lower graph) provides the applied non-linearity correction as a function of detector signal, and the standard deviations about this average give the corresponding uncertainties in the applied non-linearity correction. The corrections are measured to be < 0.1500 % for almost all pixels over an intensity range of $> 10^3$, and the reproducibility of each pixel's response is generally < 0.05 % for intensities up to 53 % (35 000 DN) of full scale, which accommodates the brightest Earth scenes viewed during Flight 2. This attenuation method can also accommodate higher-intensity scenes, allowing up to 75 % of the FPA's full scale to be utilized while maintaining uncertainties to < 0.12 %. The corresponding uncertainties are shown in Fig. 10 and tabulated in Table 6.

The greater-than-anticipated attenuation range and the lower-than-anticipated uncertainties due to this integration-time attenuation method allow flexibility in the attenuation amounts needed by the other two attenuation methods. The integration-time attenuation capabilities provided by this flight-capable FPA eliminate the need for attenuations via filters altogether, which reduces mass, cost, complexity, and power for a future flight instrument.

### 4.3.3 Filter-transmission uncertainties

Spectral filters were calibrated during Flight 2 using both the Moon and the Sun. The filter attenuation method demonstrated the desired attenuation range of $10^{-0.9}$, but with higher than the anticipated 0.05 % uncertainty due to the low lunar-signal levels at the time of this flight. As explained in Sect. 4.1.6, the narrow lunar-crescent illuminated only a few pixels on the HySICS FPA, resulting in read-noise-limiting filter-calibration uncertainties at wavelengths less than 900 nm (see Fig. 11). Since the filter calibrations are acquired while viewing stationary sources, namely the Sun and the Moon, multiple images and different integration times could be used to reduce noise in low-signal portions of the spectrum, although only single integration times were used for the calibrations during Flight 2. This operational improvement would substantially reduce the uncertainties in the visible spectral region shown in Fig. 11. Nevertheless, the fil-

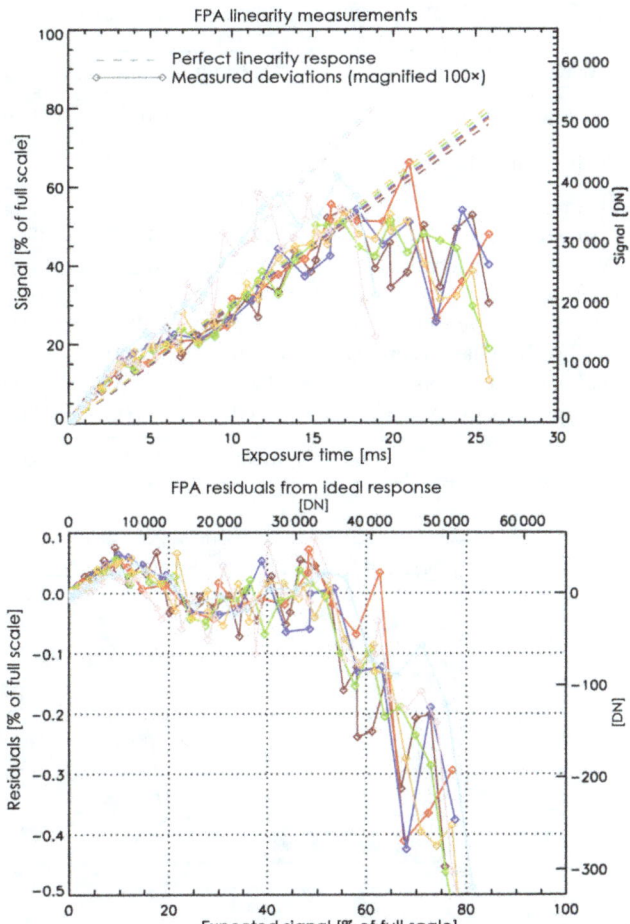

**Figure 9.** Detector response is plotted vs. exposure time over a range $> 10^3$ for seven different linearity calibrations indicated by different colors (upper plot). The differences (solid lines) between the detector response and linear fits (dashed lines), which vary between the different linearity calibrations due to intentional changes in the incident light level, have been exaggerated 100 × for visibility. These show slight sensitivity decreases at greater detector-signal levels. The residuals from the fitted linearity are shown in the lower plot as a function of signal level and provide a linearity correction for measured detector values. The repeatability of these repeated residual measurements indicates the uncertainty to which these non-linearities can be corrected and is shown in Fig. 10. These responses are determined individually for each FPA pixel.

ter uncertainties, particularly in the near-infrared, are already lower than those from the other two attenuation methods (see Table 7), although this method does not provide nearly the attenuation range of either of those other two. Fortunately, the large attenuation range provided by the integration-time attenuation method likely makes this entire filter-attenuation system unnecessary.

**Table 7.** Filter-calibration attenuation-method uncertainties.

| Parameter | Measurement uncertainty (%) | | | Measurement uncertainty (%) | | | Measurement uncertainty (%) | | |
|---|---|---|---|---|---|---|---|---|---|
| | 550 nm | 1000 nm | 2000 nm | 550 nm | 1000 nm | 2000 nm | 550 nm | 1000 nm | 2000 nm |
| Filter (Solar calibration) | NG4#2 | | | NG5#2 | | | BG25 | | |
| Shot noise | 0.94 | 0.1 | 0.047 | 0.25 | 0.065 | 0.035 | NA | 0.027 | 0.018 |
| Read noise | 0.97 | 0.097 | 0.048 | 0.25 | 0.06 | 0.035 | NA | 0.015 | 0.011 |
| Wavelength bin location | 0.027 | 0.015 | 0.009 | 0.027 | 0.015 | 0.009 | 0.027 | 0.015 | 0.009 |
| Filter-out uncertainty | 0.35 | 0.049 | 0.064 | 0.35 | 0.049 | 0.064 | 0.35 | 0.049 | 0.064 |
| Background level correction | 0.23 | 0.023 | 0.011 | 0.082 | 0.02 | 0.011 | NA | 0.005 | 0.003 |
| Blackbody radiation correction | 0.0001 | 0 | 0 | 0 | 0 | 0 | NA | 0 | 0 |
| Dark image read noise | 0.069 | 0.007 | 0.003 | 0.018 | 0.004 | 0.002 | NA | 0.001 | 0.0008 |
| Dark image shot noise | 0.003 | 0.0003 | 0.0001 | 0.0007 | 0.0002 | 0 | NA | 0 | 0 |
| Total | 1.416 | 0.150 | 0.094 | 0.505 | 0.104 | 0.082 | NA | 0.060 | 0.068 |

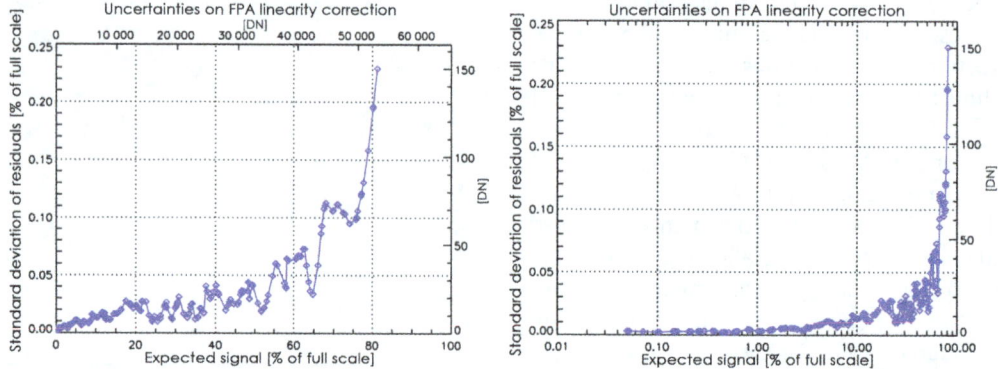

**Figure 10.** The standard deviations of the residuals from the average of the residuals shown in Fig. 9 indicate the uncertainties in the applied linearity correction as a function of signal level (left panel). The semi-log plot of the same data demonstrates the full $> 10^3$ measured intensity range (right panel).

## 4.4 Radiometric traceability to SI

### 4.4.1 Scene-reflectance uncertainties

Section 4.1 and 4.2 explain the intrinsic imaging-measurement uncertainties from Sun and Earth scenes, $S_{meas\_SSI}(\lambda)$ in Eq. (2) and $S_{meas\_obj}(\lambda)$ in Eq. (1), while Sect. 4.3 adds uncertainties from the HySICS's attenuation-system calibrations, $A(\lambda)$, which relate the relative signals of the Earth scenes to those of the Sun via the applied attenuation amount. This Earth-to-Sun ratio effectively gives the (unit-less) reflectance of the Earth scene. The uncertainty in the ratio includes uncertainties from the cross-slit solar-disk scans (Fig. 6), the Earth images (Fig. 7), and the attenuation systems applied (Fig. 8 and Table 6, since the filter-attenuation system was not utilized for the results presented here). Being independent, these uncertainties can be root-sum-squared to give the net Earth-to-Sun ratio (reflectance) uncertainties shown in Fig. 12 for both bright and dark ground scenes.

The solar cross-calibration techniques achieved a radiometric uncertainty of nearly 0.3 % across a large spectral region longward of 1000 nm from a bright ground scene, demonstrating a $\sim 6 \times$ improvement over current spacecraft uncertainties and the capability of the approach to achieve the desired radiometric accuracies. The net uncertainties shown in Fig. 12 are dominated by spectral regions where there is very little power from the reflected-Earth radiation or very low HySICS sensitivity, such as the 1800 to 2200 nm and shorter visible spectral regions, respectively, and by the increased uncertainties from the calibration of the aperture-ratio attenuations near the visible-to-NIR transition spectral region. (The latter can be improved with further calibrations and/or an improved instrument grating, as detailed in Sect. 4.3.1, and the former partially by improved flat-field calibrations, as explained in Sect. 4.1.6) At most NIR wavelengths, the uncertainties are dominated by read and shot noise in the ground scenes. The uncertainties shown are characteristic of individual pixels, and spatial or spectral binning could allow yet further reductions in uncertainties.

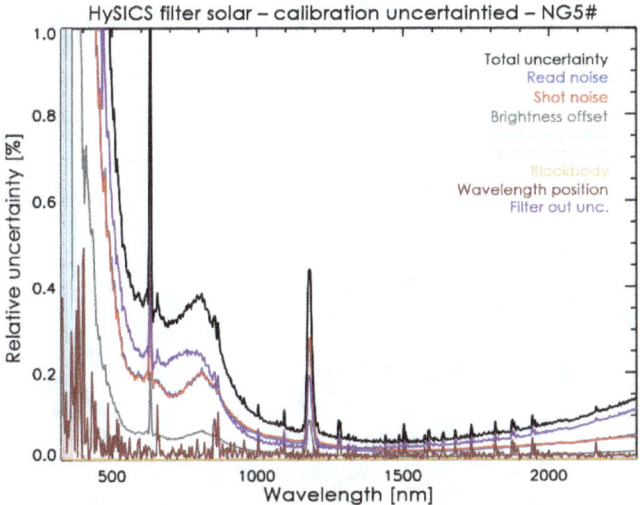

**Figure 11.** Calibration of NG5 #2 filter during Flight 2. Measurements with the filter out dominate the net uncertainty, as the spectrally flatter filter-in measurements could be done at a longer integration time to achieve a higher overall signal.

A globally averaged, all-sky estimate of Earth-reflected irradiance over the 8-year period from 2003 to 2010, based on results from observation-system simulation experiments generated using SCIAMACHY data (Y. S. Shea, personal communication, 2016), is plotted in Fig. 12 (gray) to indicate a typical, realistic reflected-solar (RS) spectrum observed by a spaceflight hyperspectral imager. Weighting the HySICS's net radiometric uncertainties by this estimated RS spectral-irradiance gives the resulting spectrally averaged radiometric uncertainties stated in the (black) figure text. These are higher than ultimately desired, which is largely caused by low instrument efficiencies and high flat-field uncertainties in the visible as well as increased aperture-ratio attenuation uncertainties near the visible-to-NIR transition. Improving these via the methods described in Sect. 4.3.1 and extending the multiple-image acquisition and a dual-scan approach to flat-field calibrations should reduce the weighted, Earth-reflected HySICS uncertainties for a future instrument by another $\sim 4 \times$ improvement over the values demonstrated here.

### 4.4.2  Conversion to physical units with SI-traceability

The high-quality data from Flight 2 with all instrument-level and attenuation-method corrections applied and with a final calibration factor, $C(\lambda)$ in Eq. (1), to provide an SI-traceable radiometric scale, enables the creation of three-dimensional spatial/spectral data cubes from ground scans that represent the end product of the HySICS solar cross-calibration technique. With sufficiently low uncertainties in the three factors in Eq. (1), this results in hyperspectral ground images with lower SI-traceable radiometric-accuracy uncertainties than existing flight instruments provide. The details of acquiring this final calibration factor are described in this sub-section.

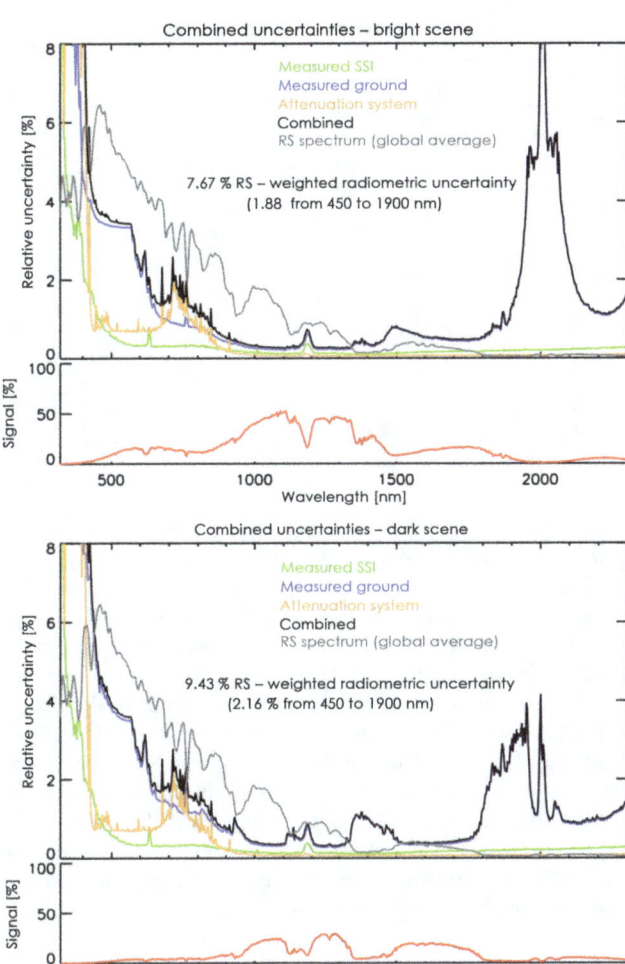

**Figure 12.** Contributions to net relative uncertainty (black) in the ratio of a bright (top) and dark (bottom) Earth scene relative to the HySICS-determined SSI during Flight 2 are shown as a function of wavelength. The small, inset lower plot (red) indicates the signal strength from each scene relative to the full scale of the instrument's FPA. Shown uncertainties are for individual pixels and could be reduced with spatial or spectral binning. Spectrally averaged uncertainties, being weighted by globally averaged reflected-solar (RS) irradiance (gray), are given for both the full (350 to 2300 nm) and partial (450 to 1900 nm) wavelength ranges. Demonstrating a minimum uncertainty of $\sim 0.3\,\%$ at wavelengths longer than 1000 nm indicates that the solar cross-calibration method used by the HySICS has promise of meeting the desired radiometric accuracies.

Multiple images as the solar disk is scanned in the cross-slit direction are spatially integrated to give a net spectral solar irradiance with corrections to account for the spectrometer's NIST-calibrated slit width as well as image overlap during the cross-slit scan. This irradiance is corrected for the diffraction and scatter described in Sect. 4.2.2 as well as other instrument effects described above. At this stage, the spectral "irradiance" is in units of instrument data numbers (DNs) and has no traceability to normal physical units. Figure 13 shows

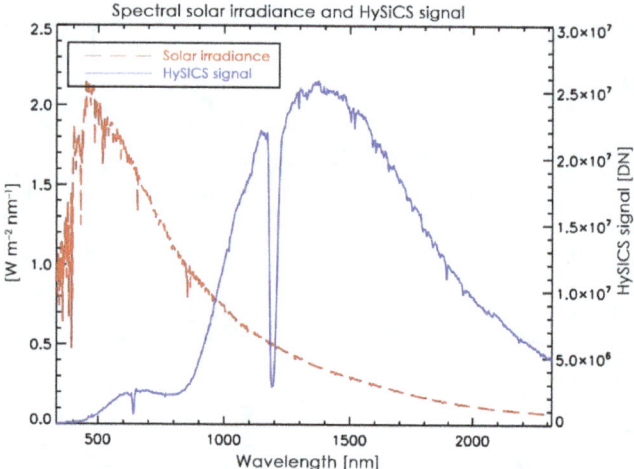

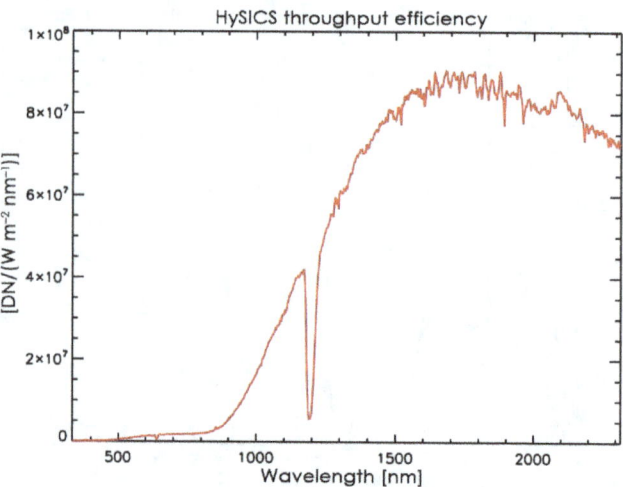

**Figure 13.** The HySICS signal from spatially integrated cross-slit solar scans with all applied instrument-level corrections gives an instrument-level spectral solar "irradiance" (blue). The values are given in instrument DNs and have no traceability to SI at this stage. The traceability and physical units are provided by scaling to the NRLSSI2 model for the day of the flight. These values, plotted in red, are adjusted to the Sun-instrument distance at the time of Flight 2 to correctly indicate the actual SSI that should be measured at the HySICS's location. These, or direct solar measurements from space-borne instruments having high absolute accuracies, enable the SI-traceable cross calibration of the HySICS-measured SSI. (Note that the DN values exceed the 16-bit maximum values from individual pixels because the plotted HySICS SSI signal is the spatially integrated sum of the entire solar disk.)

**Figure 14.** Dividing the HySICS-measured spectral solar irradiance by the modeled SSI in Fig. 13 gives the effective end-to-end sensitivity of the instrument with SI-traceability via correction factor $C(\lambda)$. This cross calibration can then be applied to Earth ground-scene observations after correcting for the HySICS's attenuations applied via factor $A(\lambda)$ to provide SI-traceable shortwave-reflected Earth radiances via Eq. (1).

the SSI determined from HySICS using a cross-slit scan of the solar disk during Flight 2.

By knowing what the actual SSI is, the HySICS instrument DNs can be converted to useful physical units and the overall instrument sensitivity can be determined. SSI values from Lean's NRLSSI2 model were applied to the HySICS Flight 2 data, since they were available prior to measurements from any on-orbit instrument. These values, which account for the solar activity state on that day, were adjusted from their as-provided 1-AU distance to the actual Earth–Sun distance on the date of the flight. They are plotted in Fig. 13 and provide the transfer to realistic physical units.

The ratio of this model-based "actual" SSI to the HySICS's measured irradiance in Fig. 13 gives the instrument sensitivity via a conversion from DNs to physical irradiance units via correction factor $C(\lambda)$, as shown in Eq. (2). This conversion factor is plotted in Fig. 14. The low instrument sensitivity in the visible is mainly the result of low efficiencies of the FPA and the grating at these wavelengths.

## 5   HySICS results: radiometrically calibrated data cubes

Applying the conversions, $A(\lambda)$ (correcting for the differences in attenuations between the solar measurements and the ground measurements) and $C(\lambda)$ (converting HySICS DNs into radiance units), to the HySICS-measured ground scenes, $S_{\mathrm{meas\_obj}}(\lambda)$ in Eq. (1), gives the resulting radiometrically calibrated ground scene shown in Fig. 15 for a single wavelength. This approach is applied at all HySICS wavelengths and provides a full three-dimensional data cube having SI-traceable radiometric accuracy, such as the example of a ground scene shown in Fig. 16 and the lunar scan in Fig. 17. Such radiometrically calibrated data cubes are the desired final products of the HySICS, and this improved-accuracy technique based on solar cross calibrations has now been successfully demonstrated via Flight 2.

## 6   Conclusions and spaceflight potential

Built under a NASA ESTO Instrument Incubator Program, the HySICS uses direct radiance measurements of the Sun to cross-calibrate hyperspectral images of other scenes, such as of the ground, Earth's atmosphere, or the Moon, with improved radiometric accuracies over similar instruments relying on indirect or diffused solar observations, on-orbit light sources, pre-launch calibrations, or measurements of vicarious ground sites. This measurement technique, utilizing three precisely characterized intensity-attenuation methods, enables direct in-flight calibrations relative to the spectral solar

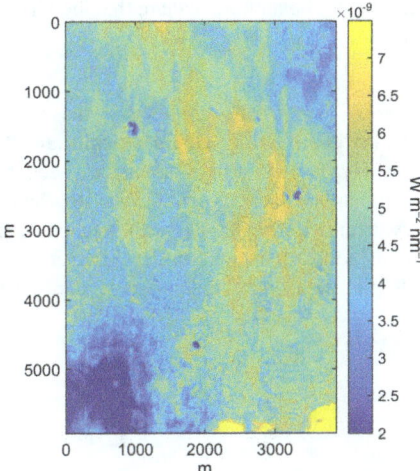

**Figure 15.** This two-dimensional spatial ground scene at 1233 nm from Flight 2 is radiometrically calibrated using the conversion values from Fig. 14.

**Figure 17.** Data cubes from HySICS's in-flight lunar scans provide spectral-radiance as well as spatially integrated irradiance measurements of the Moon (as viewed from the Earth) with improved radiometric accuracy than has as of yet been obtained from Earth-based measurements.

**Figure 16.** This representation of a three-dimensional data cube is from a ground-scene scan over mixed desert and water to show the generic HySICS data products of spatial/spectral imagery.

irradiance, which is a more stable and better-known reference than other space-based light sources, and allows SSI measurements to benchmark Earth ground scenes with radiometric accuracy and long-term precision greatly exceeding the capabilities of current space-based ground-imaging instruments. The demonstrated improvements from the HySICS were accomplished using the instrument's broadband optical design, which covers the entire reflected-solar spectrum and is based on a single flight-capable FPA. This design reduces mass, volume, power, and cost for air- or spaceflight instrumentation compared to multi-focal-plane designs intended to cover this broad spectral region.

The HySICS's solar cross-calibration methods have been applied to provide radiometrically accurate, SI-traceable spatial/spectral data cubes of ground scenes and the Moon from two high-altitude balloon flights. Using all three of its intensity-attenuation systems, the HySICS achieved net radiometric intensity reductions of $10^{-7.1}$, exceeding the $10^{-4.7}$ attenuation capability required to enable direct-view solar cross calibrations. Operating over the more limited – yet still easily sufficient – $10^{-6.2}$ attenuation-range provided by the HySICS's aperture-ratio and integration-time attenuation methods eliminates the need for the complexities of the HySICS's filter-based attenuation system for a spaceflight instrument. Although demonstrating all three intensity-attenuation methods, the second of two high-altitude balloon flights of the HySICS demonstrated a $\sim 2 \times$ improvement in radiometric accuracies over most existing spaceflight spectral imagers across a reflectance-weighted average of the visible and NIR and select spectral regions achieving a $\sim 6 \times$ improvement using only the aperture-ratio and integration-time attenuation methods. An additional radiometric-accuracy improvement of $\sim 4 \times$ could be expected for the existing instrument via better aperture-ratio lab-based calibrations and multiple in-flight flat-field calibrations. These two effects were identified as being the dominant contributors to radiometric uncertainties of ground scenes across most of the observed spectral region, and both would be relatively straightforward to improve in future measurements and calibrations. However, further improvements are expected via identified spectrometer-grating design changes.

Radiometric uncertainties based on the HySICS's solar cross-calibration approach were characterized as a function of wavelength for the balloon-flight data. The largest uncertainties were identified as being due to FPA and grating efficiencies in the visible, which cause dominant flat-field uncertainties as well as high shot and read noise, limited light sources used in laboratory calibrations of attenuations due to the aperture-ratio method, and the low lunar-signals during the times of the balloon flights. The quantified HySICS uncertainties were not limited by any intrinsic aspect of the solar cross-calibration approach, as demonstrated by the minimum pixel-level uncertainty of $\sim 0.3\%$ from a bright ground scene. This HySICS solar cross-calibration approach thus shows promise to ultimately achieve the $\sim 10 \times$ radiometric-accuracy improvements desired for future climate studies with the instrument design improvements that have been identified and will be implemented in future flight instruments.

Versions of the HySICS have been designed for accommodation on free-flyer spacecraft as well as the International Space Station, with either platform offering future instrument opportunities and the acquisition of scientifically valuable data. Studies are currently underway to manifest the HySICS on the CLARREO Pathfinder mission to improve spaceflight technology readiness, demonstrate the ability to achieve eventual CLARREO-mission climate-benchmark measurement requirements (Wielicki et al., 2013), and provide inter-calibrations of other on-orbit sensors (Roithmayr et al., 2014). The CLARREO Pathfinder/HySICS is planned for launch to the International Space Station in 2020.

*Author contributions.* All manuscript authors actively contributed to the HySICS or WASP calibrations, operations, and/or data analysis during and following Flight 2. Greg Kopp was the principal investigator of the HySICS and did the majority of the writing as the primary author of the manuscript. Second author Paul Smith completed the data analyses presented here and contributed directly to the writing, designed the flight operations software, and performed ground calibrations. Ginger Drake was the project manager and coordinated efforts between the HySICS team and the Columbia Scientific Balloon Facility for Flight 2. Joey Espejo was the optical designer and verified instrument performance prior to launch and during flight. Chris Belting, Zach Castleman, and Karl Heuerman designed the electrical and mechanical interfaces and control systems for the flight instrument, verified operation prior to launch, and performed ground calibrations. James Lanzi and David Stuchlik led the NASA/WFF WASP team that enabled the pointing accuracies needed for tracking the Sun and Moon and performed WASP operations during flight. The listed authors are most directly responsible for the results presented here, although many others at LASP and WFF contributed to the success of the HySICS balloon flights.

*Competing interests.* The authors declare that they have no conflict of interest.

*Acknowledgements.* We greatly appreciate the support of the NASA/WFF WASP team led by David Stuchlik and James Lanzi that enabled the HySICS pointing capabilities needed to demonstrated the instrument's solar cross-calibration approach.

This effort was funded by NASA's Earth Science Technology Office's Instrument Incubator Project under contract NNG04HZ05C as IIP-10-0019, and their enabling support is also greatly appreciated.

Edited by: M. Zribi

# References

Barnes, R. A. and Holmes, A. W.: Overview of the SeaWiFS ocean sensor, in: Sensor Systems for the Early Earth Observing System Platforms, edited by: Barnes, W. L., Proc. SPIE, 224–232, 1993.

Barnes, R. A. and Zalewski, E. F.: Reflectance-based calibration of SeaWiFS. II. Conversion to radiance, Appl. Optics, 42, 1648–1660, doi:10.1364/AO.42.001648, 2003.

Beiso, D.: Overview of Hyperion On-Orbit Instrument Performance, Stability, and Artifacts, aipr, p. 95, 31st Applied Imagery Pattern Recognition Workshop, Los Alamitos, CA, doi:10.1109/AIPR.2002.1182260, 2002.

Espejo, J., Belting, C., Drake, G., Heuerman, K., Kopp, G., Lieber, A., Smith, P., and Vermeer, B.: A Hyperspectral Imager for High Radiometric Accuracy Earth Climate Studies, SPIE Proc., edited by: Shen, S. S. and Lewis, P. E., 8158, doi:10.1117/12.893803, 2011.

Green, R. O., Eastwood, M. L., Sarture, C. M., Chrien, T. G., Aronsson, M., Chippendale, B. J., Faust, J. A., Pavri, B. E., Chovit, C. J., Solis, M., and Olah, M. R.: Imaging Spectroscopy and the Airborne Visible/Infrared Imaging Spectrometer (AVIRIS), Remote Sens. Environ., 65, 227–248, 1998.

Guenther, B., Barnes, W., Knight, E., Barker, J., Harnden, J., Weber, R., Roberto, M., Godden, G., Montgomery, H., and Abel, P.: MODIS Calibration: A brief review of the strategy for the at-launch calibration approach, J. Atmos. Ocean. Tech., 13, 274–285, 1996.

HyspIRI Mission Concept Team: HyspIRI Comprehensive Development Report, p. 11, prepared for NASA, Jet Propulsion Lab, California Inst. of Tech., Pasadena, CA, 2015.

Janesick, J. R.: Scientific Charge-Coupled Devices, SPIE, Bellingham, WA, 920 pp., 2001.

Kopp, G., Pilewskie, P., Belting, C., Castleman, Z., Drake, G., Espejo, J., Heuerman, K., Lamprecht, B., Smith, P., and Vermeer, B.: Radiometric Absolute Accuracy Improvements for Imaging Spectrometry with HySICS, IGARSS 2013, Melbourne, Australia, 3518–3521, 2013.

Kopp, G., Belting, C., Castleman, Z., Drake, G., Espejo, J., Heuerman, K., Lamprecht, B., Lanzi, J., Smith, P., Stuchlik, D., and Vermeer, B.: First results from the HyperSpectral Imager for Climate Science (HySICS), Proc. SPIE 9088, Algorithms and Technologies for Multispectral, Hyperspectral, and Ultraspectral Imagery XX, doi:10.1117/12.2053426, 2014.

Lukashin, C., Jin, Z., Kopp, G., MacDonnell, D. G., and Thome,

K.: CLARREO Reflected Solar Spectrometer: Restrictions for Instrument Sensitivity to Polarization, IEEE T. Geosci. Remote Sens., 53, 6703–6709, doi:10.1109/TGRS.2015.2446197, 2015.

NRC: Earth Science and Applications from Space: National Imperatives for the Next Decade and Beyond, National Academy Press, 428 pp., Washington, DC, 2007.

Pearlman, J., Segal, C., Liao, L., Carman, S., Folkman, M., Browne, B., Ong, L., and Ungar, S.: Development and Operations of the EO-1 Hyperion Imaging Spectrometer, Earth Observing Systems V, edited by: Barnes, W. L., Proc. SPIE, 4135, p. 243, 2000.

Roithmayr, C. M., Lukashin, C., Speth, P. W., Kopp, G., Thome, K., Wielicki, B. A., and Young, D. F.: CLARREO Approach for Reference Inter-Calibration of Reflected Solar Sensors: On-Orbit Data Matching and Sampling, IEEE T. Geosci. Remote Sens., 52, 6762–6774, doi:10.1109/TGRS.2014.2302397, 2014.

Shirley, E. L., Kacker, R. N., and Datla, R. V.: Diffraction Corrections in Radiometry: A Proposed Method to Estimate Uncertainties, Proceedings of the Measurement Science Conference, NIST, Gaithersburg, MD, 2002.

Smith, P., Drake, G., Espejo, J., Heuerman, K., and Kopp, G.: A Solar Irradiance Cross-Calibration Method Enabling Climate Studies Requiring 0.2 % Radiometric Accuracies, ESTF 2011, June 2011.

Stuchlik, D.: NASA Fact Sheet: Precision Arc-Second Pointing on Balloons Using WASP, FS-2015-8-329-WFF, available at: https://sites.wff.nasa.gov/balloons/docs/outreach/WASPFacts.pdf (last access: March 2017), 2015a.

Stuchlik, D.: The Wallops Arc Second Pointer – A Balloon Borne Fine Pointing System, American Institute of Aeronautics and Astronautics, Aviation 2015, BAL-02, 2150906, Session: BAL-02, Balloon Systems II, AIAA-2015-3039, 2015b.

Wielicki, B. A., Young, D. F., Mlynczak, M. G., Thome, K. J., Leroy, S., Corliss, J., Anderson, J. G., Ao, C. O., Bantges, R., Best, F., Bowman, K., Brindley, H., Butler, J. J., Collins, W., Doelling, D. R., Dykema, J. A., Feldman, D. R., Fox, N., Holz, R. E., Huang, X., Huang, Y., Jennings, D. E., Jin, Z., Johnson, D. J., Jucks, K., Kato, S., Kirk-Davidoff, D. B., Knuteson, R., Kopp, G., Kratz, D. P., Liu, X., Lukashin, C., Mannucci, A. J., Phojanamongkolkij, N., Pilewskie, P., Ramaswamy, V., Revercomb, H., Rice, J., Roberts, Y., Roithmayr, C. M., Rose, F., Sandford, S., Shirley, E. L., Smith, W. L., Soden, B., Speth, P. W., Sun, W., Taylor, P. C., Tobin, D., and Xiong, X.: Achieving Climate Change Absolute Accuracy in Orbit, B. Am. Meteorol. Soc., 94, 1519–1539, doi:10.1175/BAMS-D-12-00149.1, 2013.

Xiong, X., Che, N., and Barnes, W.: Terra MODIS on-orbit spatial characterization and performance, IEEE T. Geosci. Remote Sens., 43, 355–365, doi:10.1109/TGRS.2004.840643, 2005a.

Xiong, X., Che, N., and Barnes, W. L.: Five years of Terra MODIS on-orbit spectral characterization, Proc. SPIE – Earth Observing Systems X, 5882, doi:10.1117/12.614090, 2005b.

Xiong, X., Erives, H., Xiong, S., Xie, X., Esposito, J., Sun, J., and Barnes, W.: Performance of Terra MODIS solar diffuser and solar diffuser stability monitor, Proc. SPIE – Earth Observing Systems X, 5882, doi:10.1117/12.615334, 2005c.

Xiong, X., Butler, J., Chiang, K., Efremova, B., Fulbright, J., Lei, N., McIntire, J., Oudrari, H., Sun, J., Wang, Z., and Wu, A.: VIIRS on-orbit calibration methodology and performance, J. Geophys. Res.-Atmos., 119, 5065–5078, doi:10.1002/2013JD020423, 2014.

# 5

# Continuous wavelet transform and Euler deconvolution method and their application to magnetic field data of Jharia coalfield, India

**Arvind Singh and Upendra Kumar Singh**

Department of Applied Geophysics, Indian Institute of Technology (Indian School of Mines),
Dhanbad, Jharkhand 826004, India

*Correspondence to:* Arvind Singh (arvindsinghgps@gmail.com)

**Abstract.** This paper deals with the application of continuous wavelet transform (CWT) and Euler deconvolution methods to estimate the source depth using magnetic anomalies. These methods are utilized mainly to focus on the fundamental issue of mapping the major coal seam and locating tectonic lineaments. The main aim of the study is to locate and characterize the source of the magnetic field by transferring the data into an auxiliary space by CWT. The method has been tested on several synthetic source anomalies and finally applied to magnetic field data from Jharia coalfield, India. Using magnetic field data, the mean depth of causative sources points out the different lithospheric depth over the study region. Also, it is inferred that there are two faults, namely the northern boundary fault and the southern boundary fault, which have an orientation in the northeastern and southeastern direction respectively. Moreover, the central part of the region is more faulted and folded than the other parts and has sediment thickness of about 2.4 km. The methods give mean depth of the causative sources without any a priori information, which can be used as an initial model in any inversion algorithm.

## 1 Introduction

One of the fundamental issues in exploration geophysics is to detect differences in susceptibility and density between rocks that contain ore deposits, hydrocarbons or coal. These differences are reflected in the gravity and magnetic anomalies and also delineation of structural features, which are interpreted using several techniques (Blakely and Simpson, 1986). One of the most important objectives in the interpretation of po-

tential field data is to improve the resolution of the underlying source, delineating a lateral change in magnetic susceptibilities that provides information not only on lithological changes but also on structural trends. The edge detection techniques are used to distinguish between different sizes and different depths of the geological discontinuities (Cooper and Cowan, 2006, 2008; Perez et al., 2005; Ardestani, 2010; Hsu et al., 1996; Hsu, 2002; Holschneider et al., 2003). The derivatives of magnetic data are used to enhance the edges of anomalies and improve significantly the visibility of such features.

Gravity and magnetic signature infer that there is a dominance of sediment over Jharia coalfield (Verma et al., 1973, 1976, 1979). Thus the difference between the depths estimated using the Euler deconvolution method (EDM) (Thompson, 1982; Reid et al., 1990) and tilt depth method (TDM) (Salem et al., 2007; Cooper, 2004, 2011) may help to detect the thickness of the coal bed. Wavelet transform and EDM have been theoretically demonstrated on magnetic data. These methods provide source parameters such as the location, depth, geometry of geological bodies and interfaces in an easy and effective way. However, it may be more difficult to characterize the source properties in cases of extended sources (Sailhac et al., 2009).

Jharia coalfield in Dhanbad, India, forms an east–west trending belt of Gondwana basin, Damodar valley, in the northeastern part of India. This study region is the most coal-rich area of Gondwana basin. Analysis of Jharia coalfield suggests that the magnetic anomalies provide encouraging results which are well correlated with available gravity data and some borehole information.

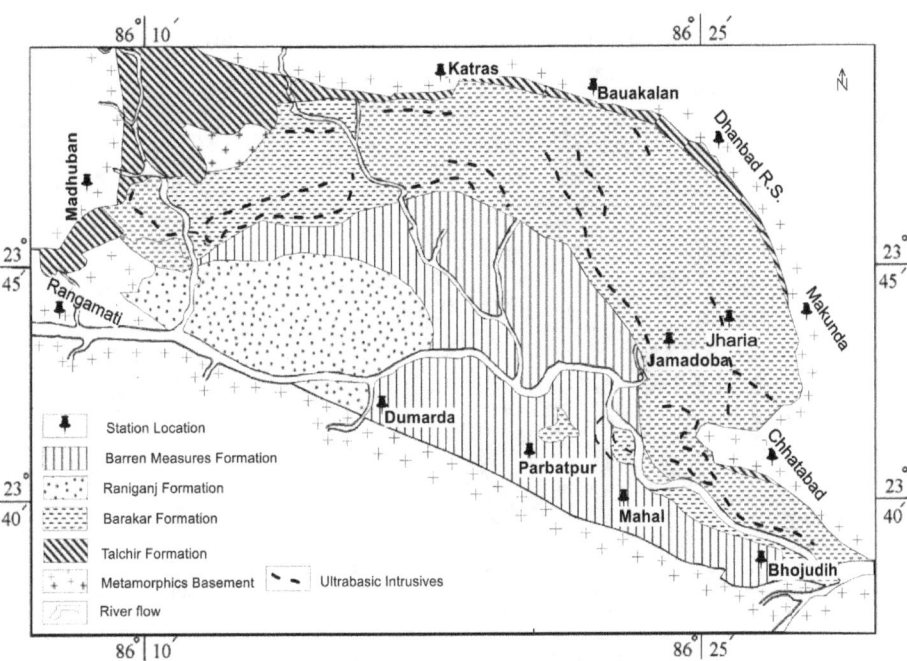

**Figure 1.** Geological map of Jharia coalfield and surrounding regions (Verma et al., 1979).

## 2 Geology of Jharia coalfield

Geology of the Jharia coal basin is shown in Fig. 1. The basin has been formed because of crustal subsidence during Gondwana periods (Fox, 1930). The coalfield has an extension along the east–west in Gondwana basin of Damodar valley in northeastern India. Gondwana basin is surrounded by crystalline gneisses of several categories from all directions. Sedimentary strata have inclination away from the gneiss contact in this region. The sedimentary strata include the rocks which belong to the Talchir series, Raniganj series, Barren Measures formation and Barakar series (Verma et al., 1979). Raniganj, Barakar and Talchir series, including Barren Measures formation, cover areas of about 58, 218 and 181 km$^2$ respectively. Various formations are shown in the Fig. 1.

Talchir and Barakar formations rest over the northern margin and dip towards the southern margin. The Barakar series covers the northern half of this coalfield. It produces one of the best quality coal in India. An elliptical outline is formed by the Raniganj formation in southwestern region of the coalfield. Geology of the Jharia coalfield has been divided into many blocks, such as the Parbatpur, Mahuda, Jarma and Moonidih blocks. There are many faults over the Jharia coalfield. A normal tensional fault exists over the southern boundary. In the southwestern part of the basin, Damodar river (Fig. 1) flows very close to the southern boundary fault (Verma et al., 1973, 1979; Verma and Ghosh, 1974).

The magnetic data were obtained from Verma et al. (1979) to study the region. We prepared the total magnetic anomaly map of magnetic data of this province as shown in Fig. 2. Magnetic anomaly variations are very smooth over the basin and irregular over Precambrian outcrops. This variation may be affected by the difference in magnetic susceptibility, weathering of the outcrop, magnetization of the outcrop by lightening, etc. At the northern portion of the basin anomalies form a semi-circular arc and are parallel to the southern boundary fault. There is no clear indication of the anomaly at the southern boundary because of uneven basement and faulting associated with Patherdih horst. So it is clear that this portion of the basin is highly folded/faulted and coal seams have been highly deformed. A noticeable part of the magnetic anomaly is the presence of major anomalous sources which are ascribed to some features within the Precambrian basement's underlying sediments.

## 3 Methodologies

### 3.1 Continuous wavelet transform (CWT)

The continuous wavelet transform is the conversion of any signal into a matrix made of sum scaler products in Fourier space. Wavelet transform method for potential field has been established by Moreau et al. (1997, 1999). This method was previously used for homogeneous, isolated and extended potential field sources (Sailhac et al., 2009). Chamoli et al. (2006), Cooper (2006), Goyal and Tiwari (2014) and Singh and Singh (2015) used wavelet transform method on various synthetic as well as field data. This method uses a Poisson group of wavelets as a mother wavelet in order to interpret the potential field data. To analyze the signal by mother wavelet, a wavelet domain signal is decomposed into the orthogonal wavelets of finite duration. The CWT coeffi-

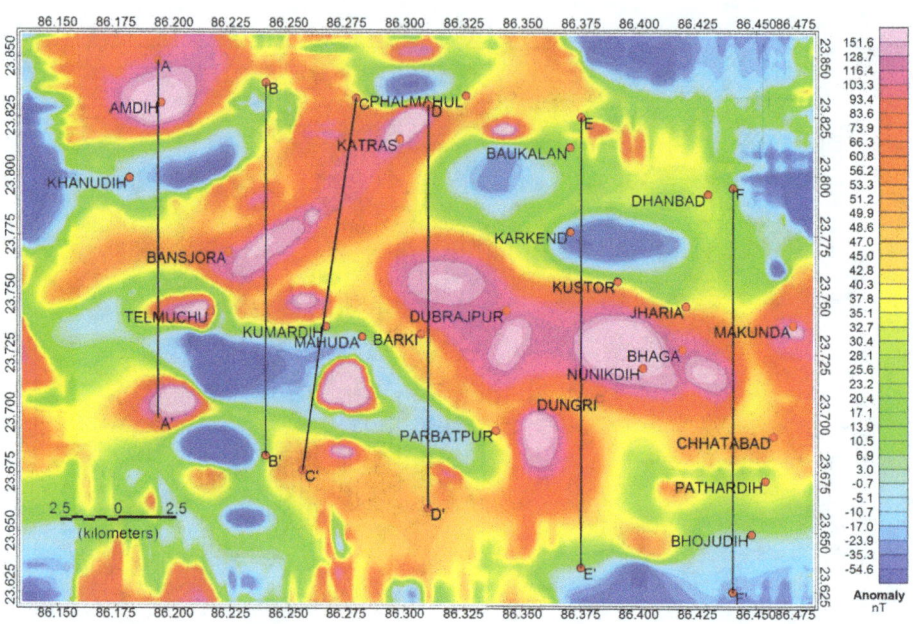

**Figure 2.** Total magnetic field anomaly (nT) map and location of the profiles over Jharia coalfield and surrounding regions (Verma et al., 1979).

cient $W_t$ of a measured potential $t(x)$ is defined as the convolution product.

$$W[\psi, t](p, o) = \int_{R^n} \frac{1}{o^n} t(x) \psi \left[ \frac{p - x}{o} \right] dx, \qquad (1)$$

$$W[\psi, t](p, o) = (D_o \psi * t)(p), \qquad (2)$$

where $\psi(x \in R^n)$ is the wavelet to be analyzed, $x$ denotes the abscissa along the particular profile line, $t(x)$ indicates the potential field (gravity or magnetic anomaly) and $(o \in R^+)$ and $p$ are the dilation and position parameter respectively. Dilation parameter allows the analyzed wavelet to act as a band pass filter. Dilation operator $D_o$ can be termed as

$$D_o \psi(x) = \frac{1}{o^n} \psi \left( \frac{x}{o} \right). \qquad (3)$$

Dilation $D_o$ fulfils two properties given below.

$$W[\psi, D_\lambda t](p, o) = \frac{1}{\lambda^n} W[\psi, t] \left( \frac{p}{\lambda}, \frac{o}{\lambda} \right) \qquad (4)$$

Equation 4 states one of the main mathematical asset of the wavelet transform, i.e., covariance of wavelet transforms with respect to the dilation. The homogeneous function $t$ of degree $\sigma \in R$ can be defined as

$$t(\lambda, x) = \lambda^\sigma t(x) \forall \lambda > 0. \qquad (5)$$

After correlation, Eqs. (4) and (5) result in the homogeneous function (i.e., by recalling $\sigma = -n$ and $\sigma = 0$ respectively)

$$(\lambda p, \lambda o) W[\psi, t] = \lambda^\sigma W[\psi, t](p, o). \qquad (6)$$

Equation (6) shows that wavelet transform of a homogeneous function is analogous to dilation and scale of any function $W(\psi, t)(p, o = \text{consant})$ of the wavelet transform. Moreau et al. (1999) suggest that the combinations of straight lines create a cone-like outline at the location where $\left( \frac{\partial^m}{\partial p^m} \right) W(\psi, t)(p, o) = 0$ and the apex of the outline is the center of homogeneity of the analyzed function. The outlines in Fig. 3 fulfils the condition $\left( \frac{\partial^m}{\partial p^m} \right) W(\psi, t)(p, o) = 0$ and are known as edges of wavelet transform or modulus maxima lines.

Potential field signal analyzed by CWT allows for estimation of depth and homogeneous distribution order of the source generating the analyzed signal. Source depth is calculated through the intersection of the converging extrema lines (Fig. 3). In addition to this, Moreau et al. (1997, 1999) established the Poisson semi-group kernel $K_o(x)$ that allows us to carry on the harmonic field $t(x, z)$ from level $z$ to the level $z + o$, which is expressed as upward continuation (Bhattacharyya, 1972).

$$P_o(x) = \frac{o}{\pi} \left( \frac{1}{o^2 + x^2} \right) \qquad (7)$$

For wavelet analysis, let us consider a local homogeneous source $x = 0$, with depth $z = z_\alpha$, of a potential field $t(x, z = 0)$. Moreau (1999) stated that the wavelet coefficients of positions and dilations that lie in the upper half plane follow a twice scaling rule with two exponent parameters. Moreau (1997) explained the relationship between wavelet coefficients at two altitudes and for any wavelets of homoge-

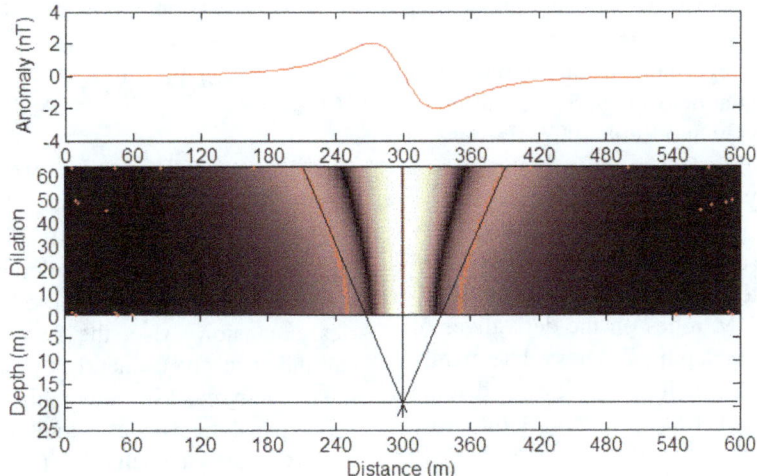

**Figure 3.** Synthetic magnetic anomaly of isolated extended source and depth estimation by wavelet transform for a Poisson wavelet for $\gamma = 1$ with mathematical expression $k(x) = -[x(2/\pi)]/(1+x^2)^2$.

neous sources as

$$W[\psi, t](p, o) = \left(\frac{o}{o'}\right)^{\gamma} \left(\frac{o' + z_\alpha}{o + z_\alpha}\right)^{\beta} W\left(p\frac{o' + z_\alpha}{o + z_\alpha}, o'\right), \quad (8)$$

where $\beta = \gamma - \sigma - 2$ indicates the holder exponent, $o$ and $o'$ denote different altitudes and $Z_\alpha$ signifies the depth of the causative source. Equations (6) is similar to Eq. (8), with the additional term $Z_\alpha$ in both the dilation and scaling factors on the right-hand side resulting in geometrical conversion. Due to geometrical conversion the cone-like outline joins at source depth because of the negative dilation $o = z_\alpha$. Therefore, the Poisson group of wavelets used on the potential field demonstrates modest assets and can be applied to find the causative source without any prior information. CWT helps to detect the edge of the formations of the extended body. Also, it offers quick and consistent results about extended and isolated source depth with location. Wavelet analysis plays a key role in depth estimation of potential field. When order of $\gamma$ increases, the obtained source depth appears shallower. For $\gamma = 1$, outlines of the cone have the point of intersection at the barycenter of the prismatic source. CWT can resolve the noisy and nonstationary dataset very well (Moreau, 1997, 1999) and magnetic data can also be analyzed without any reduction to pole.

### 3.2 Euler deconvolution method

Euler deconvolution was first developed for the interpretation of magnetic profile data by Thompson (1982), and later Reid et al. (1990) extended its approach to gridded magnetic data. Reid et al. (1990) developed the special case for the magnetic field of a contact of finite depth extent and coined the term "Euler deconvolution". Klingele et al. (1991) and Zhang et al. (2000) used it over vertical gravity gradient and tenser gravity gradient respectively. Moreover, it has been

generalized by Mushayandebvu et al. (2001, 2004), and Ravat (1996) further investigated the wider range of source nature by this method. Since then, it has been adapted and improved by Keating (1998) to interpret the gravity data. EDM makes rapid depth estimations from magnetic and gravity data in grid form using Euler's homogeneity relation (Thompson, 1982; Reid et al., 1990; Barbosa et al., 1999). Euler deconvolution is insensitive to magnetic inclination, declination and remanent magnetization and is very suitable for 3-D analyses (Keating, 1998; Mushayandebvu et al., 2004; Stavrev and Reid, 2007; Melo et al., 2013, Silva, et al., 2001).

The global acceptance of Euler deconvolution is mainly due to its simplicity of implementation and use, making it the tool of choice for a quick and reliable interpretation of potential field data (FitzGerald et al., 2004; Gerovska and Arauzo Bravo, 2003) and for finding the source information in terms of depth and geological structure. Euler deconvolution uses three orthogonal gradients of any potential quantity as well as the potential quantity itself to determine depths and locations of a source body. This method primarily responds to the gradients in the data and effectively traces the edge and defines the depth of the source body. Reid et al. (1990) and Thompson (1982) defined the 3-D Euler equation as

$$(x - x_0)\frac{\mathrm{d}F}{\mathrm{d}x} + (y - y_0)\frac{\mathrm{d}F}{\mathrm{d}y} + (z - z_0)\frac{\mathrm{d}F}{\mathrm{d}z} + NF = 0, \quad (9)$$

where $(x_0, y_0, z_0)$ is the location of magnetic source whose total magnetic field $(F)$ is observed at $(x, y, z)$. The values $\frac{\mathrm{d}F}{\mathrm{d}x}$, $\frac{\mathrm{d}F}{\mathrm{d}y}$ and $\frac{\mathrm{d}T}{\mathrm{d}z}$ are the measured magnetic gradients along the $x$, $y$ and $z$ directions. Euler deconvolution adds an extra dimension to the interpretation. It estimates a set of $(x, y, z)$ points that, ideally, fall inside the source of the anomaly. Euler deconvolution requires the $x$, $y$ and $z$ derivatives of the data and a parameter called the structural index (SI), $N$ ($N$

is a nonnegative integer). SI defines the anomaly attenuation rate at the observation point and depends on the geometry of the source. The SI is an integer number that is related to the homogeneity of the potential field and varies for different fields and source types (Stavrev and Reid, 2007; Barbosa et al., 1999; and Melo et al., 2013). For example, in the case of total field magnetic anomaly data, a dyke is represented by an SI of 1, whereas a sphere is represented by an SI of 3.

The source points that are calculated as solutions by EDM are positioned at the estimated edge of the susceptibility inhomogeneities. Thus, the EDM relies on the derivatives of the magnetic data; the resulting depth estimates relate mainly to the areas of basement heterogeneities identified as distinct sources of the field. The first vertical gradient of magnetic data is calculated by using the fast Fourier transform (FFT) method (Gunn, 1975). The vertical and horizontal derivatives of the first vertical gradient, essential for the calculation of Eq. (9), are also been calculated using the FFT method. The horizontal source locations from EDM solutions can be used to explain of lithological and structural trends. A location in the map where these solutions tend to cluster is considered to be the most probable location of the source.

Equation (9) can be explained in terms of least squares to estimate the source coordinates and structure. Since the absolute value anomalous field ($F$) is barely identified, Eq. (9) cannot be used directly over the observed data. Moreover, according to Thompson (1982) Eq. (9) does not explain the regional or background magnetic field due to adjacent source, so obtained solutions may be unreliable and may vary from their accurate location.

For the 2-D model, total magnetic field ($F$) and its derivatives at all points of observations provide the linear equation with unknown coordinates ($x_0$ and $z_0$), where $x_0$ and $z_0$ represent location and depth of the magnetic source, respectively.

Using the Taylor series, an unidentified regional field ($E$) can be described as

$$E(x, y) = E_0 + x\frac{\partial E}{\partial x} + y\frac{\partial E}{\partial y} + K(2), \tag{10}$$

where $E_0$ and $K(2)$ represent the constant background for definite window and other higher-order values in Taylor series expansion. The resultant anomalous field ($F$) can now be specified as the difference between the observed magnetic field ($O$) and regional magnetic field ($E$).

$$F = O - E \tag{11}$$

Now, after revision, modified Euler equation can be specified as

$$O \equiv (x - x_0)\frac{d(O - E)}{dx} + (y - y_0)\frac{d(O - E)}{dy}$$
$$+ (z - z_0)\frac{d(O - E)}{dz} + N(O - E) = 0. \tag{12}$$

According to Thompson (1982), Silva and Barbosa (2003) and Reid et al. (1990), Euler equation provides satisfactory results by considering the first-order term in Taylor series expansion. Also, the Euler equation becomes nonlinear and is resolved linearly by supposing tentative values of the SI (Stavrev, 1997). The higher-order term of Taylor series expansion provides the solution when singular points are closely spaced to each other (e.g., in the case of the multiple fracture or sill). In this case postulation of linear background discontinues and needs higher-order terms of Taylor series expansion for a reasonable result.

Dewangan et al. (2007) and Gerovska and Arauzo Bravo (2003) chose the second-order terms of the Taylor series expansion and favor a procedure of rational calculation in which the infinite Taylor series expansion is estimated by two polynomials (one lies in the numerator and other one in the denominator). Kopal (1961) suggested that the maximum accuracy in rational calculation may be possible when the polynomials of the numerator and denominator hold the same power. The rational function is used to calculate the background; this function can be defined as

$$E(x, y) = \left(\frac{E_0 + ax + by}{1 + cy + dy}\right), \tag{13}$$

where $a, b, c, d$ and $E_0$ are the unknown parameters. Comparison of the values of Eqs. (13) and (12) generates another nonlinear Euler equation which provides the source depth, location and structural index (Coleman and Li, 1996; Williams et al., 2003). All the variation on Euler deconvolution includes working through profiles as well as gridded datasets using a moving window (each window position is a set of linear equations which generate the solution to locate the source in plan and depth). The advantage of this method is that source magnetization direction and its result are not affected by the presence of remanence (Ravat, 1996). Moreover, it can be further used as an inversion algorithm and the design rules based on mathematical analysis proposed by Reid et al. (2014) must be considered to analyze the potential field (gravity and magnetics).

## 4 Application of CWT to synthetic magnetic anomaly

The synthetic examples demonstrate the application of the CWT technique on the magnetic anomaly due to isolated and extended homogeneous magnetic sources at 300 m, with depth about 20 m. The first analysis (shown in Fig. 3) corresponds to the magnetic anomaly of a finite length vertical

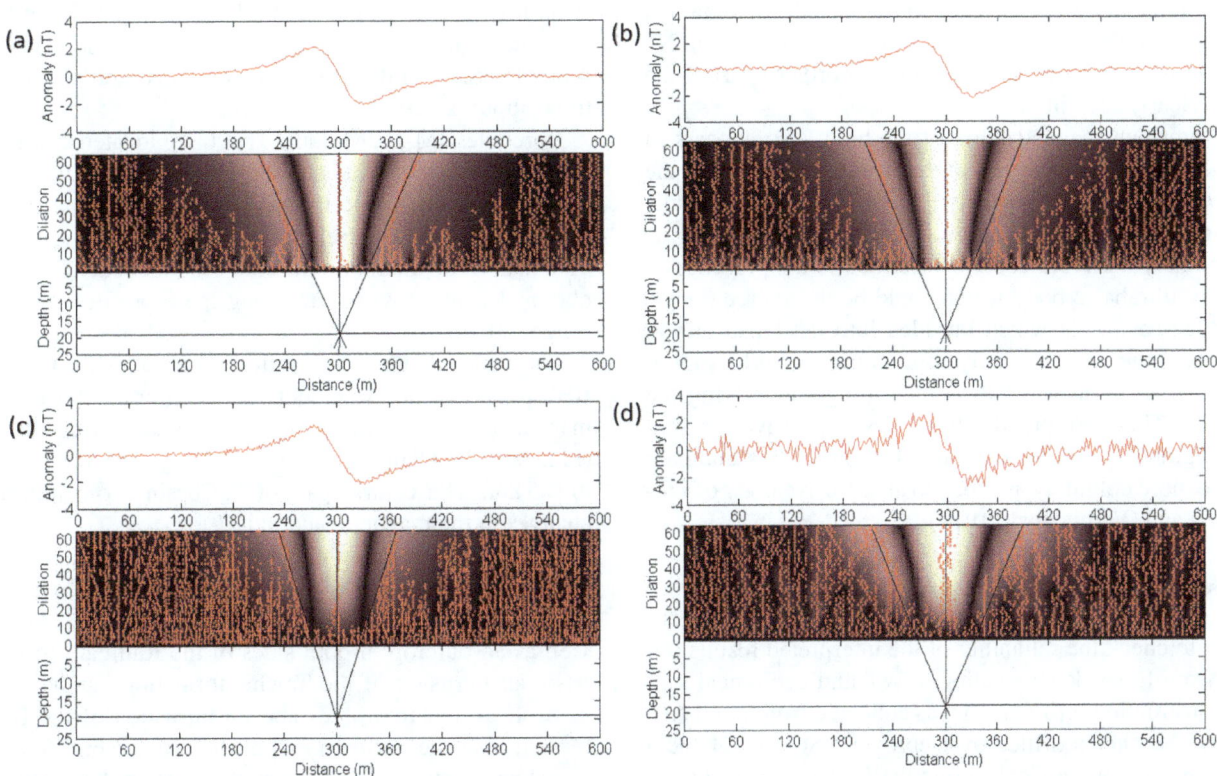

**Figure 4.** **(a)** Magnetic anomaly with 1 % random noise; **(b)** magnetic anomaly with 2 % random noise; **(c)** magnetic anomaly with 5 % random noise; **(d)** magnetic anomaly with 10 % random noise.

dipole. The wavelet coefficients of the magnetic field due to vertical dipole computed with the help of wavelet are shown in this figure (for horizontal derivative $\gamma = 1$), which shows a cone-like structure. Wavelet transform of the potential field due to homogeneous source follows a geometrical property which allows an easy estimation of source depth and location. The examples demonstrated could correspond to the zero remanent magnetization with all magnetization being induced. To understand the behavior of the modulus maxima of CWT over the magnetic anomaly due to the anomalous sources, the CWT is presented for various field examples. The converging point of ridges gives depth and location of the vertical dipole.

The wavelet coefficients are computed by applying CWT to the anomaly. Figure 4 shows the calculated values of CWT coefficients for different dilations (1–64.5) of magnetic anomaly. The maxima of modulus of CWT provide cone-like structures and are clearly shown pointing towards the position of the upper corner of the model. Whereas an approximate horizontal location has been estimated, an intersection of modulus maxima lines in the subsurface has been placed below the base line ($a = 0$) to mark the depth of the source, where $a$ is dilation.

Also, this example illustrates the application of wavelet transform to potential fields (horizontal derivative, $\gamma = 1$) where modulus maxima lines make a cone-like shape, and

ridges of the cone join below the base line or to homogeneity center of the source, where $y$ scale represents the dilation. The point where ridges join marks the depth and location of the vertical dipole. It is detected that the homogeneous source retains a geometrical possession after execution of wavelet transform on potential field. This makes a straightforward interpretation about depth and location of causative body. In order to perform wavelet analysis on field data, it has been tested on noisy data with 1, 2, 5 and 10 % random noise in the potential source data obtained because of vertical dipole (Fig. 4a–d). It is clear that wavelet analysis provides the exact depth and location of the source. When the noise level is low then it is easy to find the cone-like structure where the modulus maxima lines cross each other (Fig. 4a–b). As the noise level increases it is difficult to find the cone-like structure made by the cross section of modulus maxima line (Fig. 4c–d).

## 5   Application of CWT to magnetic field anomaly from Jharia coalfield

CWT and EDM are applied on field magnetic anomaly collected from Jharia coalfield and surrounding regions in Dhanbad, India. For CWT analysis, six profiles (AA′, BB′, CC′, DD′, EE′ and FF′) have been selected that cover the entire coalfield. These anomalies can be adequately explained by assuming an underlying body with susceptibility contrast with respect to its surroundings and which is polarized in N–S direction. The positive anomaly in the northern part of the basin is clearly seen in the profile.

The remanent magnetization of the body also appears to contribute to the anomaly. It is interesting to note that in the region of this magnetic anomaly a number of dykes and sills are found as intrusive into the sediments as shown in Fig. 1. This anomaly therefore could be ascribed to the presence of a basic or ultrabasic body which could be the source for the basic dykes and sills which intruded into the basin during Gondwana times. Alternatively, this anomaly could also represent a basic intrusive of Precambrian age underlying the sediments. There are practically no basic intrusives present in the region of positive anomaly. Therefore, this anomaly could be more definitely ascribed to an intrusive body of Precambrian age (Verma et al., 1979).

## 6   Results and discussion

In order to check the reliability of the interpreted results obtained from Euler deconvolution, CWT and geological sections, construction information was collected from published results of boreholes drilled by Geological Society of India (GSI), Bharat Coking Coal Limited (BCCL), National Coal Development Corporation (NCDC) and Central Mines Planning and Design Institute (CMPDI). Therefore, the depth to the basement configuration inferred from gravity data as well as drilled borehole information is discussed below.

Jharia coalfield and surrounding areas have been considered to estimate the source depths on the basis of technique of intersections of modulus maxima lines of CWT. The mean depths of causative sources along the profile AA′ (passes east of the Khanudih and west of the Telmuchu and Bansjora region through Amdih over the westernmost part of the Jharia coalfield, shown in Fig. 2) calculated from the CWT (Fig. 5a) and Daubechies' wavelet method (Fig. 5b) vary from 0.2 to 0.45 km. Profile AA′ shows that there is fault near the northwestern part of the basin.

Magnetic field inclination, declination and azimuth angle (clockwise from true north) of this profile are 36.44, −0.11 and 268.48° respectively. The anomaly about 77 nT between boreholes JM-4 and JK-26 has been observed because of a number of basic intrusive bodies belonging to Satpura cycle that exist over the area. Jharia coalfield consists of peridotites in the form of sills as well as dykes. Dolerite dykes are very common in the western part of this coalfield.

The central part shows a flat sedimentary region and the magnetic anomaly shows a high value on either side of the profile. Raniganj formation exists on the southern side whereas Talchir formation exists on the northern side of this profile. However, the Barren Measures and the Barakar formation lie between the Raniganj and the Talchir formations. There is an intrusion of Archean metamorphics in Talchir formation which appears as an outcrop over the surface near Amdih (Fig. 5c). Some of the boreholes provide information about the metamorphics along this profile. The maximum thickness of the sediment along this profile is observed to be about 0.8 km.

Boreholes JM-1, JM-4 and JK-26 are located close to this profile, which touches metamorphics at a depth of about 0.4, 0.55 and 0.3 km respectively. These boreholes are located west of Bansjora and Telmuchu. The depth to the basement obtained from magnetic data is nearly equal to the depth obtained from gravity data along these profiles (Singh and Singh, 2015).

The mean depths of causative sources along the profile BB′ (passes east of Telmuchu and Bansjora and west of Kumardih region, shown in Fig. 2) calculated from the CWT (Fig. 6a) and Daubechies' wavelet (Fig. 6b) vary from 1.3 to 2.5 km. The central part of the basin shows the abrupt changes in the magnetic anomaly.

Profile BB′ illustrates about the Barakar, Raniganj, Talchir formations and Barren Measures. The Barren Measures is found between the Barakar and Raniganj formations and seen as an outcrop in both sides of the Raniganj formation. Also, an intrusion of the Talchir formation has been found in Archean metamorphics and an intrustion of the Barakar formation at the northern end of the profile (Fig. 6c). There is a sloppy nature of each formation below the profile from both ends. The major portion of this area is dominated by the Raniganj and the Barakar formations. The estimated thickness of the sediments is about 2.3 km over the Raniganj formation.

Magnetic field inclination, declination and azimuth angle of this profile are 36.42, −0.11 and 268.5° respectively. This profile passes through two faults between metamorphics and sediment: one is at the southern end while the other is at the northern end of the profile. Faults are indicated by steep gradient of magnetic anomaly. The magnetic anomaly of about 103 and 162 nT southeast and east of Bansjora, respectively, represents the occurrence of Precambrian basement underlying the sediments.

The boreholes JK-7 and JM-8 are located near this profile. From borehole JM-7, it is obtained that maximum thickness of the Raniganj formation is about 0.22 km and Barren Measures lies below it. It touches the Barakar formation at a depth of about 1.2 km, east of Bansjora. From the obtained results from borehole JK-8, it is clear that sediment thickness is about 0.3 km and the borehole touches the Barren Measures at a depth of about 300 m.

The mean depths of causative sources along the profile CC′ (passes west of Mahuda and Katras through Kumardih region, shown in Fig. 2) calculated from the CWT (Fig. 7b) and Daubechies' wavelet method (Fig. 7b) vary from 1 to 2 km. The northern part of the basin shows the flatness in the basin. Most of the sedimentary formations exist along the profile CC′. Figure 7c reveals that there is a strong indication that both boundaries slope towards the central part of

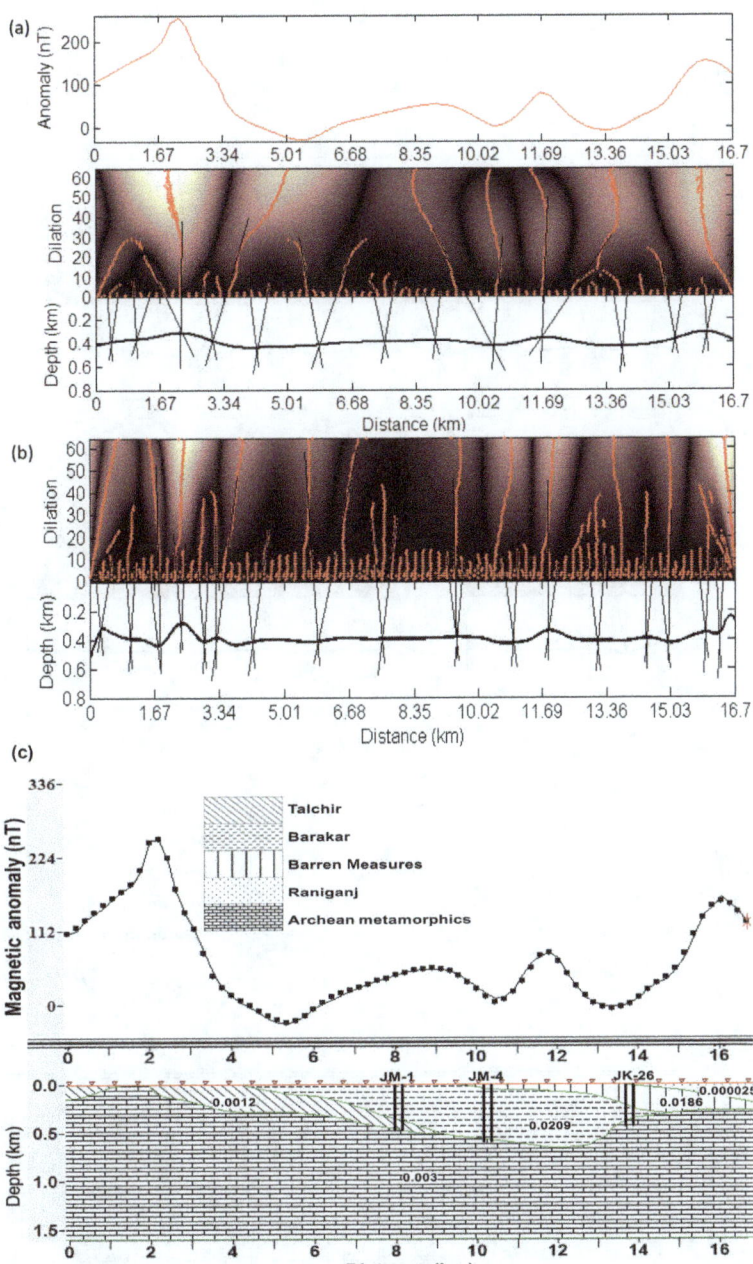

**Figure 5. (a)** Magnetic anomaly across the profile AA′ (drawn in Fig. 2) and depth estimation by continuous wavelet transform. **(b)** Depth estimation by Daubechies' wavelet method. **(c)** Geological section of the profile AA′ along with boreholes and magnetic susceptibility (shown in Table 2) of related formation.

the basin and the southern boundary is categorized by a more abrupt slope than the northern.

Magnetic field inclination, declination and azimuth angle of this profile are 36.41, −0.12 and 268.516° respectively. Gee (1932) mentioned four dykes in the memoir of this coalfield, namely Salama dyke, Sitarampur dyke, Charanpur dyke and Barakar river dyke. The flow of the Barakar river is shown in Fig. 1. It is remarkable that in the region of this magnetic anomaly profile numbers of ultrabasic dyke

(mica peridotites) and sills are found as intrusive into sediments and Barakar formation causes magnetization of the body in the present earth's field.

Similar to profile BB′, Barren Measures lies between the Raniganj and Barakar formations. Also, the Talchir formation lies between the Barakar and Archean metamorphics whose thickness varies from about 1.8 to 2.2 km at the north-central part of the basin. The thickness of sediments near Kumardih and Mahuda is about 2.4 km. Moreover, geological

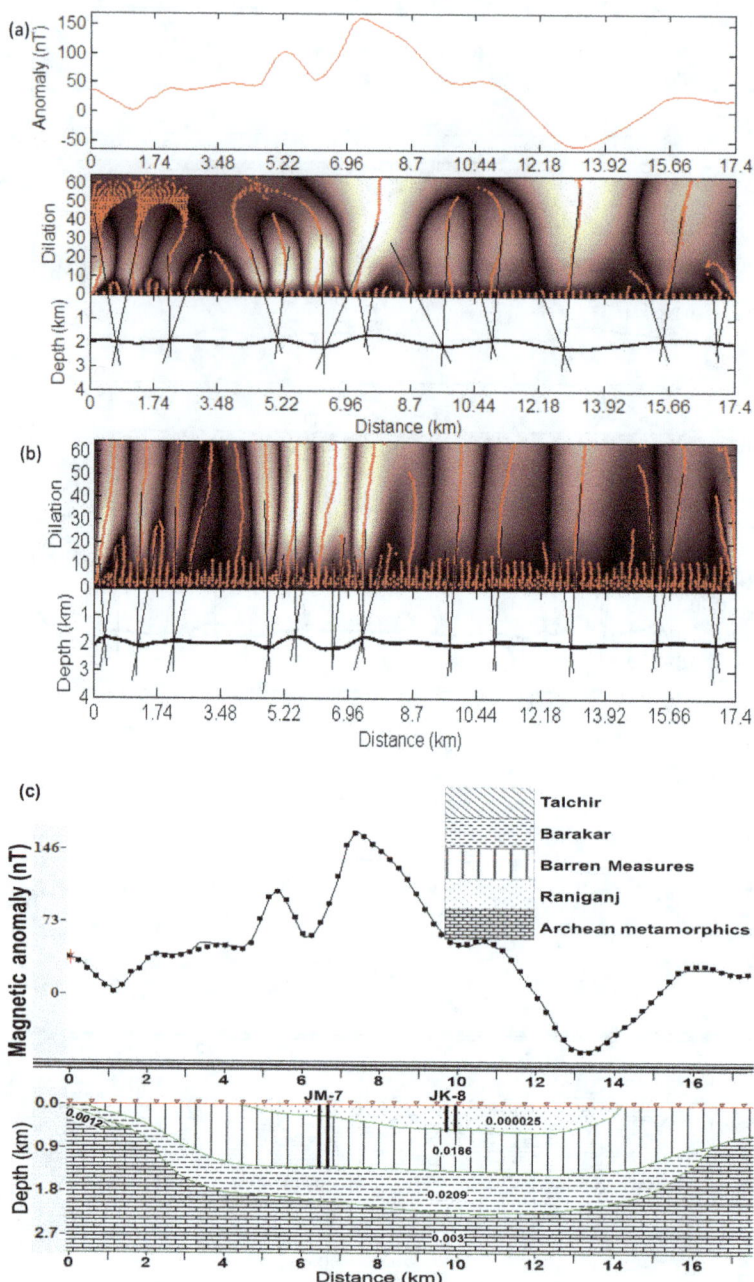

**Figure 6. (a)** Magnetic anomaly across the profile BB′ and depth estimation by continuous wavelet transform. **(b)** Depth estimation by Daubechies' wavelet. **(c)** Geological section of the profile BB′ along with boreholes and magnetic susceptibility (shown in Table 2) of related formation.

sections along the profile CC′ are also based on the results obtained from gravity data (Singh and Singh, 2015), borehole information as well as geological information. Boreholes NCJA-4, NCJA-5 and MN-11 are located near this profile. Boreholes NCJA-4 and NCJA-5 are located southwest of Katras and northeast of Kumardih. Depths of individual formations near the deepest part of the basin are about 0.4 km for Raniganj formation, 0.95 km for Barren Measures, 0.8 km for Barakar formation and 0.2 km for Talchir formation.

The mean depths of causative sources along the profile DD′ (passes east of Mahuda and Katras and west of Parbatpur and Dubrajpur through Barki region, shown in Fig. 2) calculated from the CWT (Fig. 8a) and Daubechies' wavelet method (Fig. 8b) vary from 1 to 2.4 km. Also, along this profile there are some indications of fractious contact between the Barakar formation and Barren Measures. The Barakar formation appears to pinch out close to the southern boundary fault.

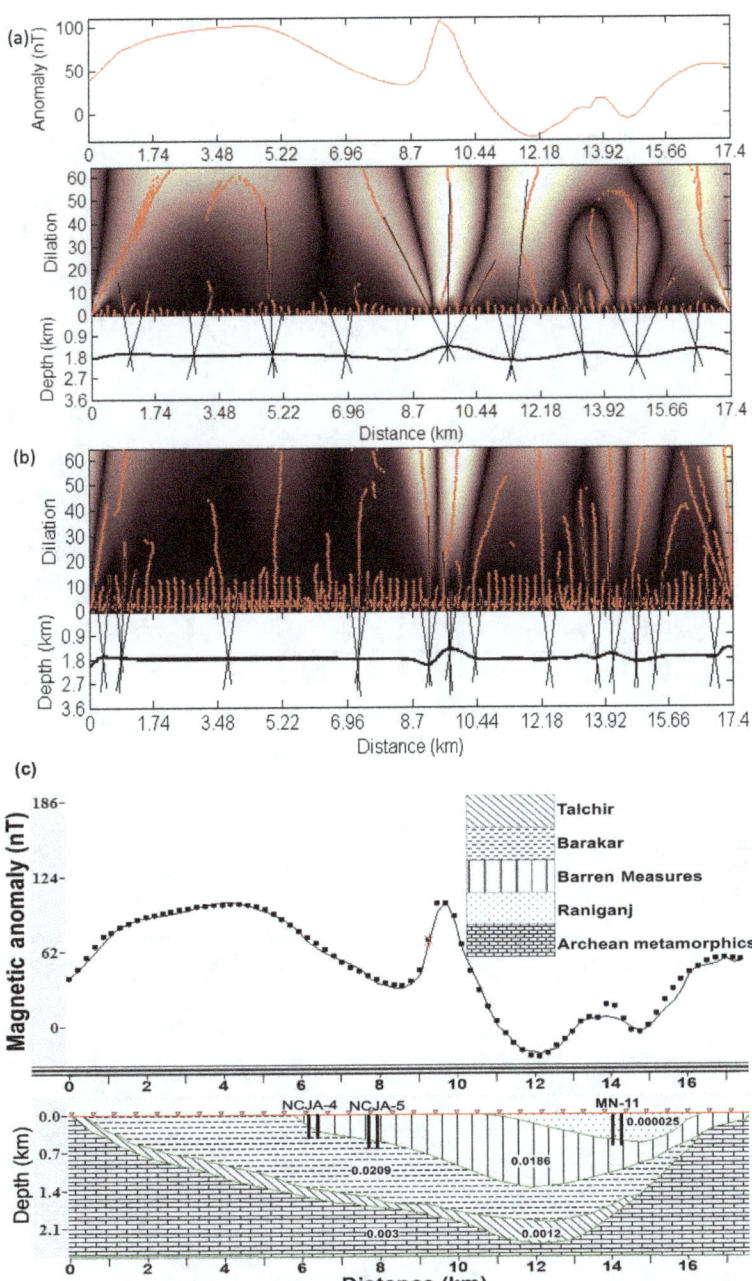

**Figure 7. (a)** Magnetic anomaly across the profile CC′ (drawn in Fig. 2) and depth estimation by continuous wavelet transform. **(b)** Depth estimation by Daubechies' wavelet. **(c)** Geological section of the profile CC′ along with boreholes and magnetic susceptibility (shown in Table 2) of related formation.

Magnetic field inclination, declination and azimuth angle of this profile are 36.40, −0.12 and 268.529° respectively. Faults between Barakar formation and metamorphics are clearly indicated by steep gradients of magnetic anomaly at the northern end of the profile. The southern end of the profile is characterized by magnetic variation that appears to be due to an uneven topography. The middle of the profile is characterized by a magnetic high of about 151 nT because of 2-D linear features and a magnetic pole which lies nearly

0.5–0.65 km below the surface in this region. The extent of the Talchir formation assumed to be underlying the Barakar formation is uncertain. Some coal seams exhibited on the surface and northern side have a steeper dip than the southern side. Approximate depth of the basement in this area estimated from a single pole was 2 km (Fig. 8c) below the surface, southwest of Parbatpur.

Geological sections along this profile were also deduced from the analysis of borehole information, gravity data

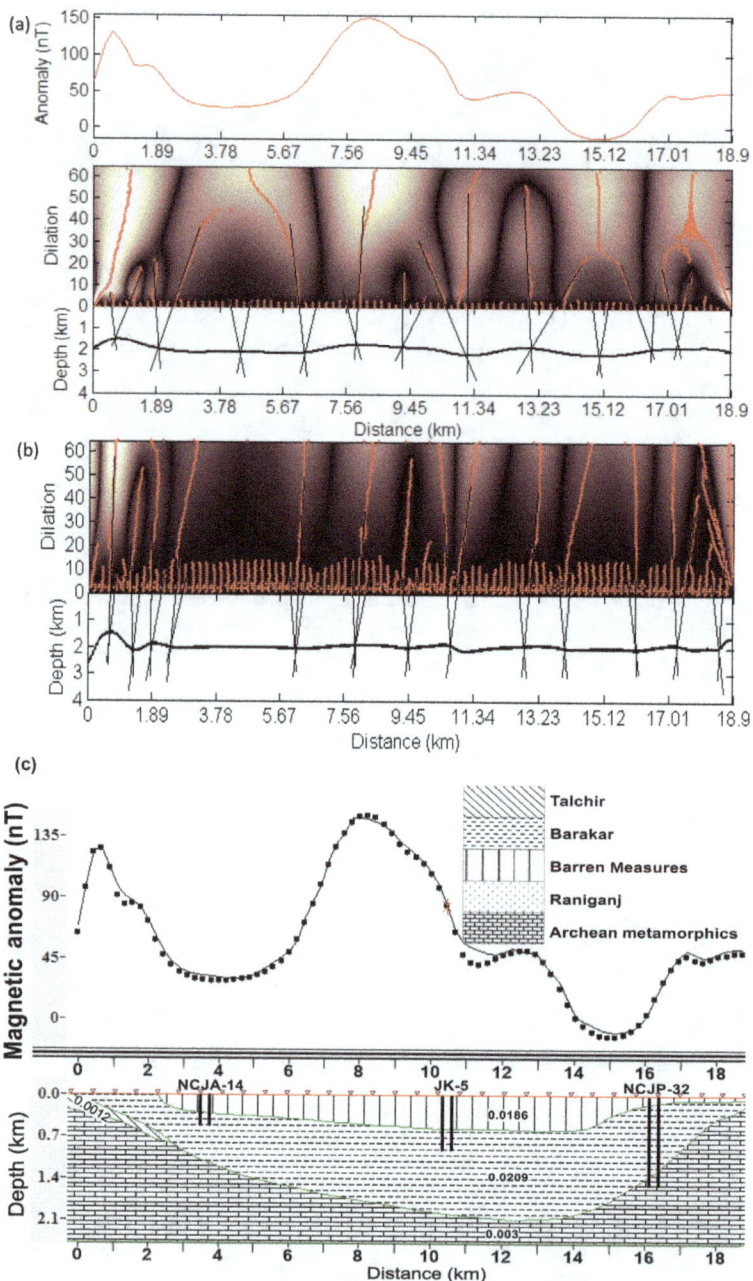

**Figure 8.** (a) Magnetic anomaly across the profile DD′ (drawn in Fig. 2) and depth estimation by continuous wavelet transform. (b) Depth estimation by Daubechies' wavelet. (c) Geological section of the profile DD′ along with boreholes and magnetic susceptibility (shown in Table 2) of related formation.

and geological information. Boreholes NCJA-14, JK-5 and NCJP-32 are located east of Katras, north west of Dubrajpur and west of Parbatpur respectively. The individual maximum thickness of various formations near the deepest part of the basin is about 0.8 km for Talchir, 0.4 km for Barren Measures and about 2 km for Barakar formation.

The mean depths of causative sources along the profile EE′ (passes east of the Parbatpur and Dubrajpur and west of Dungri, Kustore region, shown in Fig. 2) calculated from

the CWT (Fig. 9a) and Daubechies' wavelet (Fig. 9b) vary from 1.8 to 2.8 km. There is a gentle slope of the basin on the northern side, uplift of the basement in the southern part and steep slope close to the southern boundary fault, clearly indicated in this profile.

Magnetic field inclination, declination and azimuth angle (clockwise from true north) of this profile are 36.39, −0.12 and 268.556° respectively. The depth of the basement near the top pole is estimated to be about 1.5–1.6 km from the

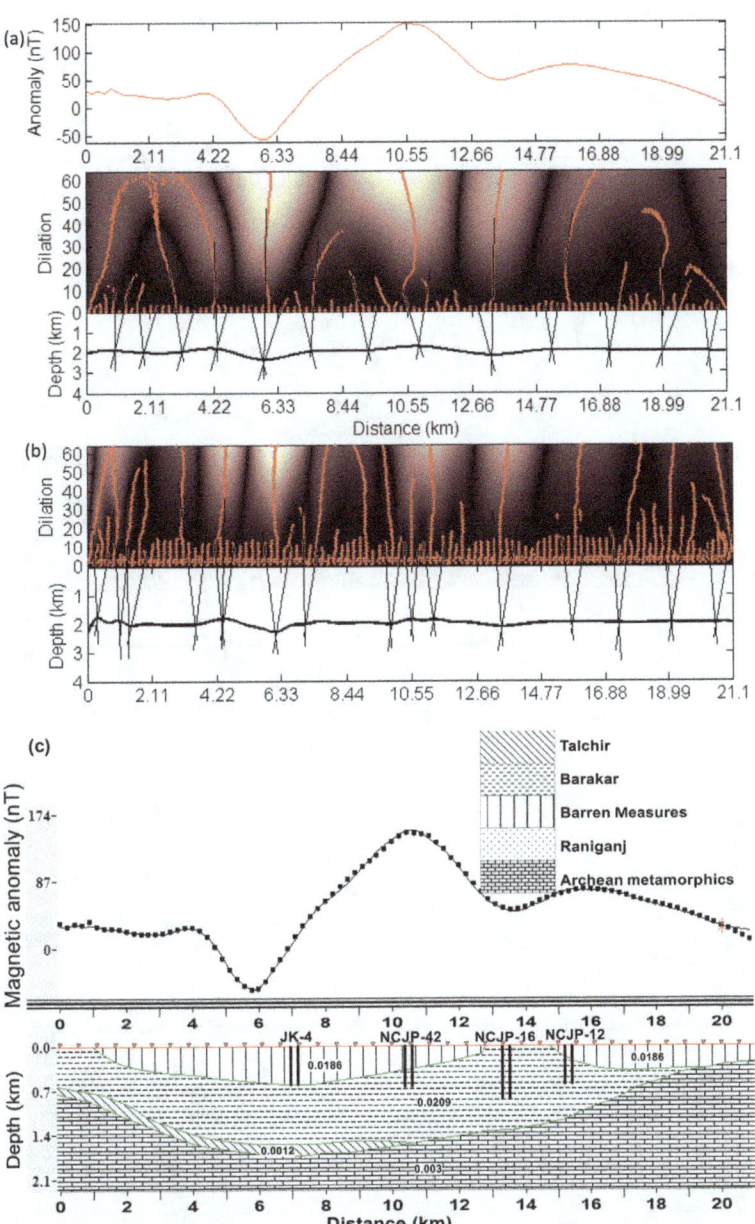

**Figure 9.** (a) Magnetic anomaly across the profile EE′ (drawn in Fig. 2) and depth estimation by continuous wavelet transform. (b) Depth estimation by Daubechies' wavelet. (c) Geological section of the profile EE′ along with boreholes and magnetic susceptibility (shown in Table 2) of related formation.

surface. The anomaly high of about 149 nT at the middle of the profile could be ascribed to the presence of basic or ultrabasic body which was a source for sills and basic dykes which intruded into the basin during Precambrian age. The south end of the underlying source is found to be at a depth of about 0.4 km and the north end at 0.7 km below the surface (Fig. 9c). The eastern margin shows the impact of the occurrence of some faults and extension of metamorphic runs under the sediments up to a distance of about 1.12 km.

Geological sections along this profile are also deduced from the gravity data, borehole information and available ge-

ological information. Individual thickness of each formation is also deduced with the help of boreholes JK-4, NCJP-42, NCJP-16 and NCJP-12, which are located southwest of Kustore, west of Nunikdih, west of Dungri and south of Dungri respectively. Maximum thickness is about 0.45 km for Barren Measures, 1.5 km for Talchir and 1.4 km for Barakar formation.

The mean depth of causative sources along the profile FF′ (passes east of the Jharia, Dhanbad and west of the Makunda and Pathardih regions, shown in Fig. 2) calculated from the CWT (Fig. 10a) and Daubechies' wavelet method (Fig. 10b)

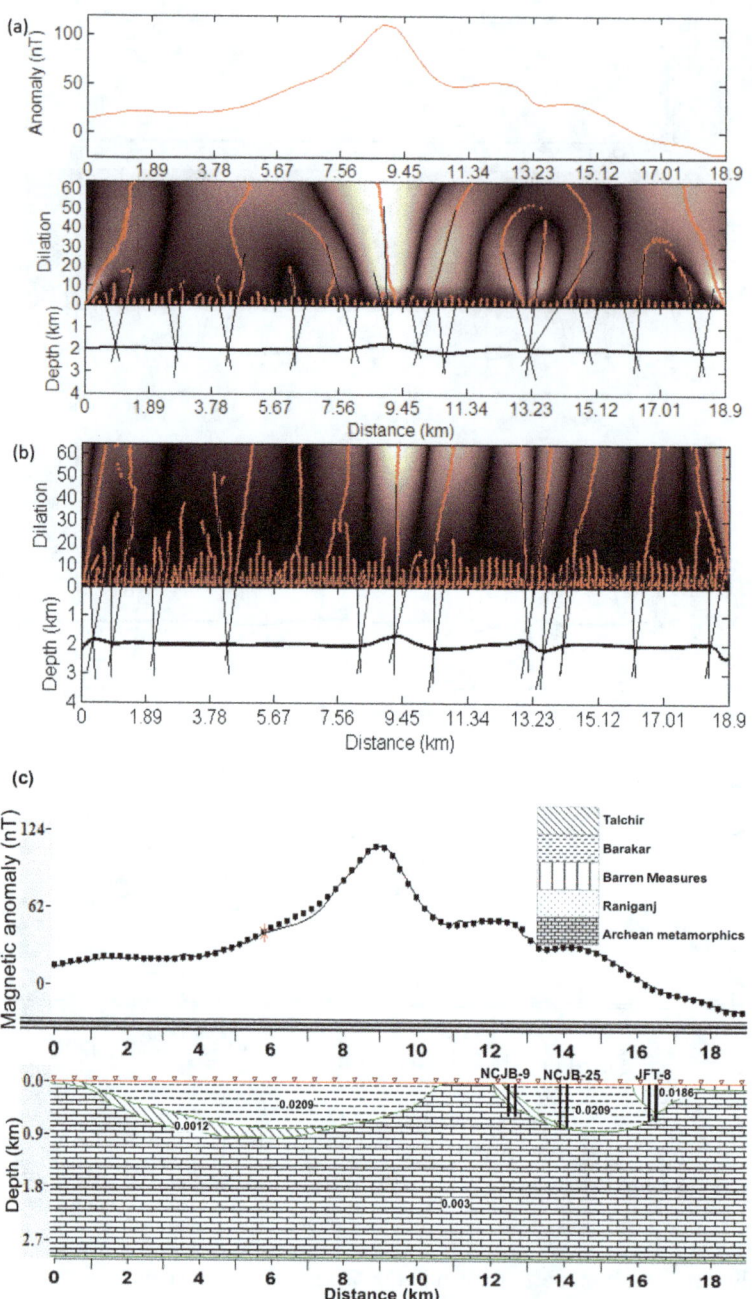

**Figure 10. (a)** Magnetic anomaly across the profile FF′ (drawn in Fig. 2) and depth estimation by continuous wavelet transform. **(b)** Depth estimation by Daubechies' wavelet. **(c)** Geological section of the profile FF′ along with boreholes and magnetic susceptibility (shown in Table 2) of related formation.

varies from 1 to 2.5 km. Also, magnetic anomaly suggests that this area is geologically highly disturbed and dips of the formations vary rapidly.

Magnetic field inclination, declination and azimuth angle of this profile are 36.33, −0.13 and 268.584° respectively. Patherdih horst, which is a tongue of gneiss, penetrates the southeast corner of this region. There are strong faults that occur at both ends of the profile. Several interesting possibil-

ities arise regarding the basic intrusives of dykes as well as schists which are normally magnetized. An anomaly of about 110 nT at the middle of the profile is due to peridotite dykes and sills having a close association with Barren Measures and Barakar formation.

It is found that in this region of magnetic anomaly remanent magnetization of the body also appears to contribute to the magnetic anomaly. A number of sills and ultrabasic

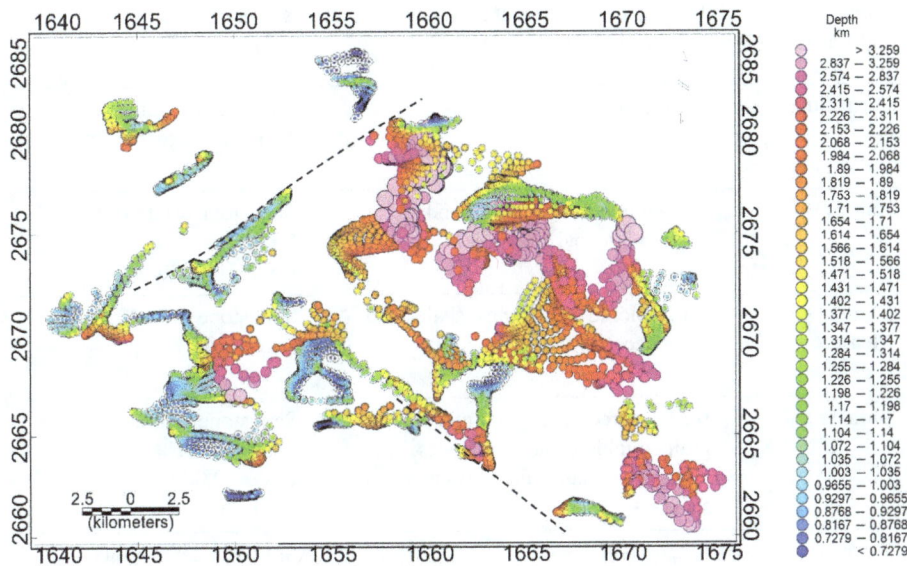

**Figure 11.** The depth estimates obtained from Euler deconvolution (SI = 2) are plotted in UTM coordinates of the study region.

**Table 1.** Mean depth of causative sources calculated from magnetic anomaly by CWT, EDM and Daubechies' wavelet along the profiles drawn over Jharia coalfield and surrounding regions.

| Names of Profiles | Distance and depth (km) | | | | | | |
|---|---|---|---|---|---|---|---|
| | Depth at 3 km from the left (km) | Depth at 6 km from the left (km) | Depth at 9 km from the left (km) | Depth at 12 km from the left (km) | Depth at 15 km from the left (km) | Depth at 18 km from the left (km) | Depth at 21 km from the left (km) |
| AA′ | 0.3 | 0.4 | 0.38 | 0.37 | 0.39 | – | – |
| BB′ | 2 | 2.4 | 2.2 | 2.5 | 1.8 | – | – |
| CC′ | 1.6 | 1.7 | 1.2 | 1.9 | 2 | – | – |
| DD′ | 2.2 | 2.8 | 1.7 | 1.8 | 2.3 | – | – |
| EE′ | 1.8 | 2.8 | 1.8 | 2 | 1.7 | 1.9 | 2.1 |
| FF′ | 2.1 | 2.2 | 1 | 1.7 | 1.8 | – | – |

dykes (mica peridotites) are found to be intrusive into the sediments. Geology over this profile could be ascribed to the presence of a basic or ultrabasic body which was the main source for the sills and basic dykes that intruded (Fig. 10c) into the basin during Gondwana times (Verma et al., 1973).

Geological strata along this profile are highly disturbed. Therefore, dips of the formations vary abruptly. The thickness of the formations is extrapolated from gravity data, boreholes NCJB-9, NCJB-25 and JFT-8 information as well as geological information. Boreholes NCJB-9, NCJB-25 and JFT-8 are located west of Chhatabad, west of Patherdih and west of Bhojudih respectively. Borehole JFT-8 has the cross contact between Barren Measures and Barakar formation and it touches the metamorphics about 0.4 km west of Bhojudih. The depth of the individual formations is approximately equal to the depth obtained from interpretation of gravity data (Singh and Singh, 2015).

The interpretation of magnetic anomaly over Jharia coalfield has been compared with some information from interpretation of gravity data (Verma and Ghosh, 1974). The mean depth of the causative sources estimated by Euler deconvolution method (Fig. 11) ranges about 0.6 to 3.2 km. The mean depth of the profiles has been shown in the Table 1.

Results from the total magnetic field of Jharia coalfield (Fig. 2) show that magnetic field anomalies are predominant due to irregular undulations of Precambrian outcrops and faults. The magnetic data are sampled at roughly 50 m along the profile direction. To enhance the signal-to-noise ratio, a high cut filter was applied in the wavenumber domain and partial derivative in the vertical direction was obtained by extending the field grid before the calculation. The SI is supposed to vary between 0 and 3, covering all plausible geological bodies. The estimates of source location and depth are obtained by minimizing the error function using the nonlinear optimization technique of Coleman and Li (1996).

**Table 2.** The following magnetic susceptibility used to prepare the geological sections. Susceptibility values are taken from the standard chart compiled by Clark and Emerson (1991) and Hunt et al. (1995).

| Formation | Litho-type | Maximum volume Magnetic susceptibility (SI units) |
|---|---|---|
| Raniganj | Fine-grained feldspathic sandstones, shales with coal seams | Sandstone = 0.0209 Shale = 0.0186 Coal = 0.000025 |
| Barren Measures | Buff-colored sandstones, shales and carbonaceous shales | Sandstone = 0.0209 Shale = 0.0186 |
| Barakar | Buff-colored coarse and medium-grained feldspathic sandstones, carbonaceous shales, fire clays and coal seams | Sandstone = 0.0209 Shale = 0.0186 Clay = 0.00025 Coal = 0.000025 |
| Talchir | Silt, carbonates Greenish shale and fine-grained sandstones | Silt/carbonates = 0.0012 Shale = 0.0186 Sandstone = 0.0209 |
| Metamorphics | Granite gneisses, quartzites, mica schists and amphibolites | Granite = 0.05 Gneisses = 0.025 Quartzites = 0.0044 Mica schists = 0.003 Amphibolites = 0.00075 |

Figure 11 shows two sets of fractures, predominantly oriented in the northeast and southeast at the northern and southern boundary respectively. The orientation of fractures sets are similar to that of the orientation obtained from regional magnetic interpretation (Verma et al., 1973). In the southern region, the depth of the Precambrian basement derived from the faults is less than that in the northern region. Furthermore, intense fracturing is detected at the center of the study area. In the western and southern regions, the basement depth is shallower compared to that of the eastern and northern region.

Profile analysis suggests that most of the basement lies below 700 m, which is reasonable as calculated by wavelet transform method. The faults and depths obtained from the Euler deconvolution, CWT and Daubechies' wavelet are related to each other according to the results obtained from the regional magnetic interpretation.

## 7 Conclusions

The present analysis demonstrates the efficiency of continuous wavelet transform to delineate the locations of causative sources of potential field. Mean depth of the causative source along the profile AA′ from across Amdih and south of Telmuchu varies from 0.2 to 0.45 km and there is a fault near the northwestern part of the study region. The magnetic anomaly of about 77 nT corresponds to the number of basic intrusive bodies belonging to the Satpura cycle. Mean depth of the profile BB′ calculated from the CWT varies from 1.3 to 2.5 km. Central part of the basin shows abrupt changes in the magnetic anomaly. The key feature of this area is that most of the major portion is covered by the Raniganj and the Barakar formations. The estimated thickness of the sediment is about 2.3 km over the Raniganj formation. Mean depth of the profile CC′ calculated from the CWT and Euler deconvolution varies from 1 to 2 km. A key feature of this profile is that sedimentary formations along the profile expose a strong indication that both boundaries have a slope towards the central part and that the southern boundary is more abrupt than the northern boundary. The mean depth of causative sources along the profile DD′ calculated from the CWT and Euler deconvolution varies from 1 to 2.4 km. Moreover, along this profile there is indication of fractious contact between the Barakar formation and Barren Measures, and the Barakar formation appears to pinch out close to the southern boundary fault. The mean depth of causative sources along the profile EE′ calculated from the CWT and Euler deconvolution varies from 1.8 to 2.8 km. There is a gentle slope of the basin on the northern end, while uplift of the basement and a steep slope close to the southern boundary fault are clearly indicated in this profile. The mean depth of causative sources along profile FF′ calculated from the CWT and EDM varies from 1 to 2.5 km. Also, magnetic anomalies suggest that the easternmost area is geologically highly disturbed and dips of the formations vary rapidly.

Thus, the wavelet transform and Euler deconvolution methods provide sufficient and relevant information necessary to find the depth and location of the causative sources. The application of the CWT methods to the synthetic and field magnetic data across Jharia coalfield demonstrates that the technique is quick, easy to use and very efficient. Continuous wavelet transform and Euler deconvolution can give the mean depth of causative sources of magnetic field data, which can be interpreted qualitatively and quantitatively to determine the cause of anomaly. Also, these methods provide a way to infer the location of causative sources without any a priori information in a very short time and can be further used as a priori models in inversion to improve accuracies.

*Competing interests.* The authors declare that they have no conflict of interest.

*Acknowledgements.* Authors are very thankful to D. C. Panigrahi, Director of IIT (ISM) Dhanbad, for providing the necessary infrastructure for this research to be successfully carried out. We are very grateful to Lev Eppelbaum, Associate Editor of this journal, who gave the initial reviews on earlier versions of the manuscript that greatly improved the final paper. We would also like to show our gratitude to Sanjay Prajapati, O. Menshov and an anonymous reviewer for their positive comments.

Edited by: L. Eppelbaum

# References

Ardestani, E. V.: Precise Edge detection of gravity anomalies by Tilt angle filters, J. Earth & Space Phys., 36, 2, 11–19, 2010.

Barbosa, V. C. F., Silva, J. B. C., and Medeiros, W. E.: Stability analysis and improvement of structural index estimation in Euler deconvolution, Geophysics, 64, 48–60, 1999.

Blakely, R. J. and Simpson, R. W.: Approximating edges of source bodies from magnetic or gravity anomalies, Geophysics, 51, 1494–1498, 1986.

Bhattacharyya, B. P.: Tectono-metamorphic effect of granite and pegmatite emplacement in the Precambrian of Bihar Mica Belt. Proc. Symp. On Metallogeny of the Precambrian, Geological Society of India, Bangalore, India, 45–56, 1972.

Chamoli, A., Srivastava, R. P., and Dimri, V. P.: Source depth characterization of potential field data of Bay of Bengal by continuous wavelet transform, Indian J. Mar. Sci., 35, 195–204, 2006.

Clark, D. A. and Emerson, D. W.: Notes on rock magnetization characteristics in applied geophysical studies, Explor. Geophys., 22, 547–555, 1991.

Cooper, G. R. J.: A semi-automatic procedure for the interpretation of geophysical data, Explor. Geophys., 35, 180–185, 2004.

Cooper, G. R. J.: The semiautomatic interpretation of gravity profile data, Computat. Geosci., 37, 1102–1109, 2011.

Cooper, G. R. I. and Cowan, D. R.: Enhancing potential field data using filters based on the local phase, Computat. Geosci., 32, 1585–1591, 2006.

Cooper, G. R. J. and Cowan, D. R.: Edge enhancement of potential field data using normalized statistics, Geophysics, 73, H1–H4, 2008.

Coleman, T. F. and Li, Y.: An interior, trust region approach for nonlinear minimization subject to bounds, SIAM J. Optimiz., 6, 418–445, 1996.

Dewangan, P., Ramprasad, T., Ramana, M. V., Desa, M., and Shailaja, B.: Automatic interpretation of magnetic data using Euler deconvolution with nonlinear background, Pure Appl. Geophys., 164, 2359–2372, 2007.

FitzGerald, D., Reid A., and McInerney, P.: New discrimination techniques for Euler deconvolution, Computat. Geosci., 30, 461–469, 2004.

Fox, C. S.: The Jharia Coal Field, Geological Survey of India, Memoir, 56, 1–255, 1930.

Gee, E. R.: The geology and coal resources of the Raniganj coalfield, Mem. Geol. Surv. India, 61, 1–343, 1932.

Gerovska, D. and Arauzo Bravo, M. J.: Automatic interpretation of magnetic data based on Euler deconvolution with unprescribed structural index, Computat. Geosci., 29, 949–960, 2003.

Goyal., P. and Tiwari, V. M.: Application of the continuous wavelet transform of gravity and magnetic data to estimate sub-basalt sediment thickness, Geophys. Prospect., 62, 148–157, 2014.

Gunn, P. J.: Linear transformations of gravity and magnetic fields, Geophys. Prospect., 23, 300–312, 1975.

Holschneider, M., Chambodut, A., and Mandea, M.: From global to regional analysis of the magnetic field on the sphere using wavelet frames, Phys. Earth Planet. In., 135, 107–124, 2003.

Hunt, C. P., Moskowitz, B. M., and Banerjee, S. K.: Magnetic properties of Rocks and Minerals, in: Rock Physics and Phase Relations – A hand book of Physical constants, AGU Reference shelf 3, edited by: Ahrens, T. J., 189–204, 1995.

Hsu, S. K.: Imaging magnetic sources using Euler's equation, Geophys. Prospect., 50, 15–25, 2002.

Hsu, S. K., Sibuet, J. C., and Shyu, C. T.: High resolution detection of geologic boundaries from potential-field anomalies: an enhanced analytic signal: technique, Geophysics, 61, 373–386, 1996.

Keating, P. B.: Weighted Euler deconvolution of gravity data, Geophysics, 63, 1595–1603, 1998.

Kopal., Z.: Numerical analysis, Chapman and Hall Ltd., London, UK, 551–553, 1961.

Melo, F. F., Barbosa, V. C. F., Uieda, L., Oliveira Jr., V. C., and Silva, J. B. C.: Estimating the nature and the horizontal and vertical positions of 3-D magnetic sources using Euler deconvolution, Geophysics, 78, J87–J98, 2013.

Moreau, F., Gibert, D., Holschneider, M., and Saracco, G.: Wavelet analysis of potential fields, Inverse Probl., 13, 165–178, 1997.

Moreau, F., Gibert, D., Holschneider, M., and Saracco, G.: Identification of sources of potential fields with continuous wavelet transform: Basic theory, J. Geophys. Res., 104, 5003–5013, 1999.

Mushayandebvu, M. F., Van Driel, P., Reid, A. B., and Fairhead, J. D.: Magnetic source parameters of two dimensional structures using extended Euler deconvolution, Geophysics, 66, 814–823, 2001.

Mushayandebvu, M. F., Lesur, V., Reid, A. B., and Fairhead, J. D.: Grid Euler deconvolution with constraints for 2-D structures, Geophysics, 69, 489–496, 2004.

Perez, C., Wijns, C., and Kowalczyk, P.: Theta map: Edge detection in magnetic data, Geophysics, 70, L39–L43, 2005.

Ravat, D.: Analysis of the Euler method and its applicability in environmental investigations, J. Environ. Eng. Geoph., 1, 229–238, 1996.

Reid, A. B., Allsop, J. M., Granser, H., Millet, A. J., and Somerton, I. W.: Magnetic interpretation in three dimensions using Euler deconvolution, Geophysics, 55, 80–91, 1990.

Reid, A. B., Ebbing, J., and Webb, S. J.: Avoidable Euler Errors- the use and abuse of Euler deconvolution applied to potential field, Geophys. Prospect., 62, 1162–1168, 2014.

Sailhac, P., Gibert, D., and Boukerbout, H.: The theory of the continuous wavelet transform in the interpretation of potential fields: A review, Geophys. Prospect., 57, 517–525, 2009.

Salem, A., Williams, S., Fairhead, J. D., Ravat, D., and Smith, R.: Tilt-depth method: a simple depth estimation method using first order magnetic derivatives, The Leading Edge, 26/12, 1502–1505, 2007.

Silva, J. B. C. and Barbosa, V. C. F.: 3-D Euler deconvolution: Theoretical basis for automatically selecting good solutions, Geophysics, 68, 1962–1968, 2003.

Silva, J. B. C., Barbosa, V. C. F., and Medeiros, W. E.: Scattering, symmetry, and bias analysis of source position estimates in Euler deconvolution and its practical implications, Geophysics, 66, 1149–1156, 2001.

Singh, A. and Singh, U. K.: Wavelet analysis of residual gravity anomaly profiles: Modeling of Jharia coal basin, India, 86, 679–686, 2015.

Stavrev, P. and Reid, A.: Degrees of homogeneity of potential fields and structural indices of Euler deconvolution, Geophysics, 72, L1–L12, 2007.

Stavrev, P. Y.: Euler deconvolution using di?erential similarity transformations of gravity or magnetic anomalies, Geophys. Prospect., 45, 207–246, 1997.

Thompson, D. T.: EULDPH: A new technique for making computer assisted depth estimates from magnetic data, Geophysics, 47, 31–37, 1982.

Verma, R. K. and Ghosh, D.: Gravity survey over Jharia coalfield, India. Geophys. Res. Bull., 12, 165–175, 1974.

Verma, R. K., Prasad, S. N., and Jha, B. P.: Magnetic Survey over Jharia Coal Field, Pure Appl. Geophys., 102, 124–133, 1973.

Verma, R. K., Majumdar, R., Ghosh, D., Ghosh, A., and Gupta, N. C.: Results of Gravity Survey over Raniganj Coalfield, India, Geophys. Prospect., 24, 19–30, 1976.

Verma, R. K., Bhuin, N. C., and Mukhopadhyay, M.: Geology, Structure and tectonics of Jharia Coal Field, India-A 3-D Model, Geoexploration, 17, 305–324, 1979.

Williams, S., Fairhead, J. D., and Flanagan, G.: Grid based Euler deconvolution: Completing the circle with 2-D constrained Euler, SEG Technical Program Expanded Abstracts, 22, 576–579, 2003.

Zhang, C., Mushayandebvu, M. F., Reid, A. B., Fairhead, J. D. and Odegard, M. E.: Euler deconvolution of gravity tensor gradient data, Geophysics, 65, 512–520, 2000.

# 6

# Seismic observations at the Sodankylä Geophysical Observatory: history, present, and the future

Elena Kozlovskaya[1,2], Janne Narkilahti[1], Jouni Nevalainen[1,2], Riitta Hurskainen[1], and Hanna Silvennoinen[1]

[1]Sodankylä Geophysical Observatory, University of Oulu, POB 3000, Oulu, 90014, Finland
[2]Oulu Mining School, University of Oulu, POB 3000, Oulu, 90014, Finland

*Correspondence to:* Elena Kozlovskaya (elena.kozlovskaya@oulu.fi)

**Abstract.** Instrumental seismic observations in northern Finland started in the 1950s. They were originally initiated by the Institute of Seismology of the University of Helsinki (ISUH), but the staff of Sodankylä Geophysical Observatory (SGO) and later geophysicists of the University of Oulu (UO) were involved in the development of seismological observations and research in northern Finland from the very beginning. This close cooperation between seismologists and the technical staff of ISUH, UO, and SGO continued in many significant international projects and enabled a high level of seismological research in Finland. In our paper, we present history and current status of seismic observations and seismological research in northern Finland at the UO and SGO. These include both seismic observations at permanent seismic stations and temporary seismic experiments with portable seismic equipment. We describe the present seismic instrumentation and major research topics of the seismic group at SGO and discuss plans for future development of permanent seismological observations and portable seismic instrumentation at SGO as part of the European Plate Observing System (EPOS) research infrastructure. We also present the research topics of the recently organized Laboratory of Applied Seismology, and show examples of seismic observations performed by new seismic equipment located at this laboratory and selected results of time-lapse seismic body wave travel-time tomography using the data of microseismic monitoring in the Pyhäsalmi Mine (northern Finland).

## 1 Introduction

Sodankylä Geophysical Observatory (SGO) was established in 1913 by the Finnish Academy of Science and Letters to perform geophysical measurements and research based on the observation results. Measurements of the Earth's magnetic field began on 1 January 1914. On 1 August 1997, the observatory became an independent research department of the University of Oulu (UO). Currently, the Sodankylä Geophysical Observatory performs continuous measurements of the Earth's magnetic field, cosmic radio noise, seismic activities, and cosmic rays. The observatory is located in central Finnish Lapland in the municipality of Sodankylä (see Fig. 1).

As described in Luosto (2001), development of seismology in Finland in the 20th century comprises several distinct periods. The initial non-instrumental period already began in the 19th century when systematic collecting of data about local seismic events in Finland started at the University of Helsinki. The second period began in 1921 when the first privately financed seismograph station in Helsinki was put in operation. This event also marks the beginning of the era of instrumental seismology in Finland.

The next period in the development of Finnish seismology started at the end of the 1950s, and it was motivated by development in seismic instrumentation worldwide. During this period, several short-period analogue seismograph stations with photo paper registration were founded in Finland, although the instrumentation had not yet been standardized and home-made seismic sensors were used (see Luosto (2001) for details). However, these seismographs were capable of recording both minor local and teleseismic earthquakes.

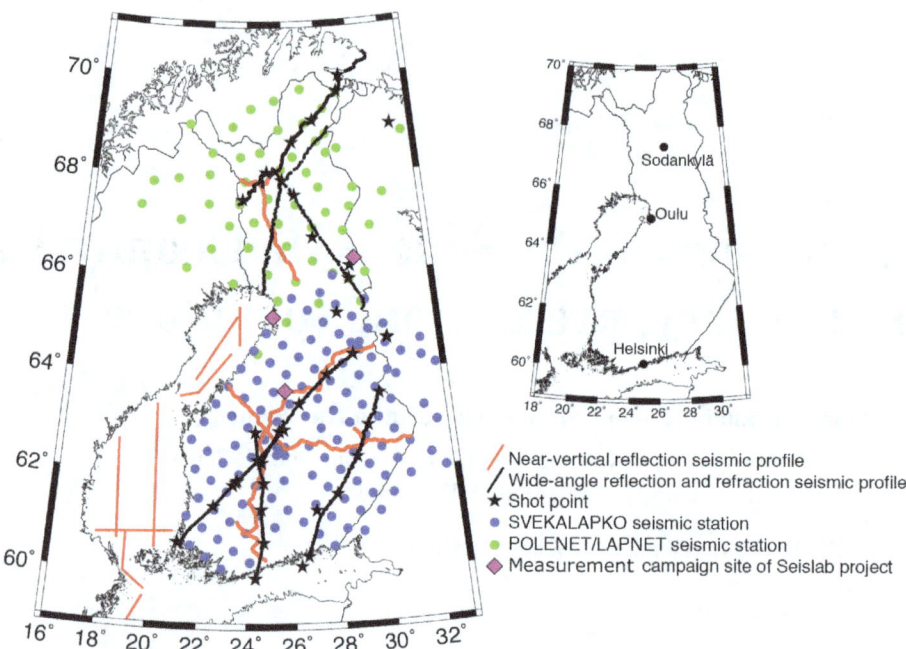

**Figure 1.** Map showing position of seismic controlled-source and passive seismic experiments in Finland, in which the seismic group of the UO and SGO has participated since the 1980s (see also Table 1 and the description of the experiments in the text). The smaller map on the right shows the locations of towns and municipalities mentioned in the text.

**Table 1.** Controlled-source seismic experiments in Finland, in which the seismic group of the University of Oulu participated with UO equipment.

| Experiment abbreviation | Year of data acquisition | References |
|---|---|---|
| FINNLAP | 1979 | Luosto et al. (1983) |
| SVEKA'81 | 1981 | Luosto et al. (1984) |
| BALTIC | 1982 | Luosto (2001) |
| POLAR | 1985 | Luosto et al. (1989) |
| BABEL | 1989 | BABEL Working Group (1993) |
| SVEKA'91 | 1991 | Luosto (2001) |
| FENNIA | 1994 | Luosto (2001) |
| FIRE | 2001–2003 | Kukkonen and Lahtinen (2006) |

The fourth development period started when the World Wide Standard Seismograph Network (WWSSN) was founded and funded by the United States of America at the beginning of the 1960s. Then several Finnish seismic stations were equipped with standard WWSSN short-period Benioff and long-period Press–Ewing seismometers, a network of seismic stations was enhanced, and the first efforts were made to transmit analogue signals from remote stations via telephone cables. The Institute of Seismology of the University of Helsinki (ISUH) was established as an independent unit in 1961. In this era, the development of seismology in Finland was strongly influenced by the massive nuclear explosion tests in Novaya Zemlya (Russia), by the International

Geophysical Year 1957–58, and by the Seventh General Assembly of International Union of Geodesy and Geophysics, held in Helsinki in 1960.

The period of digital seismology and broadband seismometry in Finland started in the 1970s–1980s. In 1981, the analogue instrumentation at WWSSN stations in Finland was upgraded to digital data acquisition systems. In the 1970s, engineer Seppo Nurminen started to design digital recorders and transmission systems at the ISUH (Nurminen, 1974, 1976). The same technique was applied in the 1980s when constructing three- or five-channel PCM-1218-80 recorders (Nurminen and Hannula, 1981), which were the first digital portable field recorders in Europe. Just at the end of the 20th century, Nurminen designed an entirely new digital seismic recorder (model DAS-98), which runs under the Linux operating system. These recorders were used both in the permanent stations and in temporary field experiments. Until the end of the century, almost the entire seismic network in Finland was operating using digital telemetric or dial-up method.

Progress in portable digital recording systems gave a start to controlled-source wide-angle reflection and refraction studies in Finland in the 1970s–1990s and the large-scale marine deep seismic reflection experiment BABEL in 1989 (Luosto, 1987, 2001; Table 1 and Fig. 1). The advanced portable instrumentation provided an opportunity for Finnish geophysicists to participate in many international controlled-source seismic experiments (Table 2).

**Table 2.** International controlled-source seismic experiments, in which the seismic group of the University of Oulu participated.

| Experiment name | Region | Year of data acquisition | Reference |
|---|---|---|---|
| EGT (European Geotraverce) | Italy, Germany | 1986 | Blundell et al. (1992) |
| LT-7 | Poland | 1987 | Guterch et al. (1994) |
| TTZ | Poland | 1993 | Grad et al. (1999) |
| EUROBRIDGE'94 | Lithuania | 1994 | Grad et al. (2006) |
| EUROBRIDGE'95 | Lithuania | 1995 | Grad et al. (2006) |
| EUROBRIDGE'96 | Belarus | 1996 | Grad et al. (2006) |
| POLONAISE | Poland, Lithuania | 1997 | Grad et al. (2006) |
| EUROBRIDGE'97 | Belarus, Ukraine | 1997 | Grad et al. (2006) |
| CELEBRATION 2000 | Austria, Germany, Poland, Hungary, Slovakia, Czech Republic, Russia, Belarus | 2000 | Grad et al. (2006) |
| ALP2002 | Austria | 2002 | Bleibinhaus et al. (2006) |
| SUDETES | Poland | 2003 | Majdanski et al. (2006) |
| DANUBE | Hungary | 2004 | Hegedus et al. (2005) |

Since the 1950s, the scientific and technical staff of SGO and the geophysical group of the UO was actively involved in the above-mentioned observatory activities and seismic projects initiated by the ISUH. Since SGO was founded much earlier than the UO (1914 and 1958, respectively) and was initially operated as an independent research institution, the seismological observations and research at these two organizations were originally developing in parallel. Presently, the research based on seismological observations is performed by the seismic group located in Oulu (Oulu unit of SGO) and comprises a significant part of the total scientific output of SGO.

The main target of our paper is to document the history of seismic observations and research, including temporary experiments, in northern Finland, both at the UO and at SGO. In the paper, we do not repeat the scientific results of seismological studies published elsewhere but mainly concentrate on such practical things as a description of instrumentation, tracing of instrument movement to alternative sites, data formats and data availability, and staff in charge. We also discuss the future of seismology at SGO in the 21st century. The future activities include participation in European Plate Observing System (EPOS) pan-European research infrastructure for solid Earth geosciences, and the development of the newly established Laboratory of Applied Seismology (SEISLAB) taking charge of campaign measurements at SGO.

## 2 History of seismic observations in northern Finland

In 1954, Eijo Vesanen, who was the head of the ISUH at the time, proposed to install a seismic station at SGO. The idea received support from the observatory administration, and the first seismic test measurements at SGO started on 11 June 1954. The instrument was a copy of the vertical component Sprengnether seismometer made at the Department of Physics at the University of Helsinki (Kataja, 2008). The measurements by the short-period vertical component Benioff seismometer at the SGO site in Tähtelä (station code SOD; Fig. 2) started on 28 June 1956. The results showed, however, that the site was not suitable for observatory level seismic measurements because of a thick (about 50 m) sand layer. The measurements at the Tähtelä site continued using a different type of analogue equipment (Table 3) until a new site (station code SDF) was found in Pittiövaara hill near Sodankylä. The seismic sensors were moved to the new site while connection and communication between sensors and recording system were established using radio link (Kataja, unpublished memoirs). In 2001, a digital seismic station with new equipment was established at the new site in an underground tunnel (station code SGF; Fig. 2, Table 3). The station recorded 3-component (3C) seismic data with a sampling rate of 50 sps in CSS data format (Anderson et al., 1990).

During the 1950s the seismographs at SGO were under the maintenance of the observatory staff. In 1959, the Finnish Academy of Sciences and Letters founded the position of seismologist at the observatory. The position was held by Airi Kataja till 1991. The seismologist was responsible for the maintenance of the seismographs and also for investigations of the seismicity in northern Finland. In 1991, this position was cancelled.

The first seismic measurements in Oulu were initiated by the University of Helsinki, and the registration at the Oulu station started on 17 Dec 1959, soon after the university and its Department of Physics were founded in 1958 and 1959, respectively. Initially, the seismic equipment was installed at the Myllytulli hydroelectric power plant, not far from the centre of Oulu. That temporary station was equipped with the Nurmia seismograph made at the University of Helsinki, and it was operated by the staff of the Department of Physics of the University of Oulu till autumn 1960.

**Table 3.** The instrumentation of the seismic stations in northern Finland prior to 2005.

| Station | Component | Type of instrument | Period T°s | Magnification at T°s | Damping ratio | Recording type | Drum speed mm·min⁻¹ | Geographical coordinates | Type of amplifier | Operation period |
|---|---|---|---|---|---|---|---|---|---|---|
| SOD Tähtelä | Z | Benioff | 1.0 | 34 000 | 15:1 | Ph. Paper | 60 | 67°22'16.2" N 26°37'44.7" E h = 181 m | Galv. | Nurmia: 28 Jun 1956–1966 |
|  | N | Nurmia | 0.5 | 35 000 | 3:1 | Ph. Paper | 30 |  | Galv. | Benioff till 14 Jun 1992 |
|  | E | Nurmia | 0.5 | 35 000 | 3:1 | Ph.Paper | 30 |  | Galv. |  |
|  | Z | Nurmia | 0.5 | 1 000 000 | 2:1 | Smoked p. | 60 |  | ElectrMech. | 1966–1973 |
|  | Microbar. | Willmore | – | – | – | Smoked p. | 5 |  | Galv. | 1968–1973 |
|  |  | Kirnos |  |  |  | Heat paper | 60 |  | Galv. |  |
| OUL Huttukylä | Z | Press–Ewing | 30 | 1500 | Inf | Ph. paper | 30 | 65°05'07" N 25°53'47" E h = 60 m | Galv. | 9 Oct 1963–1987 |
|  | N | Willmore | 0.65 | 80 000 | 4:1 | Heat paper | 30 |  | Phototube |  |
|  | N |  |  |  |  | Ph. paper | 30 |  | Galv. |  |
|  | E |  |  |  |  | Ph. paper | 30 |  | Galv. |  |
| SGF Sodankylä | Z | S-13 | 0.8 | 282k/32 | – | DAS-98 | – | 67.442° N 26.526° E $h_1$ = 180 m | – | 18 May 2001–4 Jan 2006 |
|  | N |  |  |  |  |  |  |  |  |  |
|  | E |  |  |  |  |  |  |  |  |  |
| SDF Pittiövaara | Z | Kirnos | 0.8 | 282k/32 | – | Ph.paper | 60 | 67.420° N 26.394° E h = 276.5 m | Galv. | Spring 1973 |
|  | N | S-13 |  | – |  | Ph.paper | 30 |  | Galv. | 1983–17 Jun 2000 |
|  | E |  |  |  |  | Ph.paper | 30 |  | Galv. |  |
|  | N |  |  |  |  | Lennarz ink paper recorder |  |  |  |  |
| OUL Ervasti | Z | S-13 | 1.0 | 450k/2 | – | DAS-98 | – | 65.085° N 25.842° E h = 72 m | – | Dec 1980–Jul 1996 |
| OUL Huttukylä | Z | S-13 | 1.0 | 450k/2 | – | DAS-98 | – | 65.0528° N 25.8964° E h = 60 m | – | 26 Oct 1999–31 Dec 2005 |
|  | N |  |  |  |  |  |  |  |  |  |
|  | E |  |  |  |  |  |  |  |  |  |
| MA Maaselkä MSF Maaselkä Maaselkä | Z N E | S-13 | 1.0 | – | – | DAS-98 | – | 65.9113° N 29.0402° E h = 365 m | – | Jan 1970–Jun 1998 17 May 2000–1 May 2005 |

(Column group labels: *T / dB* ; *Recorder*)

## NOISE LEVEL AT FN STATIONS IN 2014

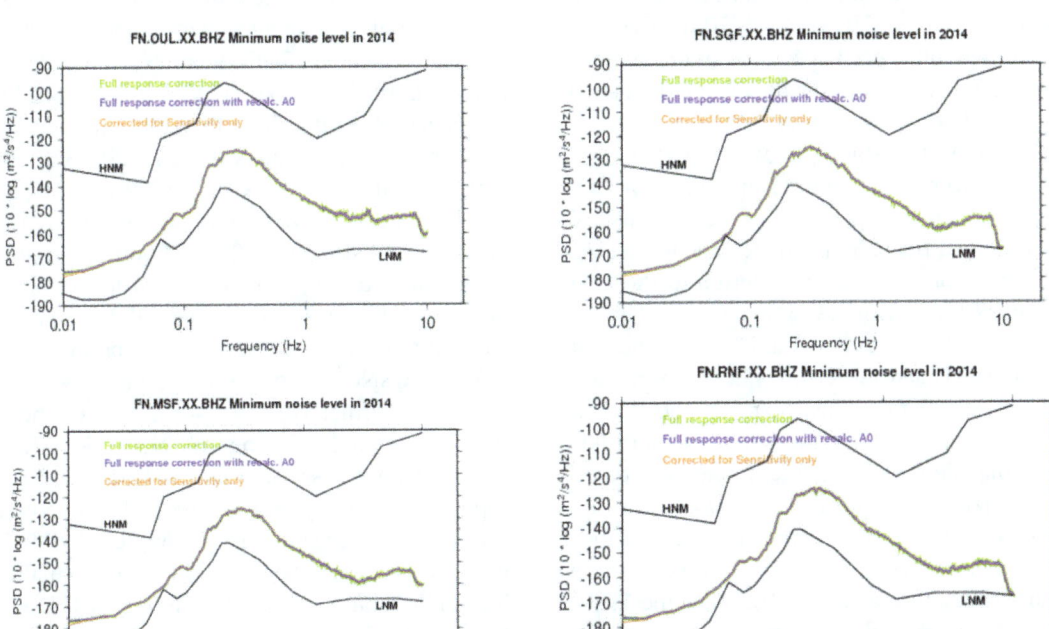

**Figure 2.** Noise level at the permanent stations of the Northern Finland Seismological Network in 2014. HNM and LNM show high noise model and low noise models, respectively (Peterson, 1993).

**Table 4.** Northern Finland Seismological Network (network code FN). Station information.

| Station name | Code | Lat. N (deg) | Long. E (deg) | Elev. (m) | Sensor | Data acquisition | | Digitizer sensitivity (microvolt/ count) | Data transfer | Data format | Start of operation |
|---|---|---|---|---|---|---|---|---|---|---|---|
| Oulu, Huttukylä | OUL | 65.085 | 25.896 | 60 | Streckeisen STS-2, 2nd generation | EarthData 24+Linux ComP | PS6-Seis- | 2.5 | Internet ADSL | mSEED | 10 Aug 2005 |
| Kuusamo, Maaselkä | MSF | 65.911 | 29.040 | 365 | Streckeisen STS-2, 2nd generation | EarthData 24+Linux ComP | PS6-Seis- | 2.5 | Internet WLAN | mSEED | 17 Oct 2005 |
| Sodankylä | SGF | 67.442 | 26.526 | 180 | Streckeisen STS-2, 2nd generation | EarthData 24+Linux ComP | PS6-Seis- | 1.0 | Internet WLAN | mSEED | 4 Jan 2006 |
| Rovaniemi | RNF | 66.612 | 26.010 | 198.1 | Streckeisen STS-2, 2nd generation | EarthData 24+Linux ComP | PS6-Seis- | 1.0 | Internet WLAN | mSEED | 6 Nov 2007 |

In 1963, the Department of Physics founded a new seismic station on the quiet site of Aarne Karjalainen Observatory in Huttukylä (about 18 km from Oulu; station code OUL; Fig. 2). The property and observatory buildings were donated to the University of Oulu by Aarne Karjalainen. The new station was equipped with the Benioff seismometer and Geotech Co. helicorder, and later by the short-period vertical Willmore seismometer and long-period vertical Sprengnether seismometer (Table 3). Since 1980, the station had operated at the Ervasti site nearby (the station code was not changed) until the equipment of the station was destroyed by a thunderstorm in summer 1996.

In 1970, the UO founded a new seismic station at the Maaselkä site (station code MSF; Fig. 2), about 10 km from town of Kuusamo in north-eastern Finland (Table 3). Digital registration using a new type of seismic equipment started at Oulu station (Huttukylä site) in 1999 and in MSF station in 2000 (Table 3) in CSS data format (Anderson et al., 1990) with a sampling rate of 50 sps. The data acquisition system was the same as the system installed in other Finnish permanent stations operated by the ISUH (Luosto, 2001). Continuous data were recorded to the hard disk drive of the station's Linux computer and transmitted to the data server located at the UO via telephone lines.

In 1968, the position of seismologist was founded at the UO together with the foundation of the Department of Geophysics. Heikki Korhonen was appointed as the first seismologist at Oulu in 1968 and Jukka Yliniemi became his successor in 1977. The position was later transferred to the Geophysical Observatory founded in 1985 at the UO. The Sodankyla Geophysical Observatory was united with the University of Oulu in 1997, and the Geophysical Observatory was merged into it in the following year. The position of seismologists was simultaneously moved to SGO, and Jukka Yliniemi was responsible for seismic measurements at SGO until 2004; Elena Kozlovskaya was his successor.

At first, studies of microseismic ambient noise and local seismicity were the main research branches of seismology in Oulu University (Korhonen et al., 1980). Since the 1980s, the geophysicists of the university have participated actively in deep seismic wide-angle reflection and refraction surveys in Finland and abroad (Luosto, 2001; Tables 1 and 2). As a result, the research direction has changed towards an interpretation of controlled-source seismic experiment data and lithosphere studies (Yliniemi, 1991, 1992).

## 3   Recent seismic observations at SGO

### 3.1   Northern Finland Seismological Network: seismology in the 21st century

In 2004, it became apparent that existing permanent seismic stations of SGO did not satisfy the requirements of the 21st century seismology. First, they were equipped with the short-period Geotech S-13 seismic sensors, while the majority of seismic network operators in Europe had already changed their equipment to broadband force-balanced seismic sensors. Another problem was that the data of the SGO stations were not open and had been used by the ISUH solely for locating local seismic events. The continuous seismic data were not archived in any international data centre, and recordings of teleseismic events were not used in seismological research.

During 2005–2007, the Oulu unit of SGO started to modernize its permanent seismic stations. During this modernization, the short-period seismic instruments were replaced by Streckeisen STS-2 broadband seismometers and the existing data acquisition system was replaced by the Earth Data PR6-24 24-bit digitizers and Linux computers with SeisComP seismic data acquisition software (SeisComP Manual, 2006). The agreement was reached with the GeoForschungsZentrum (GFZ) Potsdam on archiving and distribution of the

seismic data via the GFZ Data Archive. A new seismic broadband station in Rovaniemi (station code RNF) with the same type of equipment was established in 2008.

At the moment, SGO operates the Federation of Digital Seismograph Networks (FDSN network code FN). It is a permanent real-time broadband seismic network consisting of four real-time stations (OUL, MSF, SGF, RNF). The information about stations of the FN network is given in Table 4 and Fig. 2 shows the noise level at the vertical component of these stations in 2014. Two new stations (Oulanka – OLKF and Kolari – KLF) were installed in 2014. They are now working in test regime and will be connected to the network after testing. The network is a part of GEOFON (GEOFOrschungsNets: http://www.geofon.gfz-potsdam.de/wave) Extended Virtual Network – GEVN, of the Virtual European Broadband Seismograph Network (VEBSN) operated by ORFEUS (Observatories and Research Facilities for European Seismology) and of the global International Federation of Digital Seismograph Network (FDSN). The Oulu unit of SGO represents the University of Oulu in the Incorporated Research Institutions for Seismology (IRIS; as a Foreign Affiliate).

The continuous seismic data of the Northern Finland Seismological Network in MiniSeed format (SEED Manual, 2002) is archived in the GFZ Seismological Data Archive of the GFZ Potsdam (Germany) and at their own archive of the Oulu unit. Since 2011, the data are also archived in the European ORFEUS Data Centre (www.orfeus-eu.org) that now holds the European Integrated Data Archive (EIDA) of seismological data. The data are used for monitoring of seismic activity in northern Europe and worldwide as well as for detection of local and teleseismic events. Information about seismic events is published in several online bulletins, including the bulletin of seismic events in Fennoscandia by ISUH (http://www.helsinki.fi/geo/seismo/english/bulletins/). The data are, via the GFZ, distributed through the ORFEUS (EIDA) data distribution system.

### 3.2   Temporary seismic experiments at SGO

The seismic group of the Sodankylä Geophysical Observatory of the University of Oulu has participated with its own resources and equipment in many seismic projects in Finland (see Fig. 1) and abroad. The controlled-source seismic projects are listed in Tables 1 and 2 with references to publications introducing the projects and collaborations. See Sect. 3.2.1 for more details. In addition, SGO has participated in several passive seismic experiments that are shortly introduced in Sect. 3.2.2–3.2.4 and 3.2.6 and coordinated a passive seismic POLENET/LAPNET experiment (Sect. 3.2.5) during International Polar Year (IPY) 2007–2009.

### 3.2.1 Seismic wide-angle reflection and refraction experiments

The Seismic group of the Sodankylä Geophysical Observatory of the University of Oulu has participated with its own resources and equipment in many seismic controlled-source experiments in Finland and abroad (Fig. 1, Tables 1 and 2). The scientific results of these experiments have been published in numerous papers summarized by Luosto (2001) and Grad et al. (2006). During 2001–2005, the seismic group of SGO participated in the Finnish Reflection Experiment (FIRE) carried out by a consortium consisting of the Geological Survey of Finland, the ISUH, Department of Geosciences of the University of Oulu, and SGO. Deep seismic reflection soundings were made along four main transects with a total length of 2104 km in the central and northern parts of the Fennoscandian Shield (Kukkonen and Lahtinen, 2006). The main contractor of the project was Spetsgeofizika S.E. (Russia). The Oulu seismic group and ISUH also organized wide-angle reflection and refraction measurements along FIRE lines using its own equipment (Silvennoinen et al., 2010).

Oulu University is responsible for storing the data of several controlled-source seismic experiments. In Finland, the ISUH and the Geological Survey of Finland (GSF) are also archiving the data of a number of such experiments.

Initially, the equipment of the seismic group for controlled-source seismic experiments included Willmore vertical seismometers and PCM-1218-80 recorders developed by the ISUH. Since 1996, the equipment has consisted of 8 Reftek 72 data loggers (used in cooperation with the Department of Geophysics of UO) and eight 3C Mark Products L4C seismometers with a natural frequency of 2 Hz.

### 3.2.2 SVEKALAPKO passive seismic array research

In 1997–1999, the seismic group of SGO, together with the Department of Geophysics of Oulu University and ISUH, participated in the EUROPROBE/SVEKALAPKO Deep Seismic Tomography project (Hjelt and Daily, 1996, 2006; Bock et al., 2001). These papers also provide the information about which individual researchers and research organizations participated in the project. The project was a passive seismic array research in southern and central Finland aimed at studying the lithosphere–asthenosphere transition in the suture zone of Proterozoic Svecofennian and Archaean Karelian domains of the Fennoscandian Schield (Fig. 1). The detailed description of the experiment, including sites and equipment information, is given in Sandoval (2002). The results of the SVEKALAPKO array research dramatically changed the point of view of the structure of the mantle lithosphere beneath Finland. Prior to the experiment, it was assumed that the lithosphere there is thick, and the structure of the mantle lithosphere is relatively simple. This opinion was based on worldwide studies of upper mantle xenoliths from Archaean and Proterozoic areas that demonstrated a certain correlation between the composition of the subcontinental lithospheric mantle (SCLM) and crustal tectonothermal age (see, for example, Griffin et al., 2003). Thus prior to the SVEKALAPKO experiment, higher velocities and lower densities in Archaean domain and lower velocities and higher densities in the Proterozoic domain were expected. Instead, inhomogeneous and anisotropic upper mantle beneath the Proterozoic–Archean suture zone has been revealed (Alinaghi et al., 2003; Kozlovskaya et al., 2004; Kozlovskaya et al., 2007; Sandoval, 2002; Sandoval et al., 2003, 2004; Bruneton et al., 2002, 2004a, b; Yliniemi et al., 2004; Plomerová et al., 2006; Vescey et al., 2007; Kozlovskaya et al., 2008; Pedersen et al., 2006, 2007).

### 3.2.3 ALPASS (Alpine Lithosphere and Upper Mantle PASsive Seismic Monitoring) experiment

Leading organization of the ALPASS project was Institute of Geodesy and Geophysics, Vienna University of Technology, principal investigator E. Brückl. For more information on ALPASS, see Mittelbauer et al. (2011).

ALPASS was a passive seismic monitoring project aimed at revealing lower lithosphere and upper mantle beneath the wider eastern alpine region, and to contribute to a better understanding of the geodynamic processes at work. Participating countries were Austria, Croatia, Finland, Hungary, Poland, and the USA. The seismic group of SGO participated in the passive seismic experiment in 2005–2006 with their own field instruments. In 2009, it participated in data processing and teleseismic tomography studies (Mittelbauer et al., 2011).

### 3.2.4 PASSEQ 2006–2007: passive seismic experiment in the Trans-European Suture Zone

The primary aim of the PASSEQ 2006–2007 passive seismic array experiment was an investigation of the seismic structure of the mantle and lithosphere–asthenosphere boundary in the Trans-European Suture Zone (TESZ) in central Europe, between the young Palaeozoic platform of the western European and Precambrian eastern European platform (Wilde-Piortko et al., 2008).

SGO participated in the passive measurements in the territory of Lithuania in 2006–2007 with its own equipment. In 2009, PASSEQ research continued within the project "Investigation of local seismicity in Lithuania using the data of Passive Seismic Experiment PASSEQ 2006–2008" that was carried out by the seismic group of SGO in collaboration with the University of Vilnius and the Geological Survey of Lithuania (Janutyte, 2012). The teleseismic P-wave tomography using the PASSEQ 2006–2007 data (Janutyte et al., 2015) showed significant differences in seismic velocity structure beneath the TESZ, young Palaeozoic western Europe, and eastern European platform.

### 3.2.5 POLENET/LAPNET seismic array experiment during the International Polar Year 2007–2009

POLENET/LAPNET (Fig. 1) was a sub-project of the IPY 2007–2009 POLENET consortium related to seismic studies in the Arctic (http://ipydis.org). The main target of the project was to carry out an ambitious temporary broadband seismic array research in northern Fennoscandia (northern parts of Finland, Sweden, Norway, and Russian Karelia). The experiment was initiated by the group of scientists, who participated previously in the SVEKALAPKO experiment (Helle Pedersen, Jaroslava Plomerová, Ulrich Achauer, Eduard Kissling, Irina Sanina, Elena Kozlovskaya) and its aim was to continue the SVEKALAPKO array to the north. Equipment for the temporary deployment was provided by RESIF-SISMOB, FOSFORE, EOST-IPG Strasbourg Equipe seismologie (France), Seismic pool (MOBNET) of the Geophysical Institute of the Czech Academy of Sciences (Czech Republic), the Sodankylä Geophysical Observatory (Finland), the Institute of Geosphere Dynamics of the Russian Academy of Sciences (RAS) (Russia), the Institute of Geophysics ETH Zürich (Switzerland), the Institute of Geodesy and Geophysics, the Vienna University of Technology (Austria), and the University of Leeds (UK). For a full list of the working group see, for example, Plomerová et al. (2011) and Pedersen et al. (2013). The project was coordinated by SGO with Elena Kozlovskaya as the principal investigator. SGO also carried the responsibility of serving the stations during the data acquisition period.

The POLENET/LAPNET array, with the average spacing between stations of 70 km, was designed to solve specific tasks of polar seismology. The collected POLENET/LAPNET data set (Kozlovskaya et al., 2007) includes continuous high-frequency data (sampling rate from 50 to 100 sps) of 37 temporary stations, which were in operation during the period from 1 May 2008 to 31 September 2009, and of 21 stations of selected permanent networks in Fennoscandia. Most of the stations of the array were equipped with broadband sensors. The data of broadband stations, pre-processed into the standard seismological miniSeed format, are now deposited into the database of RESIF Data Centre at the University of Grenoble (France) (http://seismology.resif.fr). The metadata about POLENET/LAPNET stations, their coordinates, and instrumentation are also deposited into the database. The backup copy of all continuous data is stored at SGO. In addition, the data of several short-period stations are archived at SGO and Geophysical Centre RAS, Schmidt Institute of Physics of the Earth RAS, Russia.

The data of the POLENET/LAPNET array have been interpreted by different research groups at the participating institutions, using different techniques. The main results of the POLENET/LAPNET project were published in a number of papers. Plomerová et al. (2011) and Vinnik et al. (2014) estimated seismic anisotropy in the upper man-

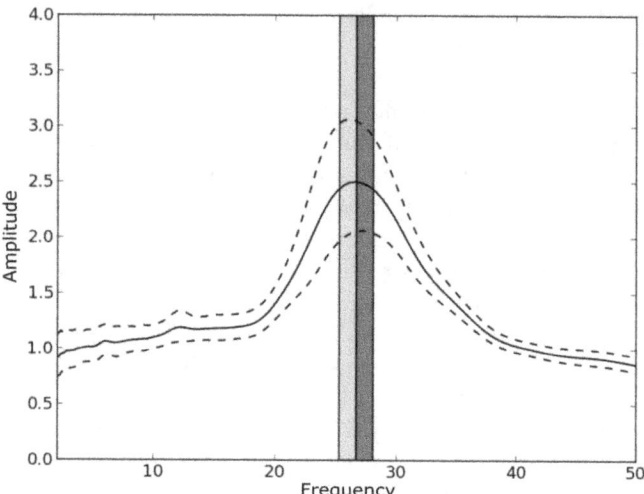

**Figure 3.** A typical H / V curve measured with MEMS-based 3-component accelerometer. The peak frequency corresponds to ~ 1 m sediment thickness.

tle beneath the LAPNET study area. Vinnik et al. (2016) estimated variations of S- and P-wave velocities, the Vp / Vs ratio and major boundaries in the upper mantle beneath the POLENET/LAPNET array using joint inversion of P- and S-receiver functions. Pedersen et al. (2013) presented results of surface wave studies. Silvennoinen et al. (2014) presented a new map of the Moho boundary for the northern part of Fennoscandia and an upper mantle P-wave velocity model estimated by teleseismic tomography (Silvennoinen et al., 2016). Krasnoshchekov et al. (2016) used the data from the array for studying of the Earth's inner core. For the first time, Poli et al. (2012, 2013) used ambient seismic noise recorded in Finland to estimate the inner structure of the Earth's crust and upper mantle. Usoltseva and Kozlovskaya (2016) presented results of local event studies, and Gibbons et al. (2015) used the POLENET/LAPNET array data to investigate the propagation of infrasound signals.

### 3.2.6 DAFNE – seismic monitoring of postglacial faults and the ICDP drilling project

The Drilling Active Faults in Northern Europe (DAFNE) project (Kukkonen et al., 2010) aims to investigate, via scientific drilling, the tectonic and structural characteristics of postglacial faults (PGFs) in northern Fennoscandia. During the last stages of the Weichselian glaciation (ca. 9000–15 000 years B.P.), reduced ice load and relaxation of accumulated tectonic stress resulted in a rapid uplift in Fennoscandia. Active faulting occurred with fault scarps up to 150 km long and up to 30 m high. Some of these faults show week seismicity even presently. That is why studying of PGFs would create information relevant for proper seismic hazard evaluation and planning and exploitation of such critical facilities as nuclear waste disposal and underground

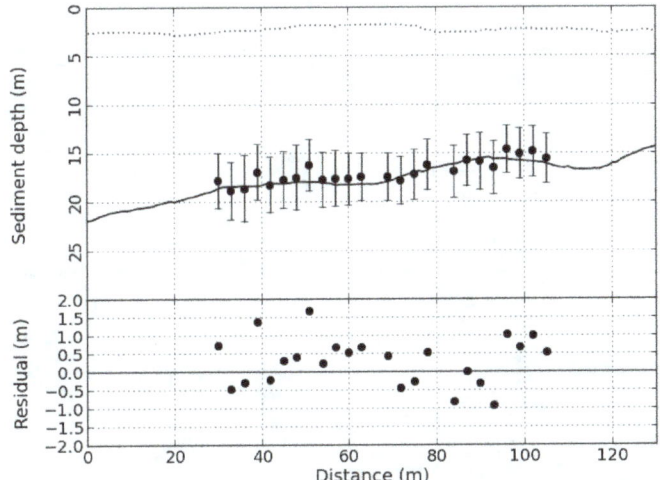

**Figure 4.** Sediment thickness extracted by the H / V method (black dots) and compared with results from ground penetrating radar (black line). The error bars correspond to the width of the H / V curve. The dotted line corresponds to water level depths.

**Figure 5.** Overview of the Pyhäsalmi mine, Finland, Photo was kindly provided by Timo Mäki, First Quantum Minerals Ltd.

mines. The main purpose of the DAFNE/FINLAND passive seismic array experiment was to characterize the present-day seismicity of the Suasselkä postglacial fault (SPGF) that was proposed as one potential target for the DAFNE project. As the fault is located far from permanent stations of regional seismic networks in Fennoscandia, no natural seismicity from the fault was reported previously. In order to check whether the fault is still active, eight short-period and four broadband 3C seismic stations were installed in the close vicinity of the fault area in September 2011. During September 2011–May 2012, we collected the data of more than 70 000 seismic events (teleseismic, regional and local ones). Recordings of the array have been analysed manually and automatically, in order to find natural earthquakes from the fault area. The detected events were located and spectral charac-

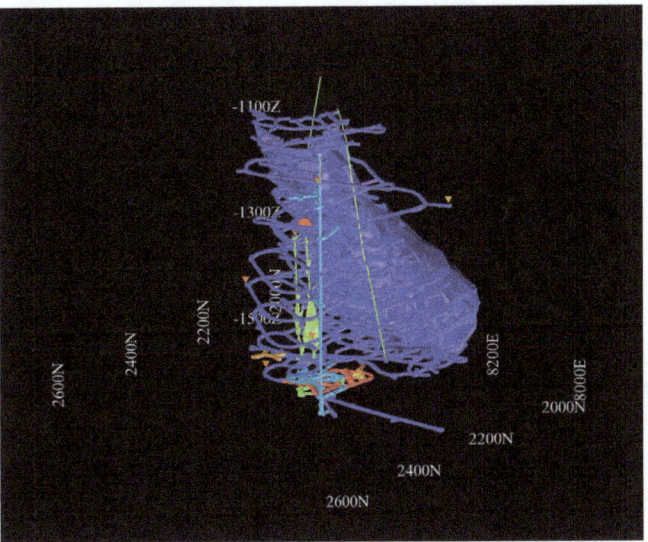

**Figure 6.** Pyhäsalmi mine deep ore body and work routes. The red ball represents seismic event and orange triangles geophones that detected it. Published with permission of First Quantum Minerals Ltd.

teristics of signals were analysed, in order to discriminate natural events originating from the fault, from both production blasts and mining induced events originating from the Kittilä Gold Mine. As a result, we found several dozens of events originating from the fault area that could be of natural origin. We also found and analysed a number of events originating from the Kittilä Gold Mine that could correspond to rock fall in the areas of production and mine development (Kozlovskaya et al., 2013).

## 4 Future of seismic observations at SGO

### 4.1 European Plate Observing System at the University of Oulu

The EPOS is the integrated open-access solid Earth Sciences research infrastructure approved by the European Strategy Forum on Research Infrastructures (ESFRI) and included in the ESFRI road map in December 2008 (European Commission, 2011). EPOS is a long-term integration plan of national existing research infrastructures (RI). The implementation phase of EPOS will be during 2015–2018. The result will be a single sustainable, permanent geophysical observational infrastructure, integrating existing monitoring networks (e.g. seismic and geodetic networks), local observatories (e.g. volcano observatories), and experimental laboratories (e.g. experimental and analytic laboratories for rock physics and tectonic analogue modelling) in Europe and adjacent regions (EPOS, 2016). Partners of the FIN-EPOS national Finnish EPOS consortium are THE Universities of Helsinki and Oulu, National Land Survey, Finnish Meteo-

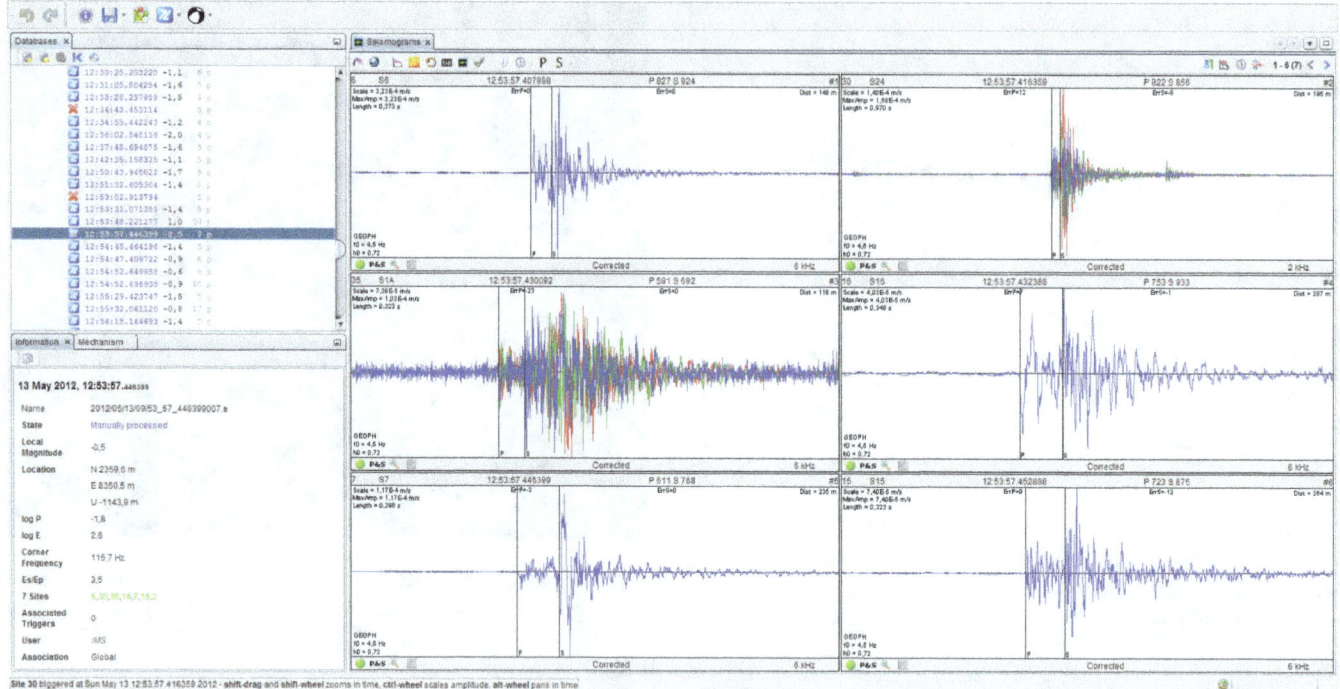

**Figure 7.** Example of seismograms of microseismic event recorded by mycroseismic monitoring network in Pyhäsalmi mine (with permission of First Quantum minerals Ltd.).

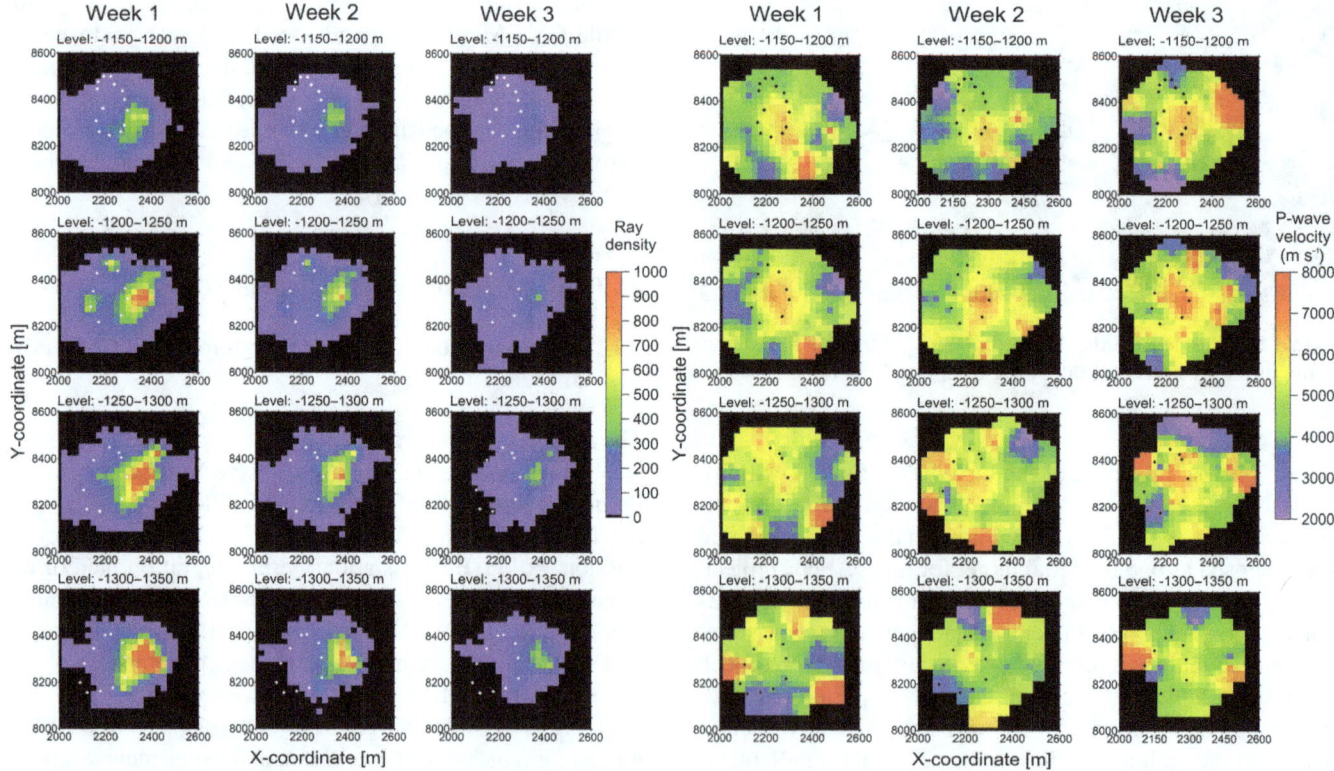

**Figure 8.** An example of results of travel-time tomography using the data of microseismic monitoring network in Pyhäsalmi Mine. Left panel: ray density images for the first 3 weeks of May 2012. Right panel: the results of seismic travel-time tomography for same time period. White and black dots show the average boundary of Pyhäsalmi deep ore body for the corresponding depths levels.

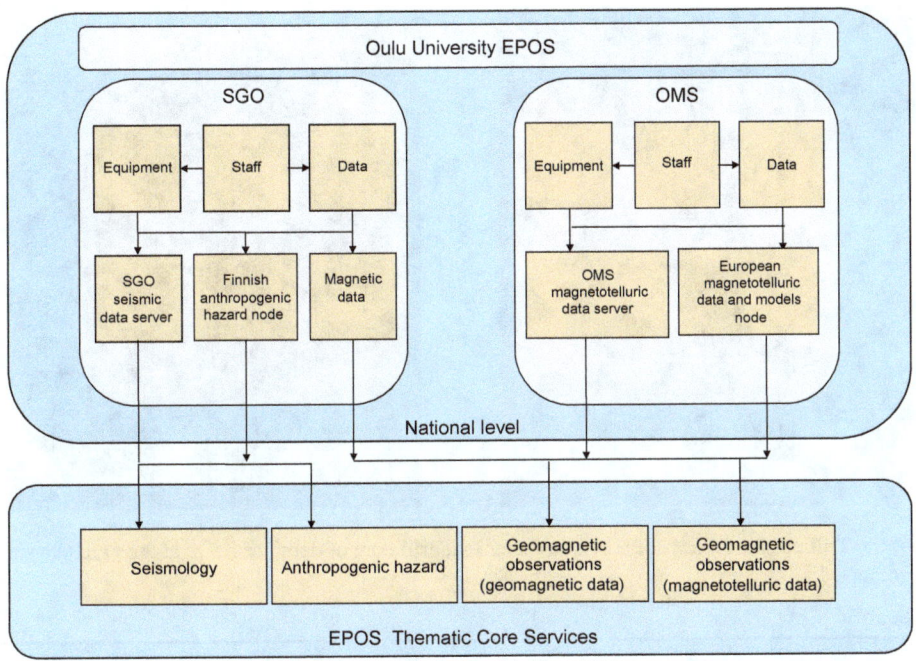

**Figure 9.** EPOS infrastructure at the University of Oulu in 2015.

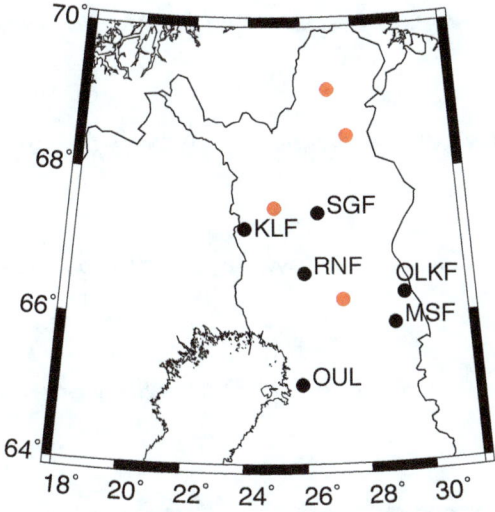

- ● Permanent seismic station of Sodankylä Geophysical Observatory
- ● Planned permanent seismic station of Sodankylä Geophysical Observatory

**Figure 10.** Location of seismic stations of the FN network. Black dots indicate position of stations that are in operation in 2015, including OLKF and KLF stations operating in the test regime. Red dots indicate position of stations that will be installed in 2016–2017.

rological Institute, Geological Survey of Finland, CSC – IT Center for Science, and VTT Technical Research Centre of Finland Ltd. The consortium is hosted by the ISUH. The national coordination office will be placed at ISUH. The con-

sortium leader/chair and principal investigator (PI) is Annakaisa Korja, Director of ISUH, and the consortium's vice-chair and co-PI is Markku Poutanen from the National Land Survey.

The EPOS will (a) build up excellent science opportunities in Earth sciences, (b) strengthen capacity building for new generations, (c) contribute to the natural hazard mitigation, (d) provide easily accessible geoscientific real-time data and data products, (e) maintaining reference frameworks for society and industry, (f) foster IT innovations related to analysis and management of large distributed data sets.

At the UO, two Departments contributing to the FIN-EPOS infrastructure are SGO and Oulu Mining School (OMS) (Fig. 9).

### 4.2  Upgrading the Northern Finland Seismological Network (FN)

The Finnish National Seismic Network (FNSN) comprises the national Helsinki University Seismological Network (HE) and the Northern Finland Seismological Network (FN) hosted by SGO. As a part of EPOS activities at the national level, both organizations started to upgrade their networks in 2015. ISUH focuses on increasing the nation-wide permanent station coverage by four stations while SGO focuses on increasing the permanent station coverage in the polar region (four stations, Fig. 10). As the networks are overlapping and complementary, additions in one network are also beneficial to the other one. The consortium project is funded by the Academy of Finland in 2015–2017. Also, SGO will establish a new national central hub for induced seismicity data. Ini-

**Figure 11.** Installation of the Trillium Posthole 120PH seismometer in a drill core of depth of 5.5 m at the Oulanka site (station code OLKF). Photo by Hanna Silvennoinen.

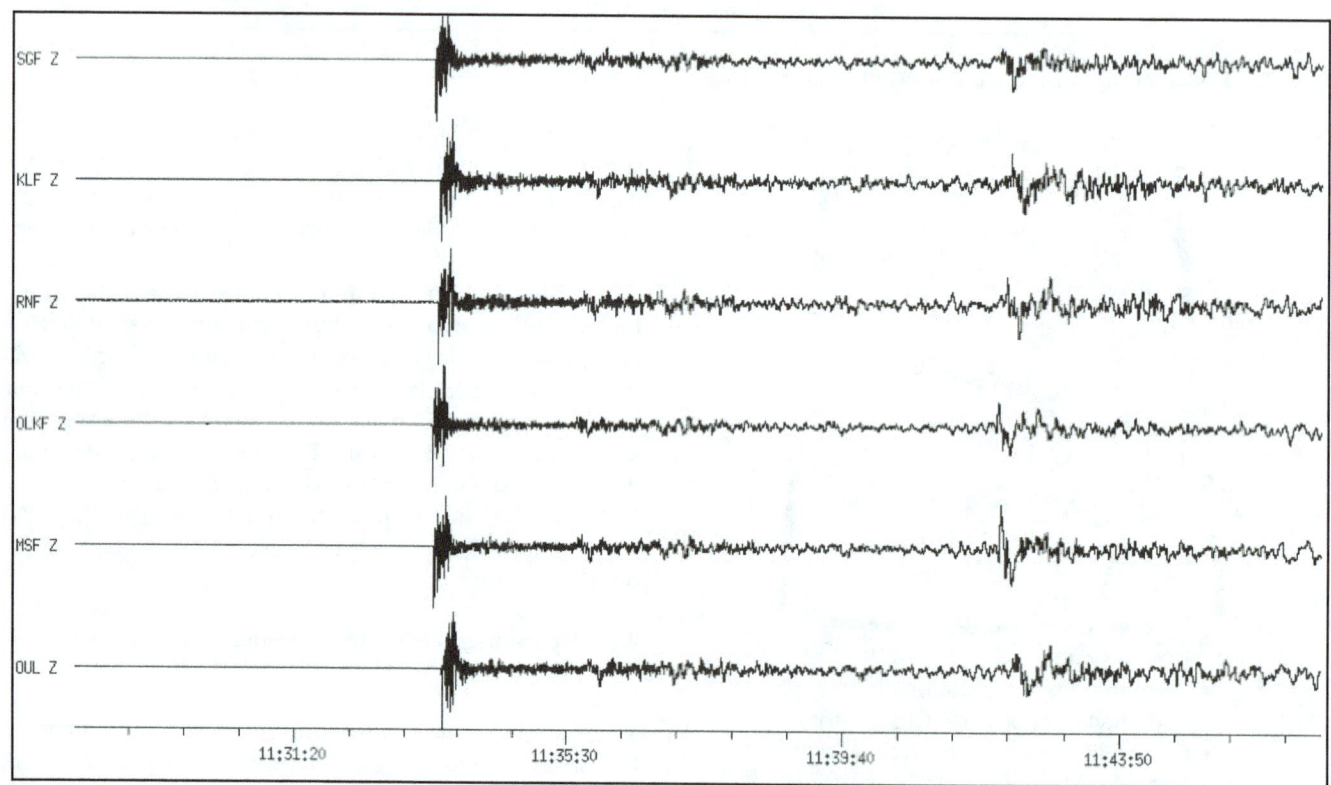

**Figure 12.** An example of unfiltered seismogram of teleseismic event on 29 May 2015 at 11:23:02 UTC from Bonin Island with $M_\mathrm{w} = 7.8$ recorded by the upgraded FN array in 2015. New stations OLKF and KLF are included.

tiation of the national induced seismicity database will contribute significantly to one of the research focus areas of the UO: the environment, natural resources and materials as well as to the mining and mineral field development area that has recently resulted in the establishment of a new mining faculty (Oulu Mining School).

The new permanent stations of SGO will be installed in boreholes and equipped with the Trillium Posthole 120PH seismometers. This type of instrument has been under testing at the OLKF station in Oulanka since 2014, where the staff of SGO is testing various materials and technical solutions for installation of equipment, insulation of sensor,

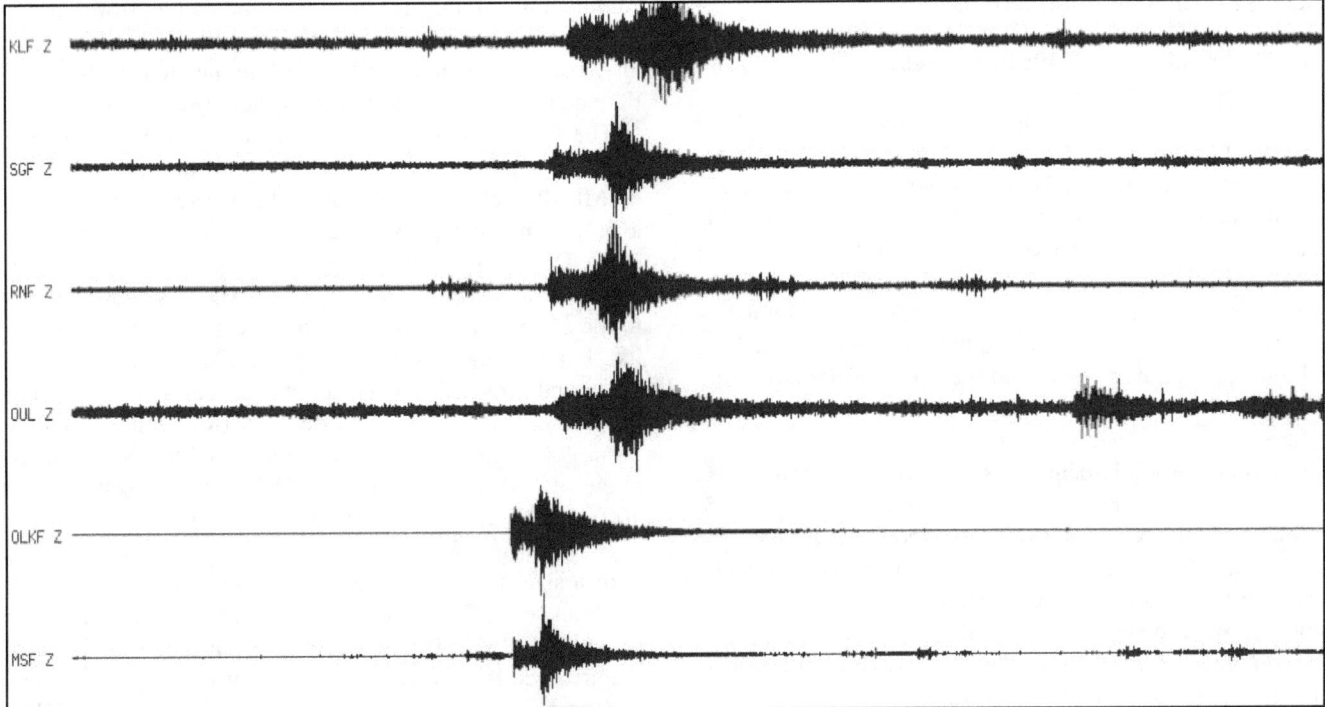

**Figure 13.** An example of seismograms of local event from northern Russia on 29 June 2015 at 13:05:08 UTC with $M = 1.9$ recorded by the upgraded FN array. Recordings are filtered by the 2–22 Hz bandpass filter. Stations OLKF and KLF are included.

power supply, and data transmission. Figure 11 shows the process of installation of the posthole seismometer at the Oulanka site. The station network of SGO is located north of 65° N, which causes some particular challenges related to the limited infrastructure and the subarctic climate of the area. For most of northern Finland, the population density is less than 1 person $km^{-2}$ (Population Register Centre of Finland), which limits the availability of necessary infrastructures such as electricity and even roads. The subarctic climate (extreme cold temperature and snow coverage) and polar night limit working during winter months. The installation of the seismometer into a borehole offers stable registration temperature throughout the year but above-ground registration electronics still require insulation and a heating system to keep operational.

Figures 12 and 13 demonstrate a comparison of recordings of existing permanent stations of FN equipped with the STS-2 seismometer with the recordings of stations working in a test regime and equipped with Trillium Posthole 120PH seismometer (OLKF station) and Trillium 120PA broadband sensor (KLF station).

## 4.3 Laboratory of Applied Seismology

### 4.3.1 General target of the Laboratory of Applied Seismology of SGO

Currently it is recognized that the Fennoscandian Shield with resources and proven potential is the most prospective ground in Europe. However, there are under-explored geological formations for a number of commodities such as base metals, gold, platinum group metals, iron ore, and diamonds. This requires a development of new geophysical methods for investigation of sub-surface structures, in particular, methods capable of mapping 3-D geological structures: metallic and non-metallic ore bodies, faults, fracture zones, overburden, intrusions, fault zones, etc., at a depth of several kilometres. In order answer to these new challenges, the seismic group of SGO decided to upgrade its portable seismic instrumentation and organize the SEISLAB. The laboratory aims to ascertain the availability of equipment and personnel to participate in both SGO's own and collaborative short-term projects. Also, to the traditional tasks of applied and engineering seismology, the target of the new laboratory is to develop monitoring techniques for mining-induced seismicity and passive seismic interferometry methods for mapping of 3-D geological targets (e.g. fault zones, intrusions, orebodies). The project was funded by the European Regional Development Fund (ERDF), Council of Oulu region and Pyhäsalmi Mine Oy in April 2012–January 2014.

### 4.3.2  SEISLAB instrumentation

**Portable broadband seismic instruments**

The portable broadband equipment consists of 11 Trillium Compact 120 broadband 3-axial seismic sensors (cut-off period of 120 s) manufactured by Nanometrics Ltd. (www. nanometrics.com), four Reftek130 24-bit 3-channel portable data loggers (www.trimble.com), three Earth Data PR6-24 24-bit 3-channel portable data loggers, and four EDR-210 24-bit portable data loggers. Power supply for autonomous operation is provided by lead acid batteries. The detailed description of each instrument and their technical characteristics are available at the web-pages of correspondent manufacturers.

The portable equipment is used in passive seismic experiments, both in Finland and in Europe (c.f. ALPASS, PASSEQ, POLENET/LAPNET, DAFNE, SEISLAB) for monitoring local earthquakes and mining-induced seismic events and for crustal and lithosphere studies using seismic tomography, receiver functions and ambient noise methods. The equipment can also be used in active source applied geophysics experiments (depth to several km).

**Sercel Unite multichannel system**

Seismic instrumentation of the laboratory also includes the Sercel UNITE multichannel seismic equipment manufactured by Sercel Ltd. (www.sercel.com). The equipment has 40 3C DSU-SA Micro-machined Electro-Mechanical Sensor (MEMS) sensors and 40 RAU ex-D data acquisition units with internal batteries, totally 120 channels. The field equipment also includes

1. wireless data harvester (Tablet PC) with cables for data harvesting and quality control

2. PFT-portable field terminal for uploading serial numbers of RAU and initiating experiment

3. a portable version of Unite LITE acquisition system with a software license for max 150 channels

4. special battery charger for 20 RAU ex-D units.

Sercel UNITE system is an autonomous recording system composed of remote acquisition units (RAU ex-D) and MEMS-based accelerometers within a digital sensor unit (DSU3SA). Previously UNITE system has been used in reflection and refraction surveys with active energy sources (Lansley et al., 2008; McWhorter et al., 2012).

RAU ex-D houses an internal Li-ion battery, a non-volatile memory (32 GB), integrated GPS and Wi-Fi in a compact IP68 rated case with weight less than 2 kg. A radio identification (RFID) enables a fast identification of recording unit. Internal battery enables 130 h autonomous operation, which can be further extended by using an external battery. Memory autonomy with 500 Hz sampling rate is more than 300 h.

Acquisition parameters can be set, and data retrieved trough Ethernet port or wireless via Wi-Fi transmission. Additionally, the licence free wireless communication enables real-time quality control (QC) of the system (Sercel Ltd.).

The DSU3SA is a 3C accelerometer that is powered by remote autonomous unit (RAU ex-D). The sensor is based on MEMS technology. These digital accelerometers provide a broadband linear response (DC to 800 Hz) (www.sercel. com). DSU3SA is a digital sensor unit in the same way that a 24-bit analogue-to-digital converter (ADC) is interconnected to the MEMS, and thus the output of the sensor unit is digital. Digital data transmission to RAUDex-D avoids pick-up noise and crosstalk related to conventional analogue transmission between sensor and digitizer (Mougenot and Thorburn, 2004). DSU3SA has a full scale of $5 \, \text{m s}^{-2}$, dynamic range of 120 dB @ 250 Hz sampling rate and self-noise on $400 \, \text{nm} /\text{s2}/\sqrt{\text{Hz}}$ (10–200 Hz) (www.sercel.com).

A typical MEMS accelerometer is a small silicon chip, with a size of $1 \, \text{cm}^{-2}$, weight $< 2 \, \text{g}$ and proof mass in microgram scale. From the application point of view, the main advantage of MEMS accelerometers over traditional electromagnetic coil-based sensors is their broadband linear phase and amplitude response that may extend from 0 (DC) to 800 Hz within 1 dB.

Additionally, MEMS resonant frequency is far above the seismic band pass (1 kHz). This makes it possible to record frequencies below 10 Hz without attenuation, including the direct current (DC) related to the gravity acceleration (Laine, 2014).

The main challenges related to MEMS technology are related to the sensitivity and self-noise affecting signal-to-noise ratio, especially at low frequencies.

The DSU3SA has self-noise of $400 \, \text{nm s}^{-2} \sqrt{\text{Hz}}^{-1}$ (between 10 and 100 Hz). However, self-noise increases toward low frequencies; below 55 Hz it becomes higher than that of a geophone-digitizer system and below 5 Hz it can exceed ambient noise (Laine, 2014), making ambient noise recording in this frequency domain impossible. As a reference, according to New Low Noise Model (NLNM) the minimum terrestrial noise to be reached is $40 \, \text{nm} \sqrt{\text{Hz}}^{-1}$ (1–100 Hz) (Peterson, 1993). At high frequencies ($> 50 \, \text{Hz}$) the floor/electric noise of the MEMS is lower than that of the equivalent geophone/station electronics (Mougenot and Thorburn, 2004).

### 4.3.3  Examples of measurements and research made during the SEISLAB project

**Passive measurements using MEMS-based sensors**

Recording of ambient seismic noise (vibration of the Earth due to natural or industrial sources) is presently used in many passive seismic methods. Ambient noise measurements can be used to extract information on geological structures or locate underground oil or gas reservoirs, or other resources. Passive seismic methods are becoming more and more important since new passive seismic methods are developed due to scientific, economic, and ecological reasons.

The suitability of a new type of seismic equipment, based on MEMS technology, to record ambient seismic was tested during the experiment in Haukipudas area near Oulu in 2013 (Fig. 1) where the MEMS seismic sensors were installed along a small-scale profile cutting known sedimentary formation. The aim was to extract information on the subsurface structure using H / V (horizontal-to-vertical) spectral ratio of ambient seismic noise. The technique originally proposed by Nogoshi and Igarashi (1971), and promulgated by Nakamura (1989), consists of estimating the ratio between the Fourier amplitude spectra of the horizontal to vertical components of the ambient noise vibrations recorded at one single station. The computation of the H / V ratio follows several steps and includes (a) recording a 3C ambient noise signal, (b) selection of the most suitable time windows (e.g. using an anti-triggering algorithm), (c) computation and smoothing of the Fourier amplitude spectra for each time window, (d) averaging the two horizontal components (using a quadratic mean), (e) computation of the H / V ratio for each window, and (f) computation of the average H / V ratio (SESAME, 2005).

In our study, we used the Geopsy software (http://www.geopsy.org) to perform the H / V analysis of the ambient noise data recorded by the Sercel multichannel seismic equipment. Figure 3 demonstrates a typical example of the H / V data analysis and interpretation. Results of the measurements along the Haukipudas profile were compared with those extracted with conventional coil-based broadband seismometers (Nanometrics Trillium Compact) and were also compared with results from other methods such as ground penetration radar (Fig. 4). The comparison showed that the new equipment of the SEISLAB can be used in passive seismic methods based on ambient noise analysis and in a number of other applied seismology tasks as well.

## Seismic travel-time tomography in Pyhäsalmi mine

During the SEISLAB project we started to investigate whether or not passive microseismic monitoring data from Pyhäsalmi mine, Finland, (Fig. 5) can be used to model seismic velocity structure within the mine. The seismicity in the Pyhäsalmi mine is driven by the changes in rock mechanic state due the ongoing mining operation, and thus it is a mine-induced seismicity. The mine-induced seismic event data in Pyhäsalmi mine have been recorded since 2002 when the passive microseismic monitoring network designed by the Institute of Mine Seismology (http://www.imseismology.org) was installed in the mine (Fig. 6). Since then over 120 000 microseismic size events have been observed (Pekka Bergström, personal communication, 2015). An example of seismogram of a microseismic event is shown in Fig. 7.

The purpose of our study was to test how the travel-time seismic tomography performs with the passive microseismic monitoring data where the source–receiver geometry is based on a non-even distribution of mine-induced events in the mine, and hence is a non-ideal one for the travel-time tomography. The tomographic inversion procedure was tested with the synthetic data and real source–receiver geometry and with the real travel-time data of the first arrivals of P-waves from the microseismic events. The synthetic modelling gave positive results as known synthetic model was retrieved by used the SIRT (simultaneous iterative reconstruction technique) method (Lo and Inderwiesen, 1994). The results showed that the travel-time tomography is capable of revealing differences in seismic velocities in the mine area corresponding to different rock types (for example, the velocity contrast between the orebody and surrounding rock can be easily distinguished). The velocity model recovered corresponds well to the known geological structures in the mine area (Fig. 8).

The second target was to apply the travel-time tomography to microseismic monitoring data recorded during different time periods in order to track temporal changes in seismic velocities within the mining area as the excavation proceeds. The result shows that such a time-lapse travel-time tomography can recover such changes (Fig. 8). In order to obtain good ray coverage and good resolution, the time interval for a single tomography iteration needs to be selected taking into account the number of events and their spatial distribution.

From our results, it can be concluded that seismic tomography is applicable to Pyhäsalmi mine passive seismic monitoring data, and the dense ore body can be detected by seismic tomography. There is also a variability between results obtained using different weekly data sets, as the number of microseismic events and correspondent ray coverage depends on ore production and changes from week to week. From the results, it can be seen, however, that there are periods of time that the distribution has been favourable for tomography even for as short a time period as 1 week.

An example of microseismic monitoring data from the Pyhäsalmi mine will be included in a database of induced seismicity episodes of the EPOS anthropogenic hazard node. At the Pyhäsalmi episode, the effect of the underground mining operation to the induced seismicity in the mine will be considered (EPOS, 2016).

## 5   Conclusions

In this paper, we have reviewed the history of seismic observation at the Sodankylä Geophysical Observatory. Also, we presented most recent and significant seismic experiments that the seismic group of the observatory has participated.

Over the years, the seismic group of SGO has gained long experience in carrying out seismological studies in the polar region of northern Fennoscandia. This experience and new seismic instrumentation of the Northern Finland Seismological Network and Laboratory of Applied Seismology can be used to initiate new projects and continue a high-level seismological research at SGO.

*Acknowledgements.* The staff of the seismic group of SGO acknowledges very much kind cooperation with the scientific and technical staff of the Institute of Seismology of the University of Helsinki during numerous common research projects and operating of permanent seismic stations. We appreciate scientific cooperation with many research organizations in Europe during temporary seismic experiments, in which the seismic group of the UO has participated. We are grateful to the technical staff of SGO, who helped a lot in installation and operation of permanent seismic stations.

The POLENET/LAPNET project is a part of the International Polar Year 2007–2009 and a part of the POLENET consortium. The study was financed by the Academy of Finland (grant no. 122762) and the University of Oulu (Finland), FBEGDY programme of the Agence Nationale de la Recherche, Institut Paul Emil Victor (France) and ILP (International Lithosphere Program) task force VIII, grant no. IAA300120709 of the Grant Agency of the Czech Academy of Sciences, and by Russian Academy of Sciences (programmes nos. 5 and 9). The SEISLAB project was funded by the European Regional Development Fund (ERDF), Council of Oulu region and Pyhäsalmi Mine Oy. Geopsy free software (www. Geopsy.org) was used for the data analysis of SEISLAB project.

DAFNE/FINLAND Working Group:

Ilmo Kukkonen, Principal Investigator (University of Helsinki, Department of Physics), Pekka Heikkinen (University of Helsinki, Institute of Seismology), Kari Komminaho (University of Helsinki, Institute of Seismology), Elena Kozlovskaya (Sodankylä Geophysical Observatory and Oulu Mining School, University of Oulu), Riitta Hurskainen, Tero Raita, Hanna Silvennoinen (all from the Sodankylä Geophysical Observatory, University of Oulu).

Edited by: J. Pulliainen

# References

Alinaghi, A., Bock, G., Kind, R., Hanka, W., Wylegalla, K., and TOR and SVEKALAPKO Working Group: Receiver function analysis of the crust and upper mantle from the North German Basin to the Archaean Baltic Shield, Geophys. J. Int., 155, 641–652, 2003.

Anderson, J., Farrell, W., Garcia, K., Given, J., and Swanger, H.: Center for Seismic Studies Version 3 Database: Schema Reference Manual, SAIC Technical Report C90-01, 1990.

BABEL Working Group: Integrated seismic studies of the Baltic shield using data in the Gulf of Bothnia region, Geophys. J. Int., 112, 305–324, 1993.

Bleibinhaus, F., Brueckl, E., and ALP 2002 Working Group: Wide-angle observations of ALP 2002 shots on the TRANSALP profile: Linking the two DSS projects, Tectonophysics, 414, 71–78, 2006.

Bock, G. and SVEKALAPKO Seismic Tomography Working Group: Seismic probing of Fennoscandian lithosphere. EOS, Trans. AGU, 82, 628–629, 2001.

Blundell, D. J., Freeman, R., and Mueller, S. (Eds.): A Continent revealed. The European Geotraverse, Structure and dynamic evolution, Cambridge University Press, 288 pp., 1992.

Bruneton, M., Farra, V., Pedersen, H. A., and the SVEKALAPKO Seismic Tomography Working Group: Non-linear surface wave phase velocity inversion based on ray theory, Geophys. J. Int., 151, 583–596, 2002.

Bruneton, M., Pedersen, H. A., Vacher, P., Kukkonen, I. T., Arndt, N. T., Funke, S., Friederich, W., Farra, V., and SVEKALAPKO STWG: Layered lithospheric mantle in the central Baltic Shield from surface waves and xenolith analysis, Earth Planet. Sci. Lett., 226, 41–52, 2004a.

Bruneton, M., Pedersen, H. A., Farra, R., Arndt, N. T., Vacher, P., Achauer, U., Alinaghi, A., Ansorge, J., Bock, G., Friederich, W., Grad, M., Guterch, A., Heikkinen, P., Hjelt, S. E., Hyvonen, T. L., Ikonen, J. P., Kissling, E., Komminaho, K., Korja, A., Kozlovskaya, E., Nevsky, M. V., Paulssen, H., Pavlenkova, N. I., Plomerova, J., Raita, T., Riznichenko, O. Y., Roberts, R. G., Sandoval, S., Sanina, I. A., Sharov, N. V., Shomali, Z. H., Tiikainen, J., Wieland, E., Wylegalla, K., Yliniemi, J., and Yurov, Y. G.: Complex lithospheric structure under the central Baltic Shield from surface wave tomography, J. Geophys. Res.-Solid Earth, 109, B10303, doi:10.1029/2003JB002947, 2004b.

European Commission: Building a pan-European Observation System for Geosciences. EU Support for research infrastructures in environmental sciences, A working document, 32 pp., http://ec.europa.eu/research/infrastructures/pdf/ last access: 2011.

EPOS: European Plate Observing System webpages, www.epos-eu.org, last access: 2016.

First Quantum Minerals: Information of Pyhäsalmi mine, Fisrt Quantum minerals Ltd., www.firstquantum.com last access: 2016.

Gibbons, S. J., V. Asming, L. Eliasson, A. Fedorov, J. Fyen, J. Kero, E. Kozlovskaya, T. Kværna, L. Liszka, S. Peter Näsholm, Tero Raita, Michael Roth, Timo Tiira, and Yuri Vinogradov: The European Arctic: A Laboratory for Seismoacoustic Studies, Seismol. Res. Lett., 86, 917–940, doi:10.1785/0220140230, 2015.

Grad, M., Janik, T., Yliniemi, J., Guterch, A., Luosto, U., Tiira, T., Komminaho, K., Środa, P., Höing, K., Makris, J., and Lund, C.-E.: Crustal structure of the Mid-Polish Trough beneath the Teisseyre–Tornquist Zone seismic profile, Tectonophysics, 314, 145–160, 1999.

Grad, M., Janik, T., Guterch, A., Sroda, P., and Czuba, W.: EUROBRIDGE '94–97, POLONAISE'97 and CELEBRATION 2000 Seismic Working Groups: Lithospheric structure of the western part of the East European Craton investigated by deep seismic profiles, Geol. Quart., 50, 9–22, 2006.

Griffin, W. L., O'Reilly, S. Y., Abe, N., Aulbach, S., Davies, R. M., Pearson, N. J., Doyle, B. J., and Kivi, K.: The origin and evolution of Archaean lithospheric mantle, Precambrian Res., 127, 19–41, 2003.

Guterch, A., Grad, M., Janik, T., Materzok, R., Luosto, U., Yliniemi, J., Lück, E., Schulze, A. & FÖrste, K.: Crustal structure of the transition zone between Precambrian and Variscan Europe from new seismic data along LT-7 profile (NW Poland and eastern Germany), C. R. Acad. Sci. Paris, 319, 1489–1496, 1994.

Hegedus, E., Brueckl, E., Csabafi, R., Fancsik, T., Grad, M., Guterch, A., Hajnal, Z., Keller, R., Kovacs, A. C., Komminaho, K., Kozlovskaya, E., Tiira, T., Torok, I., and Yliniemi,

J.: DANUBE 2004 Lithosphere Research Program, in: Eos, 86, B254 pp., 2005.

Hjelt, S.-E., Daly, S., and SVEKALAPKO colleagues: SVEKALAPKO, in: Evolution of Palaeoproterozoic and Archaean Lithosphere, 56–67, edited by: Gee, D. and Zeyen, H., EUROPROBE 1996 – Lithosphere Dynamics: Origin and Evolution of Continents, EUROPROBE Secretariate, Uppsala University, p. 138, 1996.

Hjelt, S.-E., Korja, T., Kozlovskaya, E., Yliniemi, J., Lahti, I., BEAR and SVEKALAPKO Working Groups: Electrical conductivity and seismic velocity structures of the lithosphere beneath the Fennoscandian Shield, edited by: Gee, D. and Stephenson, R., Europ. Lithos. Dynam., 541–559, 2006.

Janutyte, I., Kozlovskaya, E., Motuza, G., and PASSEQ Working Group: Study of Local Seismic Events in Lithuania and Adjacent Areas Using the Data of PASSEQ Experiment, PAGEOPH, Online First, doi:10.1007/s00024-012-0458-8, 2012.

Janutyte, I., Majdanski, M., Voss, P. H., Kozlovskaya, E., and PASSEQ Working Group: Upper mantle structure around the Trans-European Suture Zone obtained by teleseismic tomography, Solid Earth, 6, 73–91, doi:10.5194/se-6-73-2015, 2015.

Kataja, E.: Seismic station at the SGO, unpublished memoirs and presentation at the meeting of Finnish Geophysical Society, 18 March 2008, 2008.

Korhonen, H., Luosto, U., and Yliniemi, J.: Microseismic storms in December 1975 at the Finnish seismograph stations, Publ. Inst. Geophys. Pol. Acad. Sc., A-9, 183–189, 1980.

Kozlovskaya, E. and Kozlovsky, A.: Influence of high-latitude geomagnetic pulsations on recordings of broad-band force-balanced seismic sensors, Geosci. Instrum. Method. Data Syst., 1, 85–101, doi:10.5194/gid-2-107-2012, 2012.

Kozlovskaya and POLENET/LAPNET Working Group, POLENET/LAPNET, RESIF – Réseau Sismologique et géodésique Français, Seismic Network, doi:10.15778/RESIF.YV2011, http://data.datacite.org/10.15778/RECIF.XK2007, 2007.

Kozlovskaya, E., Elo, S., Hjelt, S.-E., Yliniemi, J., Pirttijärvi, M., and SVEKALAPKO Seismic Tomography Working Group: 3D density model of the crust of southern and central Finland obtained from joint interpretation of SVEKALAPKO crustal P-wave velocity model and gravity data, Geophys. J. Int., 158, 827–848, 2004.

Kozlovskaya, E., Vecsey, L., Plomerová, J., and Raita, T.: Joint inversion of multiple data types with the use of multiobjective optimization: problem formulation and application to the seismic anisotropy investigations, Geoph. J. Int., 171, 761–779, 2007.

Kozlovskaya, E., Kosarev, G., Aleshin, I., Riznichenko, O., and Sanina, I.: Structure and composition of the crust and upper mantle of the Archean-Proterozoic boundary in the Fennoscandian Shield obtained by joint inversion of receiver function and surface wave phase velocity of recording of the SVEKALAPKO array, Geoph. J. Int., 175, 135–152, 2008.

Kozlovskaya, E., Usoskina, I., and Silvennoinen, H.: Study of local seismicity in the area of Suasselkä postglacial fault (DAFNE/FINLAND project), Preliminary Technical Report, doi:10.13140/RG.2.1.4990.9600, 2013.

Krasnoshchekov, P. Kaazik, V. Ovtchinnikov, E. Kozlovskaya. Seismic structures in the Earth's inner core below Southeastern Asia, Pure Appl. Geophys., doi:10.1007/s00024-015-1207-6, 2016.

Kukkonen, I. and Lahtinen, R. (Eds.): Finnish Reflection Experiment FIRE 2001–2005, Geological Survey of Finland Special Paper 43, 2006.

Laine, J. and Mougenot, D.: A high-sensitivity MEMS-based accelerometer, The Leading Edge, 33, 1234–1242, doi:10.1190/tle33111234.1, 2014.

Lansley, M., Laurin, M., and Ronen, S.: Modern land recording systems: How do they weigh up?, The Leading Edge, 27, 888–894, doi:10.1190/1.2954029, 2008.

Lo, T. and Inderwiesen, P.: Fundamentals of Seismic Tomography, Geophysical Monograph Series, Soc. Explor. Geophys., No. 6, 1994.

Luosto, U.: Deep seismic sounding studies in Finland 1979–86, Inst. Seismology, Univ. Helsinki, Report S-15, 21 pp., doctoral dissertation, 1987.

Luosto, U.: Seismology in Finland in the Twentieth Century, Geophysica, 37, 147–185, 2001.

Luosto, U., Zverev, S. M., Kosminskaya, I., and Korhonen, H.: Observations of FENNOLORA shots on additional lines in Finnish Lapland, edited by: Bisztricsany, E. and Szeidovitz, G., Proc. 17th Assembly of the EGS Budapest, 24–29 August 1980, Elsevier, Amsterdam, 517–521, 1983.

Luosto, U., Lanne, E., Korhonen, H., Guterch, A., Grad, M., Materzok, R., and Perchuc, E.: Deep structure of the earth's crust on the SVEKA profile in central Finland, Ann. Geophys., 2, 559–570, doi:10.5194/angeo-2-559-1984, 1984.

Luosto, U., Flueh, E. R., and Lund C.-E.: Working group: The crustal structure along the POLAR Profile from seismic refraction investigation, Tectonophysics, 162, 51–85, doi:10.1016/0040-1951(89)90356-9, 1989.

McWhorter, R., Schultz, G., Clark, A., Branch, T., and Lansley, M.: 3D seismic operational optimization in the lusitanian basin, portugal. First Break, Special Topic: Land Seismic, 30, 103–108, 2012.

Majdanski, M., M. Grad, A. Guterch ja SUDETES 2003 Working Group: 2-D seismic tomographic and ray tracing modelling of the crustal structure across the Sudetes Mountains basing on SUDETES 2003 experiment data, Tectonophysics, 413, 249–269, 2006.

Mittelbauer, U., Behm, M., Brückl, E., Lippitsch, R., Guterch, A., Keller, R., Kozlovskaya E., Rumpfhuber, E.-M., and Šumanovac, F.: Shape and origin of the East-Alpine slab constrained by the ALPASS teleseismic model, Tectonophysics, 510, 195–206, 2011.

Mougenot, D. and Thorburn, N.: MEMS-based 3D accelerometers for land seismic acquisition: Is it time?, The Leading Edge, 23, 246–250, doi:10.1190/1.1690897, 2004.

Nakamura, Y.: A method for dynamic characteristics estimation of subsurface using microtremor on the ground surface, Quaterly Report of the Railway Technical Research Institue, 30, 25–30, 1989.

Nogoshi M. and Igarashi T.: On the amplitude characteristics of microtremor (part 2) (in Japanese with english abstract), J. Seismol. Soc. Jpn., 24, 26–40, 1971.

Nurminen, S.: Digital 12-bit C-MOS telemetric system for seismological use, Geophysica, 13, 89–94, 1974.

Nurminen, S.: PCM tape-recording system for seismological use, constructed of C-MOS integrated circuits. Geophysica 14, 1, 141–146, 1976.

Nurminen, S. and Hannula, A.: Deep seismic sounding equipment model PCM 1218–80, service manual, Inst. Seismology, Univ. Helsinki, Report T-7, 33 pp., 1981.

Plomerova, J., Babuska, V., Vecsey, L., Kozlovskaya, E., Raita, T., and SVEKALAPKO STWG: Proterozoic-Archean boundary in the upper mantle of eastern Fennoscandia as seen by seismic anisotropy, J. Geodynam., 41, 369–450, 2006.

Plomerová, J., Vecsey, L., Babuška, V., and LAPNET Working Group: Domains of Archean mantle lithosphere deciphered by seismic anisotropy – inferences from the LAPNET array in northern Fennoscandia, Solid Earth, 2, 303–313, doi:10.5194/se-2-303-2011, 2011.

Pedersen, H. A., Bruneton, M., and Maupin, V.: Lithospheric and sublithospheric anisotropy beneath the Baltic shield from surface-wave analysis, Earth Planet. Sci. Lett., 244, 590–605, 2006.

Pedersen, H. A., Krüger, F., and the SVEKALAPKO Seismic Tomography Working Group: Influence of the seismic noise characteristics on noise correlations in the Baltic Shield, Geophys. J. Int., 168, 197–210, 2007.

Pedersen, H. A., Debayle, E., Maupin, V., and the POLENET/LAPNET Working Group: Strong lateral variations of lithospheric mantle beneath cratons – Example from the Baltic Shield, Earth Planet. Sci. Lett., 383, 164–172, 2013.

Peterson, J.: Observations and modeling of seismic background noise, U.S. Geol. Survey Open – File Report, 93-322, 1–95, 1993.

Poli, P., Pedersen, H., Campillo, M., and LAPNET Working Group: Body-wave imaging of Earth's mantle discontinuities from ambient seismic noise, Science, 338, 1063–1065, doi:10.1126/science.1228194, 2012.

Poli, P., Campillo, M., Pedersen, H., and the POLENET/LAPNET Working Group: Noise directivity and group velocity tomography in a region with small velocity contrasts: the northern Baltic shield application to the northern Baltic Shield, Geoph. J. Int., 192, 413–424, doi:10.1093/gji/ggs034, 2013.

Sandoval, S.: The lithosphere-asthenosphere system beneath Fennoscandia (Baltic Shield) by body-wave tomography, PhD thesis, ETH, Zürich, 2002.

Sandoval, S., Kissling, E., Ansorge, J., and the SSTWGL High-resolution body wave tomo-graphy beneath the SVEKALAPKO array. I. A priori 3D crustal model and associated travel time effcets on teleseismic wave fronts, Geophys. J. Int., 153, 75–87, 2003.

Sandoval, S., Kissling, E., Ansorge, J., and SVEKALAPKO STWG: High-resolution body wave tomography beneath the SVEKALAPKO array: II. Anomalous upper mantle structure beneath the central Baltic Shield, Geophys. J. Int., 157, 200–214, 2004.

SeisComP 2.5 Configuration Manual: GeoForschungsZentrum Potsdam, 30 pp., 2006.

SESAME: Guidelines for the implementation of the H / V spectral ratio technique on ambient vibrations – measurements, processing and interpretations, SESAME European research project, deliverable D23.12, 2005.

Silvennoinen, H., Kozlovskaya, E., Yliniemi, J., and Tiira, T.: Wide angle reflection and refraction seismic and gravimetric model of the upper crust in FIRE4 profile area, northern Finland, Geophysica, 46, 21–46, 2010.

Silvennoinen, H., Kozlovskaya, E., Kissling, E., Kosarev, G., and POLENET/LAPNET working group: A new Moho boundary map for northern Fennoscandian shield based on combined controlled-source seismic and receiver function data, Geo. Res. J., 1–2, 19–32, 2014.

Silvennoinen, H., Kozlovskaya, and Kissling, E.: POLENET/LAPNET teleseismic P-wave traveltime tomography model of the upper mantle beneath northern Fennoscandia, Solid Earth, 7, 425–439, doi:10.5194/se-7-425-2016, 2016.

Standard for the Exchange of Earthquake Data: Reference Manual, SEEDFormat v2.3, FDSN, IRIS, USGS, 1992.

Usoltseva, O. A. and Kozlovskaya, E. G: Studying local earthquakes in the area Baltic-Bothnia Megashear using the data of the POLENET/LAPNET temporary array, Solid Earth, 7, 1095–1108, doi:10.5194/se-7-1095-2016, 2016.

Vecsey, L., Plomerova, J., Kozlovskaya, E., and Babuska, V.: Shear-wave splitting as a diagnostic of varying upper mantle structure beneath eastern Fennoscandia, Tectonophysics, 438, 57–77, 2007.

Vinnik, L., Oreshin S., Makeyeva, L., Peregoudov, D., Kozlovskaya, E., and POLENET/LAPNET Working Group: Anisotropic lithosphere under the Fennoscandian shield from P receiver functions and SKS waveforms of the POLENET/LAPNET array, Tectonophysics, 628, 45–54, doi:10.1016/j.tecto.2014.04.024, 2014.

Vinnik, L, Oreshin, S., Kozlovskaya, E., Kosarev, G., Piiponen, K., and Silvennoinen, H.: The lithosphere, LAB, LVZ and Lehmann discontinuity under central Fennoscandia from receiver functions, Tectonophysics, 667, 189–198, doi:10.1016/j.tecto.2015.11.024, 2016.

Wilde-Piórko, M., Geissler, W. H., Plomerová, J., Grad, M., Babuška, V., Brückl, E. Čyžienė, J., Czuba, W., Eengland, R., Gaczyński, E., Gazdova, R., Gregersen, S., Guterch, A., Hanka, W., Hegedüs, E., Heuer, B., Jedlička, P., Lazauskienė, J., Randy Keller, G., Kind, R., Klinge, K., Kolinsky, P., Komminaho, K., Kozlovskaya, E., Krüger, F., Larsen, T., Majdański, M., Málek, J., Motuza, G., Novotný, O., Pietrasiak, R., Plenefish, Th., Ržek, B., Šliaupa, S., Środa, P., Świeczak, M., Tiira, T., Voss, P., and Wiejacz, P.: PASSEQ 2006–2008: PASsive Seismic Experiment in Trans-European Suture Zone, Stud. Geophys. Geod., 52, 439–448, 2008.

Yliniemi, J.: Deep seismic sounding in the University of Oulu. Inst. Seismol., Univ. Helsinki, Report S-25, 1-6, 1991.

Yliniemi, J.: Oulun yliopisto Geofysiikan Observatorio: Seismografiasemat, in: BABEL, WABEL, SVEKA ja EUROPROBE, edited by: Manninen, J., XXVI Geofysiikan Observatoriopäivät Sodankylässä 8–10 January 1992, 79–86, 1992.

Yliniemi, J., Kozlovskaya, E., Hjelt, S.-E., Komminaho, K., Ushakov, A., and SVEKALAPKO Seismic Tomography Working Group: Structure of the crust and uppermost mantle beneath southern Finland revealed by analysis of local events registered by the SVEKALAPKO seismic array, Tectonophysics, 394, 41–67, 2004.

# 7

# Understanding of morphometric features for adequate water resource management in arid environments

**Mohamed Elhag[1], Hanaa K. Galal[2,3], and Haneen Alsubaie[2]**

[1]Department of Hydrology and Water Resources Management, Faculty of Meteorology, Environment & Arid Land Agriculture, King Abdulaziz University, Jeddah 21589, Saudi Arabia
[2]Biological Sciences Department, Faculty of Science, King Abdulaziz University, Jeddah 21589, Saudi Arabia
[3]Botany Department, Faculty of Science, Assiut University, Asyut, Egypt

*Correspondence to:* Mohamed Elhag (melhag@kau.edu.sa)

**Abstract.** Hydrological characteristics such as topographic parameters, drainage attributes, and land use/land cover patterns are essential to evaluate the water resource management of a watershed area. In the current study, delineation of a watershed and calculation of morphometric characteristics were undertaken using the ASTER global digital elevation model (GDEM). The drainage density of the basin was estimated to be very high, which indicates that the watershed possesses highly permeable soils and low to medium relief. The stream order of the area ranges from first to sixth order, showing a semi-dendritic and radial drainage pattern that indicates heterogeneity in textural characteristics, and it is influenced by structural characteristics in the study area. The bifurcation ratio (Rb) of the basin ranges from 2.0 to 4.42, and the mean bifurcation ratio is 3.84 in the entire study area, which signifies that the drainage pattern of the entire basin is controlled much more by the lithological and geological structure. The elongation ratio is 0.14, which indicates that the shape of the basin has a narrow and elongated shape. A land use/land cover map was generated by using a Landsat-8 image acquired on 10 August 2015 and classified to distinguish mainly the alluvial deposit from the mountainous rock.

## 1 Introduction

Soil studies that include interpolation techniques focus mainly on either inverse distance weighting (IDW) or Kriging methods, followed by accuracy assessment. Gotway et al. (1996) found that the IDW method generated more ac-

curate results for mapping soil organic matter and soil $NO_3$ levels. Wollenhaupt et al. (1994) compared these two interpolation techniques and concluded that IDW was more accurate for mapping P and K levels soil. Mueller et al. (2004) observed that for the optimal parameters of the method, the accuracy of IDW interpolation generally equaled or exceeded the accuracy of Kriging at each scale of measurement. However, other scholarly works observed Kriging to be more accurate for the interpolation of soil attributes. Leenaers et al. (1990) found the Kriging interpolation method to be more accurate in comparison to IDW for mapping soil Zn content.

Further studies have compared Kriging, IDW and radial basis function interpolation techniques in soil science. Schloeder et al. (2001) observed that ordinary Kriging and inverse distance weighting were similarly accurate and effective methods, while thin-plate smoothing splines with tension were not. Weller et al. (2007) concluded not only that the predications for Kriging were not satisfied by the Kriging method but also that it was as good as any other radial base function interpolation.

Hydrological parameters are essential for adequate water resource management plans. Morphometric characteristics are used to investigate watershed delineation, site selection in water recharge and discharge, runoff modelling and other geomorphological studies (Sreedevi et al., 2013; Elhag, 2015). GIS helps with a wide variety of basin characterization and evaluation applications under different terrain conditions (Pankaj and Kumar, 2009; Magesh et al., 2011).

Digital elevation model (DEMs), such as the ASTER global digital elevation model (GDEM) (USGS, USA), are of key importance in various extractions of geohydrological parameters of a watershed. Several parameters including slope, aspect, stream network, and upstream flow areas can be retrieved from the DEM characterization (Grohmann et al., 2007; Elhag, 2015). Reliable results of implementing remote sensing and GIS-based morphometric evaluation using ASTER GDEM data have been reported in numerous scholarly works on watershed characterization (Farr and Kobrick, 2000; Panhalkar, 2014; Elhag, 2015).

The main aim of the present study is to identify and investigate various drainage attributes for geometrical evaluation of the Yalamlam Basin for sustainable rainwater harvesting management and conservation of water resources.

## 2 Materials and methods

### 2.1 Study area

The Wadi Yalamlam Basin is located about 125 km southeast of the city of Jeddah and is bounded by latitudes 20°26′ and 21°8′ N and longitudes 39°45′ and 40°29′ E (Fig. 1). The Wadi Yalamlam Basin drains a large catchment area of about 180 000 ha. The boundary of the lower part of the basin is enlarged to include nearly all of the flat area in the downstream part. The Wadi Yalamlam Basin is initiated from the high-elevation Hijaz escarpment, with a mean annual rainfall of about 140 mm. The basin elevations varies greatly from upstream and downstream parts and range between 2850 and 25 m (a.s.l.) respectively. The main course of Wadi Yalamlam crosscuts highly fractured granitoids and gabbroic and metamorphic rocks until the coastal plain of the Red Sea. The upper and middle parts of the Wadi Yalamlam Basin are covered by dense natural vegetation. The lower part is covered mainly by Quaternary deposits and sand dunes with sparsely scattered, highly altered granitoids and metamorphosed basaltic hills. Several basic dykes are recorded in the lower part of the Wadi Yalamlam Basin. The thickness of Quaternary wadi deposits increases in the lower part (Elhag and Bahrawi, 2017). Regional groundwater flow drains toward the south and southwest, following the general trend of the main wadi channel. The mean monthly maximum temperature is 38 °C and the mean monthly minimum temperature is 20 °C. The annual rainfall is about 120 mm, falling mainly in the winter season (Elhag, 2016). The thickness of the saturated zone within the aquifer varies from less than 1 m upstream of Yalamlam to about 30 m in the Sa'diyah area. The aquifer is generally unconfined, especially in the upper parts of the wadi. Semi-confining conditions may exist in the lower parts, where layers of clay exist. There are about 31 wells in the basin of Wadi Yalamlam, of which 23 are hand-dug wells and the others are drilled.

### 2.2 Soil sampling

Map accuracy and quality depend on the sampling method scale, analytical laboratory errors and prediction errors. Sampling approaches depend on the objectives of the study, which are highly correlated with scale. Random stratified sampling was the adopted sampling design. First, the landscape was divided into smaller areas, named strata, and afterwards 150 random samples were taken from the designated study area (Johnson and Riess, 1982).

### 2.3 Physical and chemical soil analysis

Each individual sample was analyzed separately, and each measurement was repeated three times for the same extract. Thus, the final values of the measured attributes are represented by the mean value of three measurements. Soil samples were analyzed in order to estimate physical analyses (clay, silt and sand).

For standard particle size measurement, the soil fraction that passes through a 2 mm sieve is considered. Laboratory procedures normally estimate the percentage of sand (0.05–2.0 mm), silt (0.002–0.05 mm) and clay (<0.002 mm) fractions in soils. Soil particles are usually cemented together by organic matter; this has to be removed with $H_2O_2$ treatment. However, if substantial amounts of $CaCO_3$ are present, actual percentages of sand, silt or clay can only be determined by prior dissolution of the $CaCO_3$.

### 2.4 Interpolation techniques

Geostatistical interpolations are based on the assumption that all values of a variable that is measured are the result of a random process. The phrase "random process" does not indicate that all events are independent. More specifically, geostatistical techniques are based on random processes with dependence, otherwise called autocorrelation, and rely on some notion of replication. Repeated observations in nature can result in understanding the variation and uncertainty of natural phenomena, and furthermore in estimating their sequence in space and time. Three interpolation techniques were used for the generation of the prediction maps (interpolation). Inverse distance weighting (IDW), radial basis function (RBF) and ordinary Kriging (OK) methods were compared according to the accuracy of the results (Bahrawi et al., 2016). The following equations were used:

– Spatial distribution (Weibel, 1997):

$$\gamma_{(k)} = \frac{1}{2 \times n(k)} x \sum_{i=1}^{n(k)} [Z(X_i) - Z(X_{i+k})]^2, \tag{1}$$

where $n(k)$ is the number of pairs of observation and $Z(x_i)$ the soil property measured in point $x$ and in point $x + k$.

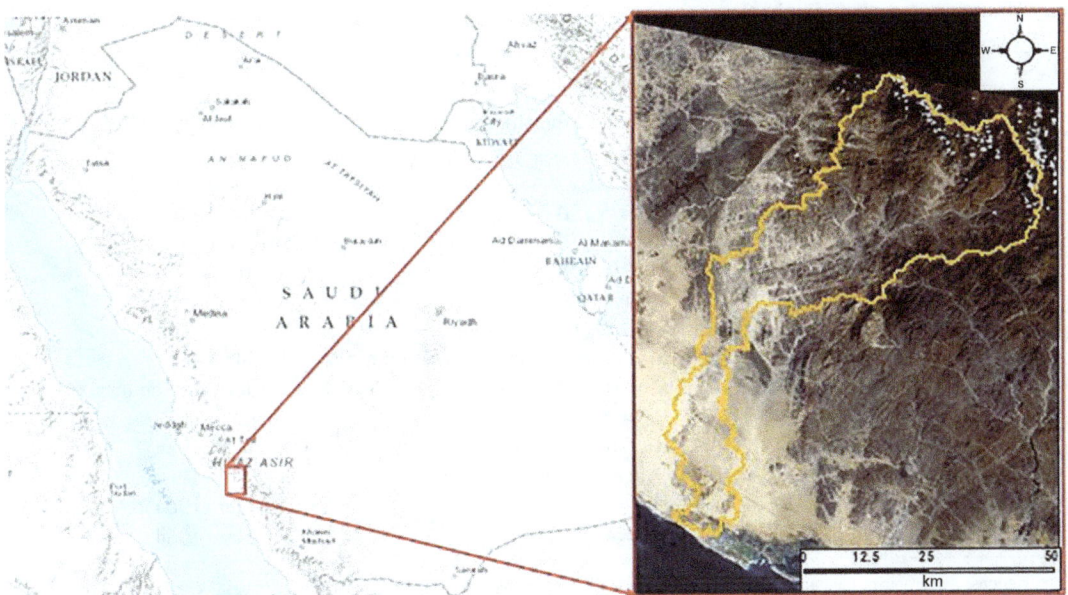

**Figure 1.** Location of the study area (Elhag, 2016).

– Interpolation (Stoer and Bulirsch, 1980):

$$Z_x\left(X_o\right) = \sum_{i=1}^{n} \lambda_i \, x \, Z\left(x_i\right), \tag{2}$$

where $Z_x(x_o)$ is the interpolated value of variable $Z$ at location $X_o$, $Z(x_i)$ are values measured at location $x_i$, and $\lambda i$ are weighed coefficients calculated based on the semivariogram.

– Trend and random error (Johnson and Riess, 1982):

$$Z\left(S\right) = \mu\left(S\right) + \varepsilon\left(S\right) \tag{3}$$

where $S$ stands for the location of the prediction location, $Z(S)$ is the variable being predicted (total extractable heavy metal concentration), $\mu(S)$ is the deterministic trend, and $\varepsilon(S)$ is the spatially autocorrelated random error.

## 2.5 Morphometric parameters

Based on the scholarly works of Horton (1945), Schumm (1963), Strahler (1964) and others, several morphometric parameters were used and computed utilizing the ASTER GDEM in a GIS environment. Consequently, watershed delineation, stream network identification, drainage frequency, drainage density, shape, elongation ratio, circularity ratio and form factor were computed and evaluated using the ASTER GDEM at 30 m spatial resolution. The methodologies adopted for the evaluation and computation of morphometric features are given in Table 1.

## 2.6 Supervised classification

Remote sensing data, acquired on 10 June 2013, were obtained from Landsat Operational Land Imager (OLI-8). Typical atmospheric and radiometric corrections and spatial resolution enhancement were implemented for each band individually. Furthermore, supervised classification was implemented using a support vector machine (SVM) classifier for better classification of results (Psilovikos and Elhag, 2013). The final step in the digital image analysis was the evaluation of the accuracy of the computer-derived classification results. These results are often expressed in tabular form, known as a confusion matrix (Elhag et al., 2013). The SVM classifier is implemented as

$$K\left(x_i, x_j\right) = \tanh\left(g x_i^T x_j + r\right), \tag{4}$$

where $g$ is the gamma term in the kernel function for all kernel types except linear and $r$ is the bias term in the kernel function for the polynomial and sigmoid kernels.

## 3 Results and discussion

Quantitative evaluation of the watershed through the analysis of morphometric parameter can provide significant information about the hydrological characteristics of rocks which are exposed within the basin. The nature of drainage of a basin reveals reliable information about the permeability of the rocks and the yield of the basin.

Evaluation of drainage characteristics and other morphometric parameters of the Yalamlam Basin has been undertaken to calculate the parameters and construct the topology

**Table 1.** Summary of the implemented morphometric features.

| Item | Morphometric feature | Equation | Citation |
|---|---|---|---|
| 1 | Stream length ($L_u$) | Length of the stream | Horton (1945) |
| 2 | Stream length ratio ($R_L$) | $R_L = L_u/(L_u+1)$ | Horton (1945) |
| 3 | Form factor ($F_f$) | $F_f = A/L^2$ | Horton (1945) |
| 4 | Drainage frequency ($F_d$) | $F_d = N_u/A$ | Horton (1945) |
| 5 | Drainage density ($D_d$) | $D_d = L_u/A$ | Horton (1945) |
| 6 | Drainage texture ($T$) | $T = D_d * F_d$ | Smith (1950) |
| 7 | Bifurcation ratio ($R_b$) | $(R_b) = N_u/(N_u+1)$ | Schumm (1956) |
| 8 | Elongation ratio ($R_e$) | $R_e = D/L$ | Schumm (1956) |
| 9 | Mean bifurcation ratio ($R_{bm}$) | $R_{bm}$ = average of bifurcation ratios | Strahler (1957) |
| 10 | Relief ($R$) | $R = H - h$ | Hadley and Schumm (1961) |
| 11 | Relief ratio ($R_r$) | $R_r = R/L$ | Schumm (1963) |
| 12 | Stream order ($S_o$) | Hierarchical rank | Strahler (1964) |
| 13 | Stream number | Order-wise no. of streams | Strahler (1964) |
| 14 | Mean stream length ($L_{sm}$) | $L_{sm} = L_u/N_u$ | Strahler (1964) |
| 15 | Circularity ratio ($R_c$) | $R_c = 4\pi A/P^2$ | Strahler (1964) |

Abbreviations: $A$, the area of the basin (km$^2$); $D_d$, drainage density; $F_f$, form factor; $F_s$, stream frequency; $L$, basin length (km); $L_{sm}$, mean stream length; $L_u + 1$, the total stream length of its next higher order $u$; $L_u$, the total stream length of order $u$; $N_u + 1$, number of stream segments of the next higher order; $N_u$, the number of stream segments of order $u$; $P$, perimeter (km); $R_b$, bifurcation ratio; $R_c$, circularity ratio; $R_e$, elongation ratio; RL, stream length ratio; and $T$, drainage texture; $\pi = 3.14$

of the basin. Different types of areal and linear aspects and their characteristics have been calculated, such as basin area (A), basin length (L), basin perimeter (P), bifurcation ratio ($R_b$), elongation ratio ($R_e$), circularity ratio ($R_c$), drainage frequency ($F_d$) and drainage density ($D_d$).

### 3.1 Stream order ($S_o$) and stream number

The lower Yalamlam Basin encompasses the basin mega-fan, which is formed by ancient and modern radial drainage patterns in the study area. The channel of this area is characterized by higher sinuosity, decreased widths and lesser discharge capacity than numerous traverse paleo-alluvial channels (Bahrawi et al., 2016). Therefore, the stream ordering of the study area has been ranked based on the Strahler (1964) method and demonstrated in Table 2.

### 3.2 Bifurcation ratio

The bifurcation ratio was calculated as the number of streams of an order to the number of the streams of the next higher order. The values vary from 2.0 to 4.42 for the Yalamlam stream basin, which also indicates the maximum structural influences (Strahler, 1964). After the calculation of the bifurcation ratio, the average value is calculated; the mean bifurcation ratio is 3.84 for the basins. The value also indicates that the drainage pattern has been affected by structural disturbances within the basin. The obtained number for the bifurcation ratio varies from one order to another. Such variation is interpreted as the result of irregularities in the lithological and geological development within the watershed. The values of the bifurcation ratio and mean bifurcation ratio are shown in Table 3.

### 3.3 Drainage texture and drainage density

Drainage density is an expression of spacing and the distribution of channels as proposed by Horton (1932), measuring the total length of the streams of all orders as calculated per unit area. The relative relief and slope gradient of the river basin primarily control the stream density. The stream density of the watershed has been calculated and is shown in Table 3. The drainage density value is 0.92 in the study basin. The drainage density has been classified into five kinds of drainage texture as proposed by Smith (1950).

Drainage density values are classified as follows: more than 8, very fine drainage texture; 8–6, fine; between 6 and 4, moderate; 4–2, coarse; and less than 2, very coarse. The observer drainage texture is 0.138, which indicates resistant permeable rock with a moderate infiltration rate and moderate relief (Bahrawi et al., 2016). The value of the variation of drainage texture depends on different types of natural factors, i.e., rainfall and other climatic characteristics, rock type, soil type, vegetation characteristics, permeability, relief, and infiltration capacity within the watershed. The relationship between the hydrological features and the geological structures is estimated to be with a high drainage density caused by the mountainous relief in the basin. The lower drainage density value reveals that the region is composed of permeable subsurface material and low-relief, dense vegetal cover, which results in an increase in infiltration capacity in the basin. The high drainage density value indicates mountainous relief, thin vegetation and impermeable subsurface material, and highly resistant rock types in the river basin.

**Table 2.** Stream network order based on the Strahler method.

| Strahler | Cnt_Strahler | $R_b$ | Nu-r | $R_b \cdot$ Nu-r | Sum_Length |
|---|---|---|---|---|---|
| 1 | 598 | | | | 872.847 |
| 2 | 135 | 4.42963 | 733 | 3246.919 | 452.488 |
| 3 | 22 | 6.13636 | 157 | 963.409 | 237.306 |
| 4 | 6 | 3.66667 | 28 | 102.667 | 112.047 |
| 5 | 2 | 3.00000 | 8 | 24.000 | 54.635 |
| 6 | 1 | 2.00000 | 3 | 6.000 | 58.259 |
| Sum | 764 | 19.232659932 | 929 | 4342.99427609428 | 1787.582302 |
| Mean | | 3.847 | | 4.675 | |

**Table 3.** Wadi Yalamlam morphometric features.

| Parameters | Descriptions | Remarks |
|---|---|---|
| Basin area (km$^2$) | 1940.3 | The basin area is too large |
| Basin length (km) | 60.56 | Basin length is very high |
| Basin perimeter (km) | 417 | High basin perimeter |
| Elongation ratio | 0.14 | Elongated |
| Form factor | 0.06 | Elongated shape and flatter peak flow |
| Circularity ratio | 0.08 | Strongly elongated and heterogeneous geological structure |
| Drainage frequency | 0.34 | Low stream frequency |
| Drainage density | 0.92 | Drainage density is considerably high |
| Drainage texture | 0.138 | Highly resistant permeable rock with moderate infiltration rate |

## 3.4 Drainage frequency

Drainage frequency, or stream frequency, is calculated as the total number of streams per unit area of all stream orders proposed by Horton (1932). The correlation value of drainage density and stream frequency plays a positive role in the basin, which suggests that the number of streams and population increase with the increase in drainage density. The observed value of stream frequency is about 0.34 for the watershed, showing a highly positive connection with stream density; see Table 3.

## 3.5 Elongation ratio

The elongation ratio is calculated as the ratio between the maximum length of the basin and the diameter of a circle which is fitted in the same basin area, as proposed by Schumm (1956). The elongation ratio value generally varies between 0.6 and 1.0, with a wide variety of geological conditions and climatic characteristics. According to Strahlar (1964), values close to 1.0 represent a region of very low relief with few structural influences, and values ranging from 0.8 to 0.6 are generally associated with a much steeper slope and high relief. The values of the elongation ratio can be categorized into three groups: less than 0.7 indicates an elongated basin shape, values of 0.8 to 0.9 represent an oval shape, and values more than 0.9 represent a circular shape. Thus, the elongated ratio of the study area is 0.14, which suggests that

the basin shape is a much more elongated type (Table 3) of basin over which there is a considerable structural influence.

## 3.6 Circularity ratio

According to Miller (1953), the circularity ratio is the ratio between the area of a circle which is fitted in the basin perimeter and the total basin area. The circularity ratio is much more influenced by geological structure, relief, slope, climate, frequency and length of stream, and land use/land cover within the basin. The basin circularity ratio is 0.08, which shows that the basin is strongly elongated and has heterogeneous geological structure and materials. The observed values also indicate high runoff capacity and low permeability capacity of subsoil and subsurface soil along the basin area (Table 3).

## 3.7 Form factor

Horton (1932) defines the form factor as the ratio between the square of the basin length and basin area. The values of form factor represent the flow intensity of the study area. Generally, the elongation shape and the values of form factor have a negative relationship, which means that a smaller value indicates a more elongated shape of the basin. The values should always be lower than 0.7854; a higher value of form factor represents higher peak flows of a higher period. The observed value of form factor is 0.06 for the Yalamlam

**Table 4.** Wadi Yalamlam land cover classifications.

| Land cover category | Area (km$^2$) | Percentage (%) |
|---|---|---|
| Vegetation | 82.7 | 4.26 |
| Alluvial deposit | 803.3 | 41.4 |
| Bare rocks | 1054.3 | 54.3 |
| Total | 1940.3 | 100 |

watersheds, which indicates an elongated shape of the basin (Table 3). Therefore, the lower values and elongated basin shape indicate that the watershed has a flatter peak flow of shorter duration.

### 3.8 Relief and relief ratio of the watershed

Relative relief is the difference between the highest and lowest elevation of the watershed. The relief ratio is the ratio between relative relief and the maximum length of the basin as proposed by Schumm (1956). It can analyze the steepness of the basin and evaluate the intensity of the erosion process in the study area. Here the relief ratio is 4.17, which indicates that most of the designated basin is situated along the rough, mountainous slope and is much narrower in the lower areas.

### 3.9 Slope map

Slope is the ratio between horizontal and vertical surface of a region, which can be expressed by the percentage and degree. It was found that the most of the area (i.e., the upper middle part) of Yalamlam watershed is classified as steep, very steep and very high steep slopes, which indicates that the area has very mountainous topography. The main channel slope of the basin has a gentle slope (0.042), meaning flat topography that is excellent for groundwater management through favoring infiltration.

### 4   Geohydrological inferences from morphometric evaluation

The classification of remote sensing data was to quantify the area of all alluvial deposits in the bare rock area within the designated study area as illustrated in Fig. 2. Table 4 indicates that the area of the alluvial deposits is roughly equal to the bare rock area, which means that the watershed is likely to be used for rainwater harvesting (Elhag, 2014 and Elhag and Bahrawi, 2014a).

The quantitative morphometric evaluation has considerable utility for watershed delineation, water and soil conservation, and their management for future sustainability. The morphometric analysis of the Yalamlam Basin shows that the watershed has a narrow, elongated shape and very high mountainous relief. The planning for runoff and artificial recharge of the area has been chosen based on small-scale

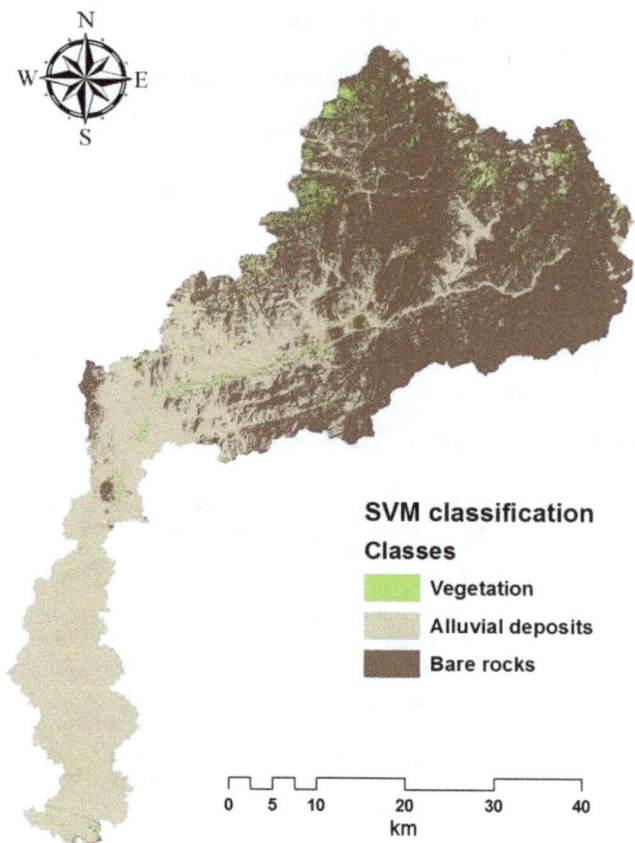

**Figure 2.** Supervised classification of the Yalamlam Basin.

topographical maps and low relief (gentle slope) in the lower area of the basin. Also, drainage morphometry plays a positive role in GIS and remote sensing techniques by selecting artificial recharge sites and creating a demand storage point in the mountainous region with the basin. In addition, if the morphometric information is integrated with other hydrological parameters of the river basin, the strategy for water harvesting and recharging measures gives a better plan for groundwater management and development for the future.

The drainage pattern of the basin is sub-dendritic and radial in nature. The pattern was affected by more or less heterogeneous structural and lithological characteristics. In addition, high drainage density is observed all over the watershed along with very high relief and impermeable subsoils and land rock substratum, as well as mountainous terrain slope. Nevertheless, lower riparian areas have low drainage density, which is favorable for the identification of water storage areas and potential groundwater zones. However, slope plays a significant role in determining the relationship between infiltration rate and runoff velocity, where infiltration rate is inversely controlled by regional slope. Therefore, all evaluated parameters are more important for the analysis of future water availability in the study region.

Results obtained from previous scholarly work of Şen (1995) showed that the average transmissivity values calculated within the study area range from 91 to $147\,m^2\,day^{-1}$. However, the transmissivity values increase sharply in the downstream area, ranging between 267 and $731\,m^2\,day^{-1}$ (average of $500\,m^2\,day^{-1}$). Such findings support the hypothesis that the aquifer is of high potential therein. Nonetheless, the hydraulic conductivity values calculated for the Yalamlam Basin attain high hydraulic conductivity ($16\,m\,day^{-1}$) due to more permeable alluvial deposits (Sen, 1995; Elhag and Bahrawi, 2014b, c).

## 5 Conclusions

Our evaluation of hydrological characteristics of the Yalamlam watershed confirms that the area has high relief and its shape is elongated in nature. The stream network of the watershed is essentially of dendritic type in the lower portion, which indicates a lack of structural influences and homogeneity of textural characteristics, but the upper portion of the watershed is highly influenced by tectonic and structural activity due to the parallel pattern of the drainage network. The drainage characteristics of the basin help to understand the different types of terrain parameters, i.e., runoff, infiltration capacity and the nature of the bedrock etc. The drainage density and frequency of the drainage basin are low, which indicates a high permeability rate and well-drained capacity of the subsurface formation. A variety of basic and derived parameters reveal the importance of the water recharge areas, and measures can be undertaken for the soil conservation structures and water resource management. Therefore, watershed analysis using remote sensing data, GDEM data and GIS techniques is an efficient, precise tool for the understanding of any terrain attributes such as surface runoff, the nature of bedrock, and infiltration capacity; this helps in better understanding drainage evolution and management of groundwater potential, the status of landforms, and their characteristics for watershed management and planning. The study should be useful for water as well as natural resource management of any terrain at the watershed level. For sustainable water resource management and watershed development, decision makers and planners can use these kinds of hydrological analyses.

*Competing interests.* The authors declare that they have no conflict of interest.

*Acknowledgements.* This article was funded by the Deanship of Scientific Research (DSR) at King Abdulaziz University, Jeddah. The authors therefore thank DSR for technical and financial support.

Edited by: Andrea Benedetto

## References

Bahrawi, J. A., Elhag, M., Aldhebiani, A. Y., Galal, H. K, Hegazy, A. K., and Alghailani, E.: Soil Erosion Estimation Using Remote Sensing techniques in Wadi Yalamlam basin, Saudi Arabia, Advances in Materials Science and Engineering, 2016, 9585962, https://doi.org/10.1155/2016/9585962, 2016.

Elhag, M.: Remotely sensed vegetation indices and spatial decision support system for better water consumption Regime in Nile Delta. A case study for rice cultivation suitability map, Life Science Journal, 11, 201–209, 2014.

Elhag, M.: Characterization of a Typical Mediterranean Watershed Using Remote Sensing Techniques and GIS Tools, Hydrology: Current Research, 6, 197–204, 2015.

Elhag, M.: Evaluation of Different Soil Salinity Mapping Using Remote Sensing Techniques in Arid Ecosystems, Saudi Arabia, Journal of Sensors, 2016, 96175–96175, 2016.

Elhag, M. and Bahrawi, J.: Cloud Coverage Disruption for Groundwater Recharge Improvement Using Remote Sensing Techniques in Asir Region, Saudi Arabia, Life Science Journal, 11, 192–200, 2014a.

Elhag, M. and Bahrawi, J.: Conservational Use of Remote Sensing Techniques for a Novel Rainwater Harvesting in Arid Environment, Environ. Earth Sci., 72, 4995–5005, 2014b.

Elhag, M. and Bahrawi, J.: Potential Rainwater Harvesting Improvement Using Advanced Remote Sensing Applications, The Scientific World Journal, 2014, 806959, https://doi.org/10.1155/2014/806959, 2014c.

Elhag, M. and Bahrawi, J. A.: Realization of daily evapotranspiration in arid ecosystems based on remote sensing techniques, Geosci. Instrum. Method. Data Syst., 6, 141–147, https://doi.org/10.5194/gi-6-141-2017, 2017.

Elhag, M., Psilovikos, A. and Sakellariou, M.: Detection of Land Cover Changes for Water Recourses Management Using Remote Sensing Data over the Nile Delta Region, Environment, Development and Sustainability, 15, 1189–1204, 2013.

Farr, T. G. and Kobrick, M.: Shuttle radar topography mission produces a wealth of data, American Geophysical Union, 81, 583–585, 2000.

Gotway, C. A., Ferguson, R. B., Hergert, G. W., and Peterson, T. A.: Comparison of kriging and inverse distance methods for mapping soil parameters, Soil Sci. Soc. Am. J., 60, 1237–1247, 1996.

Grohmann, C. H., Riccomini, C., and Alves, F. M.: SRTM-based morphotectonic analysis of the Pocos de Caldas alkaline Massif, southeastern Brazil, Comput. Geosci., 33, 10–19, 2007.

Hadley, R. F. and Schumn, S. A.: Sediment sources and drainage basin characteristics in the upper Cheyenn River Basin, US Geological Survey, Water supply paper, 1531-B, 137–197, 1961.

Horton, R. E.: Erosional development of streams and their drainage basins; hydrophysical approach to quantitative morphology, U.S. Geological Survey Professional, p. 282A, 1945.

Horton, R. E.: Drainage basin characteristics, Trans. Amer. Geophys. Union, 13, 350–361, 1932.

Johnson, L. W. and Riess, R. D.: Numerical Analysis, 2nd edn., Reading, MA: AddisonWesley, Chapter 1.3., 1982.

Leenaers, H., Okx, J. P., and Burrough, P. A.: Employing elevation data for efficient mapping of soil pollution on floodplain, Soil Use Manage., 6, 105–113, 1990.

Magesh, N., Chandrasekar, N., and Soundranayagam, J.: Morphometric evaluation of Papanasam and Manimuthar watersheds, parts of Western Ghats, Tirunelveli district, Tamil Nadu, India: a GIS approach, Environmental Earth Sciences, 64, 373–381, 2011.

Miller, V. C.: A Quantitative Geomorphologic Study of Drainage Basin Characteristics in the Clinch Mountain Area, Virginia and Tennessee, Project NR 389042, Tech Rept 3. Columbia University Department of Geology, ONR Geography Branch, New York, 1953.

Mueller, T. G., Pusuluri, N. B., Mathias, K. K., Cornelius, P. L., Barnhisel, R. I., and Shearer, S. A.: Map Quality for Ordinary Kriging and Inverse Distance Weighted Interpolation, Soil Sci. Soc. Am. J., 68, 2042–2047, 2004.

Panhalkar, S. S.: Hydrological modeling using SWAT model and geoinformatic techniques, Egypt, The Egyptian Journal of Remote Sensing and Space Science, 17, 197–207, 2014.

Pankaj, A. and Kumar, P.: GIS based morphometric analysis of five major sub-watershed of Song River, Dehradun district, Uttarakhand with special reference to landslide incidences, Journal of the Indian Society of Remote Sensing, 37, 157–166, 2009.

Psilovikos, A. and Elhag, M.: Forecasting of Remotely Sensed Daily Evapotranspiration Data over Nile Delta Region, Egypt, Water Resour. Manag., 27, 4115–4130, 2013.

Schloeder, C. A., Zimmerman, N. E., and Jacobs, M. J.: Comparison of methods for interpolating soil properties using limited data, Soil Sci. Soc. Am. J., 65, 470–479, 2001.

Schumm, S. A.: Evolution of drainage systems and slopes in badlands at Perth Amboy, New Jersy, National Geological Society of America Bulletin, 67, 597–646, 1956.

Schumm, S. A.: Sinuosity of alluvial rivers in the Great Plains, Bull. Geol. Soc. Am., 74, 1089–1100, 1963.

Sen, Z.: Wadi Hyrdology, CRC Press, Taylor and Francis, New York, 1995.

Smith, K. G.: Standards for grading texture of erosional topography, Am. J. Sci., 248, 655–668, 1950.

Sreedevi, P. D., Sreekanth, P. D., Khan, H. H., and Ahmed, S.: Drainage morphometry and its influence on hydrology in a semi-arid region: using SRTM data and GIS, Environ. Earth Sci., 70, 839–848, 2013.

Stoer, J. and Bulirsch, R.: Introduction to Numerical Analysis, New York, Springer-Verlag, Chapter 2, 1980.

Strahler, A. N.: Quantitative Geomorphology of drainage basins and channel networks, in: Handbook of Applied Hydrology, edited by: Chow, V. T., Newyork; Mc Graw hill, Section 4–11, 1964.

Strahler, A. N.: Quantitative analysis of watershed geomorphology, Trans. Am. Geophys. Union, 38, 913–920, 1957.

Weibel, R.: A Typology of Constraints to Line Simplification, in: Advances in GIS Research II (7th International Symposium on Spatial Data Handling), edited by: Kraak, M. J. and Molenaar, M., London, Taylor and Francis, 533–546, 1997.

Weller, U., Zipprich, M., Sommer, M., zu Castell, W., and Wehrhan, M.: Mapping clay content across boundaries at the landscape scale with electromagnetic induction, Soil Sci. Soc. Am. J., 71, 1740–1747, 2007.

Wollenhaupt, N. C., Wolkowski, R. P., and Clayton, M. K.: Mapping soil test phosphorus and potassium for variable-rate fertilizer application, J. Prod. Agric., 7, 441–448, 1994.

# 8

# Martian magnetism with orbiting sub-millimeter sensor: simulated retrieval system

Richard Larsson[1], Mathias Milz[2], Patrick Eriksson[3], Jana Mendrok[3], Yasuko Kasai[1], Stefan Alexander Buehler[4], Catherine Diéval[5], David Brain[6], and Paul Hartogh[7]

[1]National Institute of Information and Communications Technology, Tokyo, Japan
[2]Luleå University of Technology, Kiruna, Sweden
[3]Chalmers University of Technology, Gothenburg, Sweden
[4]University of Hamburg, Hamburg, Germany
[5]Lancaster University, Lancaster, UK
[6]University of Colorado, Boulder, USA
[7]Max Planck Institute for Solar System Research, Göttingen, Germany

*Correspondence to:* Richard Larsson (ric.larsson@gmail.com)

**Abstract.** A Mars-orbiting sub-millimeter sensor can be used to retrieve the magnetic field at low altitudes over large areas of significant planetary crustal magnetism of the surface of Mars from measurements of circularly polarized radiation emitted by the 368 GHz ground-state molecular oxygen absorption line. We design a full retrieval system for one example orbit to show the expected accuracies on the magnetic field components that one realization of such a Mars satellite mission could achieve. For one set of measurements around a tangent profile, we find that the two horizontal components of the magnetic field can be measured at about 200 nT error with a vertical resolution of around 4 km from 6 up to 70 km in tangent altitude. The error is similar regardless of the true strength of the magnetic field, and it can be reduced by repeated measurements over the same area. The method and some of its potential pitfalls are described and discussed.

## 1 Introduction

In the past decades, there have been several proposals to fly a sub-millimeter sensor on a satellite mission to Mars. One such proposal is to fly the Far-InfraRed Experiment, presented by Kasai et al. (2012). In their work, Kasai et al. (2012) show that molecular oxygen, carbon monoxide, water (even heavy water), ozone, isotopologues of carbon dioxide,

hydrogen peroxide, and various other hydrogen radicals all should have strong signals in the spectrum of Mars which they propose to observe. Additionally, wind speed parameters along the line of sight should have measurable readings. With this work we aim to develop the idea presented by Larsson et al. (2013) for remote measurement of magnetism by utilizing the Zeeman effect (Zeeman, 1897) on molecular oxygen in its ground state: $X(^3\Sigma_g^-)$. It is possible to combine this work with the idea of Kasai et al. (2012) into a single instrument that is capable of measuring and mapping both meteorological parameters and crustal magnetic structures, but this work will only focus on the magnetic aspects of flying such an instrument.

The Martian magnetic field is thought to be a remnant of a past global dipole that disappeared about 3.5 billion years ago (Acuña et al., 1998). All that remains of the past dipole is several magnetic sources in the crust, which were first measured in situ by the Magnetometer–Electron Reflectometer on board the Mars Global Surveyor (MGS) orbiter (Acuña et al., 1999) down to an altitude of 100 km. The strongest sources are located in the Southern Hemisphere, with strengths of up to 2000 nT at 100 km altitude. The ongoing Mars Atmosphere and Volatile Evolution mission will provide further coverage of the magnetic field down to $\sim 150$ km altitude in normal mode, but during week-long campaigns its periapsis will be even lower at $\sim 125$ km altitude, improving the potential to map the crustal magnetism (Jakosky et al., 2015). There are to our knowledge two planned landers, InSight (http://insight.jpl.nasa.gov) and ExoMars 2020, that will carry magnetometers to the equatorial surface of Mars.

The shape and distribution of the present field is informative for the crustal evolution of Mars (Nimmo and Tanaka, 2005). There are two main ideas about the magnetisms' formation: either a large impact demagnetized the north, leaving the south magnetized, or there was a southward migration of crustal material after the global dipole disappeared (see, e.g., Connerney et al., 2004, and Citron and Zhong, 2012, for further discussions). Regardless of the reason, the strongest crustal field should be associated with the oldest intact crustal material since these regions are linked to the times when Mars still had an effective global dipole. Identifying the reasons why the Northern Hemisphere has a lower average elevation, is younger, and also has a less magnetized crust than the Southern Hemisphere is important for questions related to, e.g., how similar Mars and Earth were in their early years. The method we propose for measuring the crustal magnetic field is useful in this regard, as it allows the determination of the magnetic field strength at different altitudes in the lower atmosphere. This means that it would help in the creation of profiles of magnetic field data that in turn can be used to estimate the depths, nature, and locations of the crustal field sources. The estimation of these parameters is beyond the scope of the paper; the interested reader may check the work by Brain et al. (2003), Connerney et al. (2004), and references therein. However, we note from Connerney et al. (2004) that there is more information about the structure of the magnetic field that is revealed at 100 km but is hidden at 400 km altitude (see their Fig. 6). If there are more structures at even lower altitudes, then these cannot be sampled by satellite magnetometers. Deep dips below 200 km altitude cost significant amounts of fuel due to increased air drag and therefore would shorten the lifetime of a satellite mission. Finding the crustal sources' characteristics accurately using only satellite data is thus difficult – which is why the available Martian magnetic field models differ in some regions by up to 2000 nT (cf. Cain et al., 2003; Morschauser et al., 2014) and the range of the strongest surface field is from just above 10 000 nT up to potentially 20 000 nT (Brain et al., 2003). The characteristics of the crustal field sources strongly limit the possible processes that led to the disappearance of the past dipole and to the north–south dichotomy, making the crustal sources interesting targets for geological exploration.

## 2   Method

This work is based only on simulations. We perform limb simulations around the molecular oxygen absorption line at 368 GHz using a measurement scenario that achieves several measurements around the same latitude–longitude tangent

profile during successive satellite revolutions. These simulated measurements are fed into a retrieval toolbox that estimates the errors of the magnetic field components in the tangent profile.

Section 2 is divided as follows. The first subsection summarizes the ideas behind the approach in generic terms. The second subsection goes over the basic aspects of the modeling theory. The last subsection describes the data required for the simulations and our practical design choices. Then Sect. 3 describes and discusses the results. Finally Sect. 4 is our conclusion.

### 2.1   Measurement idea

The magnetic field influences the absorption of molecular oxygen through the Zeeman effect. Molecular oxygen exhibits the clearest Zeeman effect in the sub-millimeter region of the molecules available in the Martian atmosphere. Other molecules have either a weaker Zeeman effect or a lower volume mixing ratio and are therefore less suitable for magnetic field retrievals. The Zeeman effect changes the energy states of the molecule to split an otherwise singular absorption line into several closely separated lines as a linear function of the strength of the magnetic field. The direction of the magnetic field is important as emission and absorption are polarized by this splitting. The Martian crustal magnetic field is strong, but not strong enough to cleanly separate the split lines from the temperature and pressure broadening of the line shape. Measuring intensity peaks and valleys of the radiation on a frequency resolved grid to directly get the magnetic field strength by peak-to-peak frequency separation is not possible. Instead, the split lines act to broaden or shift the absorption profile by a few, up to hundreds of, kilohertz (kHz). Such frequency broadening or shifting also happens from increased temperatures and from greater wind velocities.

So, an important question to ask is how we can distinguish the atmospheric effects from the magnetic effects on the absorption line. The most obvious way is to simply measure the polarization state of the radiation. Neither temperature nor wind polarizes the radiation, so the level of polarization in the split/broadened spectra is from the magnetic field in clear-sky limb view. A full sampling of the polarization state of the radiation is therefore the best way to retrieve the magnetic field using the Zeeman effect. However, it is more expensive and more difficult to build a sensor capable of measuring the different polarization components (so that the components then can be combined for the total polarization state of the radiation) than it is to build a sensor capable of measuring just one polarization component. Assuming we can only measure one polarization component, is it still possible to get distinct magnetic information? Our approach has been to set up an observational strategy that observes the same limb tangent profile multiple times from several directions during a few satellite revolutions. Temperature and pressure broaden the absorption line shape the same regardless of observational direction through the limb. Wind shifts the frequency along just one axis. The total signal strength in a transparent passively emitting atmosphere is from the number density of the emitting molecule. What remains of the signal after these effects are accounted for is therefore the polarization state and frequency shift due to the Zeeman effect.

## 2.2 Theoretical considerations

We use the Atmospheric Radiative Transfer Simulator (ARTS; Buehler et al., 2005; Eriksson et al., 2011) to simulate measurements. ARTS is a multi-purpose radiative transfer simulator, originally developed for limb sounding of Earth's atmosphere in the microwave spectral range. It currently handles microwave to infrared radiation and can simulate the full Stokes vector in arbitrary viewing geometry, with or without scattering. Applications for Earth's atmosphere have been as diverse as simulating detailed millimeter-wave three-dimensional radiative transfer through clouds (Davis et al., 2007) and simulating the infrared outgoing long-wave radiation leaving Earth to space across the full spectrum (Buehler et al., 2006). Recently, ARTS was adapted to handle other solar system planets, such as Mars. The retrieval toolbox Q-package (Qpack; Eriksson et al., 2005) is used to determine the magnetic field component error. Together, these code bases set up our retrieval system. The source codes of both software packages are freely available under the General Public License (GNU) at http://www.radiativetransfer.org.

The approach to radiative transfer taken by ARTS is to calculate the monochromatic pencil-beam polarized radiative transfer equation in Stokes formalism along the path of the radiation through a three-dimensional inhomogeneous atmosphere and magnetic field. Antenna size, sensor characteristics, and polychromatic signal averaging are considered (as described by Eriksson et al., 2006). The Zeeman effect module of ARTS (Larsson et al., 2014) is applied for these calculations to simulate the left circular polarization component as observed by the simulated sensor for a 20 MHz passband of 201 channels of 100 kHz Gaussian shape surrounding the central absorption line at 368 GHz. The ARTS Zeeman effect module has been validated using ground-based measurements (Navas-Guzmán et al., 2015) and by comparisons with meteorological satellite measurements (Larsson et al., 2016), in applications considering the radiation in different states of polarization. Circular polarization is noticeably more influenced by the magnetic field than linear polarization, and 10 MHz on both sides of the line captures most of the information given by the Zeeman effect. For circular polarization, the magnetic signal is strongest when the local magnetic field vector points directly at or is pointing directly away from the sensor (along the track of the measured radiation). So in limb-viewing geometry radiative transfer, only the horizontal magnetic field components are important at the tangent point. We expect some sensitivity to the radial compo-

nent at higher altitudes than the tangent altitude (since part of the radial component is then along the line of sight), but the sensitivity to the radial component is expected to be low since most of the signal is from around the tangent point. If we measure linear instead of circular polarization, then these measurements are sensitive to both the radial component and the horizontal components of the magnetic field. However, the magnetic signal is significantly weaker for linear than for circular polarization, so the retrieved magnetic field would be noisier. The ARTS simulations also give the Jacobian matrix for specified retrieval quantities (in our case the magnetic field components and, for testing, temperature, wind components, and the molecular oxygen volume mixing ratio).

We use a moderately nonlinear error characterization method, as presented by Rodgers (2000), to estimate the errors associated with a simulated measurement on a retrieval quantity. This error characterization is from

$$
x = x_a + S_a K^\top \left( K S_a K^\top + S_\epsilon \right)^{-1} \\
\left[ y - F(x) + K(x - x_a) \right],
\tag{1}
$$

where $x$ is the derived atmospheric variables and magnetic field, $x_a$ is the a priori atmospheric variables and magnetic input at different retrieval grid points (atmospheric pressure level, latitude, longitude), $S_a$ is the covariance matrix of the a priori, $K$ is the Jacobian matrix, $S_\epsilon$ is the covariance matrix of the instrument error, $y$ is the simulated measurement vector (made up of several individual measurements), and $F(x)$ is the forward-model simulations. With measurements $F(x) \neq y$, however, with pure simulations the terms are identical. Therefore, the error itself is found from $F(x) = y - \epsilon$, where $\epsilon$ is the random noise of the observation – this "noise" encompasses every error not accounted for by the simulated measurement, especially defined instrumental errors. Thus the retrieval error of Eq. (1) is $S_a K^\top \left( K S_a K^\top + S_\epsilon \right)^{-1} \epsilon$. We also estimate the smoothing of the calculations by the matrix $S_a K^\top \left( K S_a K^\top + S_\epsilon \right)^{-1} K$, which is called the averaging kernel. This matrix gives information on the measurement response of the system and the vertical resolution of the measurements. The method above gives a linear error estimate; even if a problem is inherently nonlinear, the error is often still linear (Rodgers, 2000).

About the retrieval grid, as explained in the previous subsection, several measurements of radiation from the same tangent profile can be combined to find the magnetic field components. The observational geometry is important as the individual measurements will observe different parts of the atmosphere at higher altitudes but the same parts of the atmosphere at lower altitudes. We therefore use a retrieval grid that takes the three-dimensional inhomogeneous atmosphere and magnetic field into account by setting a 3-by-3 grid of latitudes and longitudes, with the central grid point at the latitude and longitude of the tangent profile. The latitude grid positions are separated by a change of 2°, with the same horizontal distance separating the longitude grid positions. We use a vertical retrieval grid separation of 2 km from 0 up to 100 km. So $x_a$ is a $3 \times 3 \times 51$ long vector of inputs per retrieval quantity. We note that the measured magnetic field for the altitude range we have considered (below 100 km) is essentially the planetary crustal magnetism (Brain et al., 2003). External fields (interplanetary magnetic fields draping around the conductive ionosphere and induced ionospheric magnetic fields) play a role at higher altitude, where the solar wind interacts with the upper atmosphere. In areas of significant crustal magnetism, the influence of external fields starts at altitudes around 500 km (Fig. 11 of Brain et al., 2003); therefore they are not a concern here.

For $S_a$, we assume that there is no correlation between the different retrieval quantities but that there is some correlation in spatial distance. For the tests we have performed on the wind, temperature, and volume mixing ratio retrievals, we use covariance matrices that work for Earth. For the magnetic field components, since the crustal magnetic field is a mostly static variable at the altitudes below 100 km where we simulate the measurements, its covariance matrix should also be mostly static. However, in a Bayesian framework, the covariance matrix should describe the knowledge we have of the magnetic field at the time of measurement. As mentioned earlier, the strongest difference between the models compared in Morschauser et al. (2014) was 2000 nT, but many sources differ less. So different models give different results. For simplicity we set the uncertainty of the magnetic field to 1000 nT at all altitudes, and the correlation in altitude is $e^{-1}$ after 1 order of magnitude of atmospheric pressure change. We also set a small correlation between horizontal grids of $e^{-1}$ after a horizontal distance equivalent to 1.5° of latitude change.

## 2.3 Model inputs

### 2.3.1 Orbit and sensor considerations

A circular orbit at 330 km altitude with around a 2 h period, 97° inclination, and ascending node at 0° is used in all simulations. This orbit was selected ad hoc, such that it is able to cover Mars in a short time frame with measurements covering almost a full circle in azimuthal resolution. Other orbital parameters can be used to retrieve the magnetic field as well but require different considerations. An elliptical orbit, for instance, will have different vertical resolutions depending on true anomaly, and a less inclined orbit means that a different measurement scheme has to be adopted and that fewer poleward latitudes can be sounded – most likely with increased measurement noise. With enough measurements the measurement noise could be reduced regardless of orbit. However, the quantitative details of the change of the measurement noise to any specific orbit change have to be worked out on a per-orbit basis, so how much the noise can

be reduced is not possible to tell without doing specific simulations for a given set of orbital parameters.

With this choice of orbital parameters, we model a sensor with a 30 cm diameter antenna for a vertical resolution of about 4 km at the tangent points sampled in limb sounding by an orbiter at 330 km altitude. The antenna size was chosen for these simulations because it is reasonably small. We will only simulate left circular polarized radiation, and we assume 1500 K single-sideband system noise temperature. This number is a rough extrapolation to lower frequencies from the expected system noise temperature of the Jupiter Icy Moons' Sub-millimeter Wave Instrument's 600 GHz band (see Sobis et al., 2011, 2014, and Treuttel et al., 2016, for more details on the JUICE/SWI receiver). This extrapolation simply scales the SWI number by the frequency ratio of the channels and rounds the results up for good measure. The engineering has to be done to give a more exact number. We enforce 2 s as the time of a measurement, dedicating one-third to calibration efforts and the remaining $\sim 1.3$ s as integration time.

The position of the tangent profiles tracks this orbit as a pseudo-orbit with the same orbital parameters but with an ascending node offset by half the longitude drift of a full satellite revolution. These tangent profiles can be observed in limb geometry four times within about two revolutions: for each revolution, once before passing the tangent profile and once after passing the tangent profile. The tangent altitudes we select to observe this tangent profile extend from 6 to 78 km altitude, with a 3 km vertical separation for 25 tangent altitudes. So a total of 100 individual measurements are considered (25 measurements × 4 observation sets) in the error characterization. We call the grouping of all these individual measurements for a given tangent profile a measurement block. The error characterization of this work is from these measurement blocks. The results in the next sections track the tangent profile pseudo-orbit through seven complete revolutions at a distance of 15° of true anomaly between two retrieval profiles. This is a sufficient number of revolutions so that we can give the global error characterization.

An example of observation geometry and orbit can be seen in Fig. 1 (the Supplement provides additional details). As mentioned, we simulate 201 channels for the sensor. With 100 individual measurements in a measurement block, $y$ is a 20 100 long vector, and **K** is a 20 100 × $(3 \times 3 \times 51 \times n)$ large matrix, where $n$ is the number of retrieval quantities (usually three for the magnetic field components) and the other numbers are from the retrieval grid. The position of all 100 of these individual measurements in the example of Fig. 1 is shown in panel a. Panel b of the same figure shows the grouping of individual measurements for one set of observations of the tangent profile. We call this smaller grouping a measurement cluster. A satellite moving towards a tangent profile will, by geometrical necessity, see the tangent altitude increase the closer it gets to the profile, and vice versa a satellite moving away will see the tangent altitude decrease the farther away it gets. Finally, panel c shows that there is a

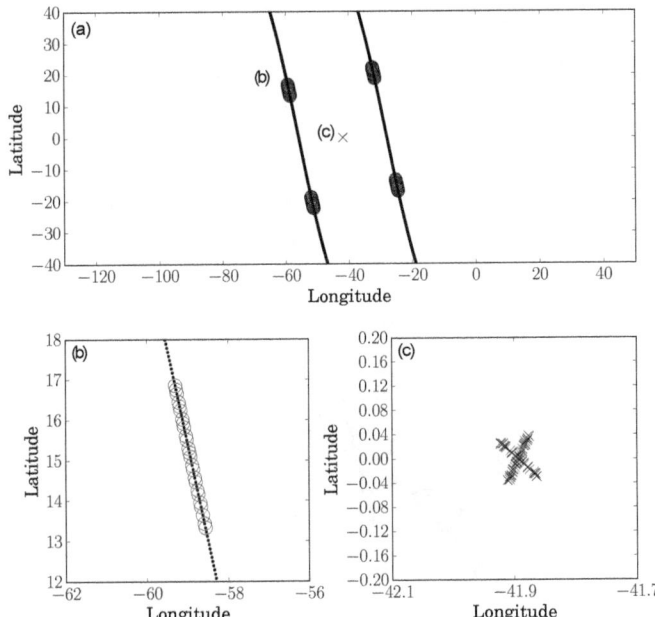

**Figure 1.** Example of the geometry of a measurement block. Panel (**a**) shows the orbit of a satellite as two thin lines indicating two satellite revolutions: the satellite positions during the two observation sets during each revolution in the measurement block as larger circles, and the mean tangent position of the measurement block as a cross. This example retrieval point is above a location of moderate modeled magnetism strength near the equator (750 nT at surface altitudes crustal magnetism model by Cain et al., 2003). Panel (**b**) zooms in to show the satellite positions at one of the clusters of satellite positions. Panel (**c**) zooms in to show the tangent positions. The original intent of the measurement block was to observe in the center of this block of individual measurements.

slight drift in tangent point positions between different measurements. It is thus expected that the tangent position will drift slightly; the greater circle of positions that can observe a tangent altitude does not strictly intersect the satellite orbit at discrete time intervals. The drift is accounted for in the calculations of **K** and **y**.

The only spectroscopic effect included in our radiative transfer calculations is the contribution of the 368 GHz molecular oxygen absorption line. We select this absorption line because the Zeeman effect line splitting frequency is linear in magnetic field strength. So the lower the frequency, the stronger the signal from the magnetic field becomes as Doppler broadening is also linear with absorption line frequency. The reason we are not simulating one of the 60 GHz molecular oxygen band lines is that we consider the antenna diameter required to have a decent vertical resolution in limb view as too large to reasonably fly the sensor. Our line data are from a recently compiled planetary toolbox, which gives line pressure broadening as a function of atmospheric composition for mixtures of common planetary

species[1] by Mendrok and Eriksson (2014), who assume that the molecular oxygen pressure broadening by carbon dioxide is $\sim 20$ kHz Pa$^{-1}$ at 296 K for our absorption line. This is directly taken from the 118 GHz line value presented by Golubiatnikov et al. (2003). This spectroscopic toolbox allows for direct calculations in Mars' carbon-dioxide-rich atmosphere. Note that the collision-induced absorption between pairs of carbon dioxide molecules is also important for the total absorption in our frequency range. Using the model by Ho et al. (1971), we estimate that it can be as important as adding a 20 K baseline signal for a pencil beam tangential to the surface. This added absorption is not sufficient to make the atmosphere opaque. We can anyways ignore this effect because at 6 km altitude – our lowest tangent altitude – the collision-induced absorption is much weaker than at the surface. Still, we note this as a lacking feature to be added in the future.

To summarize the orbit and sensor constraints, we present Fig. 2 to give an overview of how a simulation of a measurement block looks with the assumptions outlined above. Panel a contains peak brightness temperature measurements for two clusters (looking ahead and behind the tangent point) for one revolution, and again measurements for two clusters for another revolution, with each cluster having 25 measurements at tangent altitudes between 6 and 78 km. As expected, the main contribution to the signal strength is the tangent altitude (through which the column number densities are regulated). At 6 km measurement tangent altitude, the radiation signal has a peak brightness temperature of about 140 K, and at 78 km the same number is about 10 K. The sinusoidal pattern of measurements in each cluster comes from the resolution of the spectra: the peaks are at the line centers, and the valleys are away from the line centers. Only a very small part of this signal is influenced by the magnetic field. This part of the signal is not clear from the brightness temperature in panels a–b but is instead shown in the Jacobian panel c. We see that the magnetic signal is of the order of about 0.2 mK nT$^{-1}$ as strongest per 2 km altitude level per 100 kHz spectrometer channel bin. The resolution of the antenna is indicated from the coverage of the circular central disk of the Jacobian panel. Again, the simulated full-width vertical resolution is around 4 km for a 30 cm antenna. The covariance matrix is shown for completeness in panel d. The figure shows neither the full **K** nor the full $S_a$ matrices but is zoomed in on a single sub-matrix. Nevertheless, the figure presents the basic retrieval setup. The only missing entry required for the error characterization is $S_\epsilon$. This matrix is not shown because it is simply filled with a constant describing the noise of the sensor. This constant is the square root of the system noise equivalent temperature divided by the integration time times

---

[1]Our atmospheric and spectroscopic data are available in Extensible Markup Language files designed for ARTS via http://www.radiativetransfer.org.

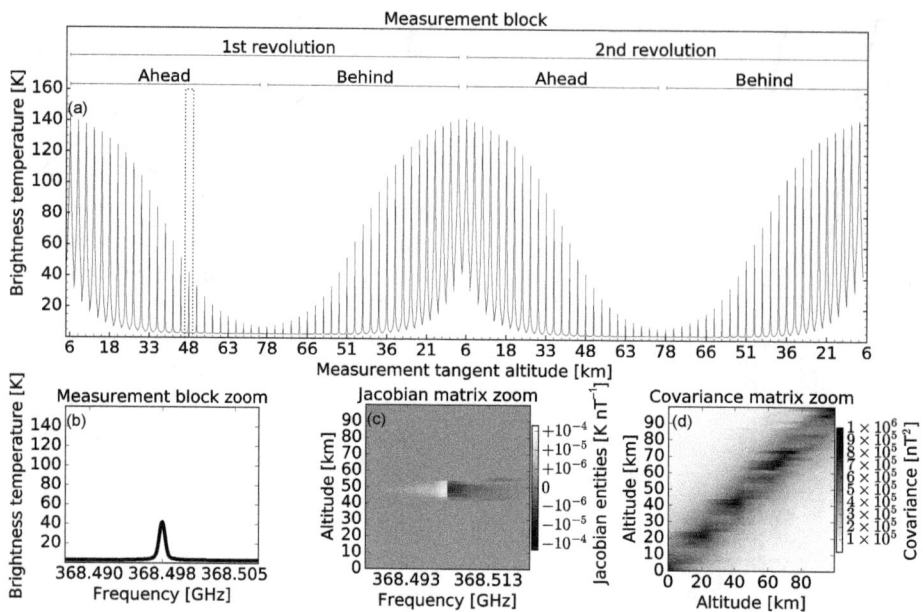

**Figure 2.** Example retrieval scenario. This is $y$, part of $\mathbf{K}$, and part of $\mathbf{S}_a$ from the example in Fig. 1. Panel **(a)** is the simulated measurements $y$ of ARTS with individual clusters marked by their revolution number and looking direction. Each peak corresponds to an individual measurement – the horizontal axis gives the tangent altitudes. The bottom row zooms in on the marked region of $y$ to show an individual simulation at 48 km tangent altitude **(b)**, a zoom on the transpose of the $\mathbf{K}$ sub-matrix at the tangent point latitude–longitude grid point for one magnetic component **(c)**, and its corresponding $\mathbf{S}_a$ sub-matrix **(d)**.

the spectrometer channel full width – which turns out to be just below 0.1 K².

### 2.3.2 Atmosphere and magnetic field models

The molecular oxygen profile is assumed constant at a volume mixing ratio of 1400 parts per million volume, which follows the profile reported by Hartogh et al. (2010b). This profile was derived by observations of the Heterodyne Instrument for the Far Infrared (HIFI) on Herschel as part of the Herschel Solar System Observations program (Hartogh et al., 2009) by using a temperature profile derived at the same day from carbon monoxide observations (Hartogh et al., 2010a). At altitudes above 90 km, this constant mixing ratio is not valid (as shown by measurements presented by Sandel et al., 2015), and it is likely not valid at our lower altitudes either. However, since we can see the molecular oxygen radiometric signal even at 78 km altitude, and since a changed mixing ratio only affects the total signal strength, the magnetic signal – that is, the relative polarization and splitting caused by the magnetic field – is not affected by this different mixing other than lowering or increasing the sensitive altitude range. With this selection of volume mixing ratio, we are not sensitive to molecular oxygen at altitudes much higher than 78 km. This is seen by the low signal strength at this altitude in Fig. 2a and the reduction compared to lower tangent altitudes. Additionally, panel c of the same figure shows that most of the signal is from around the tangent point, which is a property of limb observation geometry. Higher-altitude

molecular oxygen content is therefore not important for our results.

We use the northern spring (solar longitude $L_s = 0°$) carbon dioxide volume mixing ratio, temperature profile, wind profile, and pressure profiles from the ARTS planetary toolbox by Mendrok and Eriksson (2014), who base it on the Laboratoire de Météorologie Dynamique's global circulation model (LMD GCM) by Forget et al. (1999). These atmospheric profiles only provide global averages during the season. We thus ignore most effects of time since we are interested in demonstrating the feasibility of the technique of magnetic field measurements on a general basis. Presented errors are hence of average character. The a priori atmospheric profiles can be constrained by the measurements, but we have not taken this into account in any of our simulations shown in this work, though we give an example of what adding the atmospheric components to the error characterization means for the errors of the magnetic components. If the atmospheric parameters are not stable over the 2 h of a revolution, then the evolution of the temperature and wind profiles must be accounted for in the retrieval setup. This is still possible without changing the theoretical formalism of the retrieval setup if the problem remains moderately nonlinear, but it complicates the preparation of the atmospheric data and the simulations. From a theoretical point of view, the error characteristics of the magnetic field will remain moderately nonlinear even if the temperature, wind, and volume mixing ratio are accounted for in the retrieval problem because there

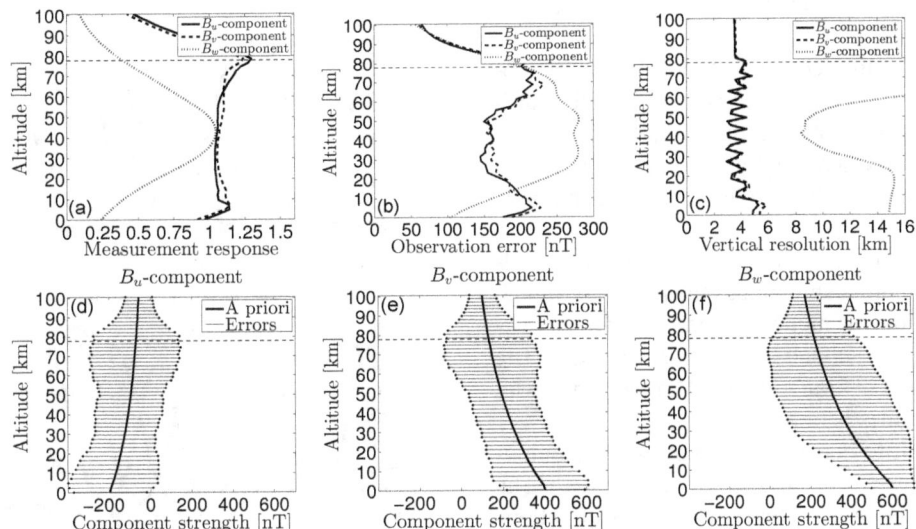

**Figure 3.** Retrieval information for Fig. 1. The first row shows Qpack output following Eq. (1). Panel (**a**) shows the response function per magnetic field component, (**b**) shows the observation error per magnetic field component, and (**c**) shows the vertical resolution per magnetic field component (zigzag is from discrete number of vertical limb measurements at 3 km spacing). The lower row (**d–f**) compares this observational error to a priori magnetic field components. The horizontal dashed line shows the highest tangent altitude.

should be no correlation between the atmospheric parameters and the magnetic field.

The primary magnetic field in our altitude range is the crustal magnetic field (Brain et al., 2003). Our model magnetic field is from the spherical harmonics fit to a selection of Mars Global Surveyor crustal magnetic field vector data by Cain et al. (2003). The magnetic field in ARTS is an extraction of the fit of Cain et al. (2003) that has been gridded at a global resolution of 0.5° latitude, 0.5° longitude, and 5 km altitude from the surface up to 100 km altitude. This means that the magnetic field is allowed to change along the path of the radiative transfer in our simulations. If the magnetic field had changed more dramatically through the radiative transfer than it does, then this might have reduced the magnetic signal by changing the polarization of the signal propagation. Since this is not a problem that we encounter, we will not pursue any ideas on how to deal with it. The model by Cain et al. (2003) is based on a 90th-order Legendre polynomial, so the spatial resolution is limited to $\sim 130$ km. As mentioned, other models, such as that of Morschauser et al. (2014), differ from that of Cain et al. (2003) by about 2000 nT at most (see Fig. 9 of Morschauser et al., 2014) but are otherwise close. This 2000 nT difference is therefore an estimation of the accuracy of the magnetic field components with the present data. Some authors (Brain et al., 2003) speculate that the strongest field strength at the surface is up to 20 000 nT. Cain et al. (2003) give the strongest field as 12 000 nT, so 8000 nT serves as an estimate of the potentially largest discrepancy today of the crustal magnetic field strength of Mars.

## 3 Results and discussions

The structure of this section is that we begin by presenting the results of running the measurement block of Fig. 1. This example is used as a generic measurement block, and noise levels are lower or higher depending on the geometry of observations. We then present an average for many observations spread out globally. There is a final summary at the end of this section showcasing where our results suggest that the proposed measurements would be useful given the expected noise levels of individual measurement blocks. Note that we define the magnetic field components as follows: $B_u$ is the east–west component of the magnetic field, $B_v$ is the north–south component, and $B_w$ is the radial component.

### 3.1 Example measurement

Figure 3 shows the results of running the retrieval system for the tangent profile example of the measurement block in Fig. 1 with the simulated measurements and Jacobian as in Fig. 2. It shows that the measurement response is good (i.e., around unity) for the $B_u$ and $B_v$ components (horizontal components) from the ground up to about 70 km for 16 mostly independent measurements of the magnetic field in the tangent profile. The $B_w$ component (vertical component) has a poorer response than the other two components, but the measurements are in this case sensitive to $B_w$ at altitudes around 40 km. The vertical resolution is about 4 km for $B_u$ and $B_v$ over the entire sensitive altitude range because it is perpendicular to the line of sight in the limb geometry, which reflects the achievable resolution for the simulated antenna size and tangent profile altitude spacing. The vertical resolution is much worse for $B_w$ between 8 and 15 km. The observational error that $B_u$ and $B_v$ experience varies between around 150 to 200 nT in the sensitive altitude range. The observational error that $B_w$ experiences is much worse, but it settles close to 300 nT at sensitive altitudes. Note that the observation errors are low where the measurement response is low; without measurements there are no errors due to observations.

As for vertical structures, the main limitation at lower altitudes is that pressure broadening hides the magnetic signal, and the main limitations at higher altitudes is that there is very little molecular oxygen due to the lower pressure. The optimal magnetic signal is from around 40 km, which is why the noise at these altitudes is lower (panel b).

$B_w$ is more difficult to detect than the other two components because of the characteristics of the Zeeman effect. The multiple measurements from different azimuthal directions allow the angle of the magnetic field in the horizontal plane to be tested almost directly, but $B_w$ acts only to broaden the absorption line in limb view. It is therefore difficult to measure it without more constraints. In fact, if we add retrievals of wind, temperature and molecular oxygen volume mixing ratio to the retrieval problem – thus reducing the constraints on the retrieval problem – the measurement response to $B_w$ is no longer good at any altitude. So we will ignore $B_w$ from here on since it is not measurable by the presented method. The horizontal magnetic components are not affected by the lessened constraints on the atmospheric parameters.

## 3.2 Suggested operational setup

The average observation errors expected for our suggested operational setup scenario are presented in Fig. 4. We simulate only seven revolutions, or about half a Martian day, of measurements, since this gives a good coverage for the whole Martian globe with our ad hoc orbit.

We find that the global average profile observation error for the $B_u$ component is 170 nT and that the same average observation error for the $B_v$ component is 200 nT. Near the equator, the two components have similar observation errors (almost identical to what is shown in Fig. 3). Closer to the poles, $B_u$ has better observation errors, and $B_v$ has worse observation errors. The reason for this is the azimuthal angles of the observation geometry of a measurement block. With the selected orbit and observational strategy, the tangent profiles are observed from the northeast/west and southeast/west azimuthal angles at the equator. Near the poles, however, the azimuthal angles are almost aligned with the east–west direction. On one hand, this indicates that it is possible to improve $B_v$ observation errors at polar latitudes to the same levels as $B_u$ by more observations from directly north or south of the tangent profiles. Similarly, it also indicates that it is possible to improve the equatorial observation errors for both horizontal components to the levels of $B_u$ at polar latitudes by focusing on east–west or north–south observation geometry

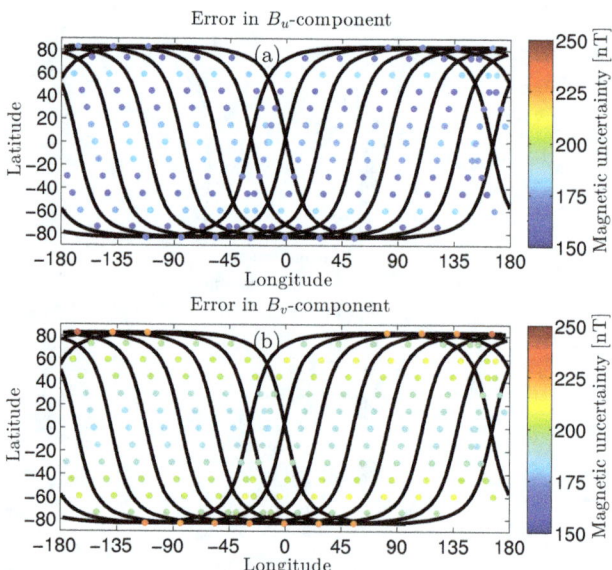

**Figure 4.** Suggested operational setup scenario. The black lines represent the orbit through seven revolutions. The maps show the estimations for errors per component averaged between 6 and 78 km altitude.

depending on the component of interest. On the other hand, if the inclination of the orbit is such that there will be fewer measurements at higher latitudes, then the errors in the $B_v$ component will be worse than shown in Fig. 4. Clearly, a highly inclined orbit is beneficial for the suggested measurements when compared to a less inclined orbit.

## 3.3 Simple estimation of measurable area and a comparison to satellite magnetometer

We prepared Fig. 5 to show the areas over the Martian surface where these measurements would be useful. This plot shows the modeled magnetic field strength extracted at different altitudes from Cain et al. (2003). At 40 km altitude, an area covering about 36 % of Mars total area has magnetic field strengths above 200 nT, with the southern sources being stronger than the northern sources. The areas with more than 1000 nT field strength cover about 4 % of the surface at this altitude. So our suggested measurements are estimated to provide information on the Martian magnetic field over somewhere between 4 and 36 % of the area of Mars at 40 km altitude, i.e., in areas where the crustal magnetism is significant. The same area percentages at 20 km are 9 % above 1000 nT and 49 % above 200 nT, and at 60 km the numbers are 2 and 27 %, respectively. For the rest of the area of Mars, the method is only capable of confirming the sub-sensitivity surface field strengths of the Cain et al. (2003) model.

Comparing the 200 nT per 4 km noise attainable by submillimeter measurements to satellite-borne magnetometer measurements by Mars Global Surveyor offers some additional information into the limitations and possibilities of the

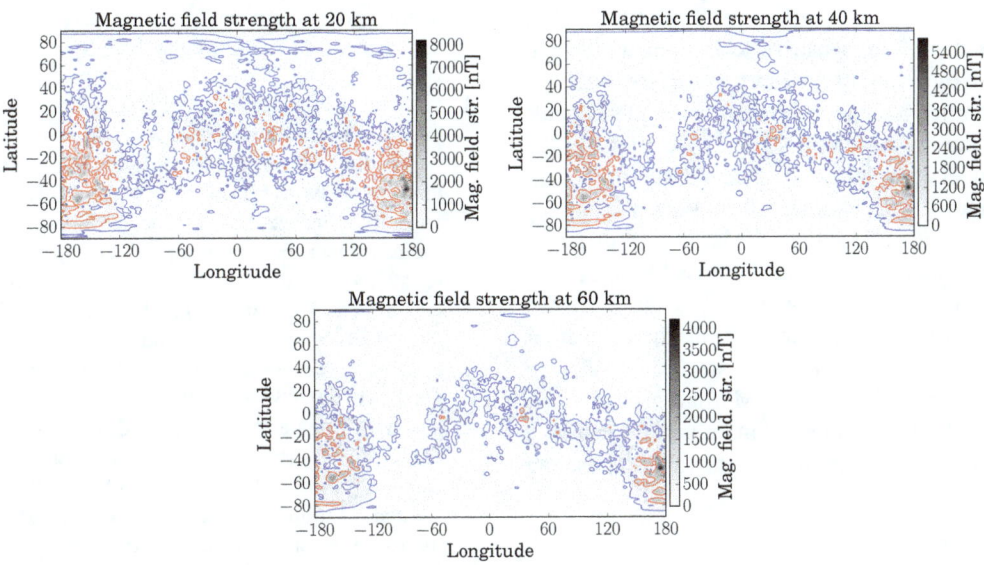

**Figure 5.** The modeled magnetic field strength of Mars at 20, 40, and 60 km altitude from Cain et al. (2003). The red contour marks 1000 nT, and the blue contour marks 200 nT.

sub-millimeter measurements. The MGS magnetometer had noise levels of only 3 nT during its deep dips to $\sim$ 100 km (Acuña et al., 1998). At 100 km, almost 99 % of the area of Mars has a magnetic field strength stronger than 3 nT, with 76 % of the area having a magnetic field strength over 15 nT (Cain et al., 2003). How much area a satellite-borne magnetometer can cover at 100 km altitude during a mission is therefore mostly limited by the amount of fuel brought with the satellite to maintain a good orbit after the deep dips. Measurements far above 100 km altitude contribute less to understanding the structures of the magnetic field necessary to estimate the strengths, shapes, and locations of the crustal sources. This is because the structures of the magnetic field are hidden by the added distance (as seen in, e.g., Fig. 6 by Connerney et al., 2004). The higher altitude of these more stable orbits are thus less interesting in regards to crustal sources but nevertheless offer much better noise levels. The noisier sub-millimeter measurements are able to sense lower altitudes, and, in contrast to the magnetometer, this can be done from a stable orbit. We consider that the measurements of the satellite-borne magnetometers and the sub-millimeter sensors are complementary to one another, with satellite magnetometers able to more strictly limit the orientation and strength of the field and sub-millimeter spectrometers more capable to sample low-altitude structures.

Finally, we remind the reader that the magnetic field strength is mostly static and will smoothly decrease in magnitude with increasing altitude. These potentials have not been used in the retrieval system presented in this work. Starting with the altitude, we find 16 data points of errors of about 200 nT per component from 6 km altitude up to 70 km altitude. Using these to fit to a model of smoothly decreas-

ing magnetic field magnitude, the noise should reduce by the square root of the number of independent data points, potentially fitting such a model to an error of about 50 nT per tangent profile. As also mentioned, we only consider measurements taking place during less than 24 h in this paper. Reasonably, a mission with a stable orbit can survive substantially longer. Throughout the lifetime of the entire mission it will be possible to measure the static magnetic field over the same areas multiple times. The error of 200 nT per 4 km per horizontal component is thus just a beginning and can be reduced continuously during the mission via proper post-processing techniques. It is beyond the scope of this work to analyze such post-processing in detail. We will leave such details to the future, if or when the sensor concept described in this work flies to Mars.

## 4   Conclusions

We have described an idea that makes measurements of the Martian near-surface crustal magnetic field profiles possible using remote-sensing techniques by combining several limb observations. We have performed radiative transfer simulations for radiation measurements from the atmosphere of Mars for one orbiting sensor. These radiative transfer simulations have been fed into a retrieval toolbox to find sensitivities to the magnetic field. Our work shows that limb observations of the sub-millimeter radiation of one absorption line of molecular oxygen can be used to measure two of the three components (the horizontal components) of the magnetic field with around 200 nT accuracy between 6 and 70 km altitude. The vertical resolution of such measurements will be about 4 km for a 30 cm diameter antenna measuring the 368 GHz line from an orbit at 330 km altitude. The described measurements are sensitive to the magnetic field strength over about one-third of the Martian surface, which is the area where significant localized planetary magnetic fields exist. We have made few assumptions on correlations in the retrieval system to let our method enhance measurement sensitivities rather than a priori sensitivities. With reasonable assumptions on the magnetic source fields or with different orbit/observation geometry, it should be possible to reduce the noise further with proper post-processing techniques. We suggest flying a sensor capable of measuring the molecular oxygen absorption line at 368 GHz on a satellite mission to Mars to make a detailed map of the Martian crustal magnetic field and thereby help – in combination with available satellite magnetometer data – to find the depths, shapes, and positions of the crustal field sources.

*Acknowledgements.* We want to acknowledge all those in the ARTS user community who have contributed to the ARTS development.

Edited by: L. Vazquez

# References

Acuña, M. H., Connerney, J. E. P., Wasilewski, P., Lin, R. P., Anderson, K. A., Carlson, C. W., McFadden, J., Curtis, D. W., Mitchell, D., Rème, H., Mazelle, C., Sauvaud, J. A., d'Uston, C., Cros, A., Medale, J. L., Bauer, S. J., Cloutier, P., Mayhew, M., Winterhalter, D., and Ness, N. F.: Magnetic Field and Plasma Observations at Mars: Initial Results of the Mars Global Surveyor Mission, Science, 279, 1676–1680, 1998.

Acuña, M. H., Connerney, J. E. P., Ness, N. F., Lin, R. P., Mitchell, D., Carlson, C. W., McFadden, J., Anderson, K. A., Rème, H., Mazelle, C., Vignes, D., Wasilewski, P., and Cloutier, P.: Global Distribution of Crustal Magnetization Discovered by the Mars Global Surveyor MAG/ER Experiment, Science, 284, 790–793, 1999.

Brain, D. A., Bagenal, F., Acuña, M. H., and Connerney, J. E. P.: Martian magnetic morphology: Contribution from the solar wind and crust, J. Geophys. Res., 108, 1424, doi:10.1029/2002JA009482, 2003.

Buehler, S. A., Eriksson, P., Kuhn, T., von Engeln, A., and Verdes, C.: ARTS, the atmospheric radiative transfer simulator, J. Quant. Spectrosc. Ra. 91, 65–93, doi:10.1016/j.jqsrt.2004.05.051, 2005.

Buehler, S. A., von Engeln, A., Brocard, E., John, V. O., Kuhn, T., and Eriksson, P.: Recent developments in the line-by-line modeling of outgoing longwave radiation, J. Quant. Spectrosc. Ra., 98, 446–457, doi:10.1016/j.jqsrt.2005.11.001, 2006.

Cain, J. C., Ferguson, B. B., and Mozzoni, D.: An $n = 90$ internal potential function of the Martian crustal magnetic field, J. Geophys. Res., 108, 5008, doi:10.1029/2000JE001487, 2003.

Citron, R. I. and Zhong, S.: Constraints on the formation of the Martian crustal dichotomy from remnant crustal magnetism, Phys. Earth Planet. Int., 212–213, 55–63, 2012.

Connerney, J. E. P., Acuña, M. H., Ness, N. F., Spohn, T., and Schubert, G.: Mars Crustal Magnetism, Space Sci. Rev., 111, 1–32, 2004.

Davis, C. P., Evans, K. F., Buehler, S. A., Wu, D. L., and Pumphrey, H. C.: 3-D polarised simulations of space-borne passive mm/sub-mm midlatitude cirrus observations: a case study, Atmos. Chem. Phys., 7, 4149–4158, doi:10.5194/acp-7-4149-2007, 2007.

Eriksson, P., Jiménez, C., and Buehler, S. A.: Qpack, a general tool for instrument simulation and retrieval work, J. Quant. Spectrosc. Ra., 91, 47–64, doi:10.1016/j.jqsrt.2004.05.050, 2005.

Eriksson, P., Ekström, M., Melsheimer, C., and Buehler, S. A.: Efficient forward modelling by matrix representation of sensor responses, Int. J. Remote Sens., 27, 1793–1808, doi:10.1080/01431160500447254, 2006.

Eriksson, P., Buehler, S. A., Davis, C. P., Emde, C., and Lemke, O.: ARTS, the atmospheric radiative transfer simulator, Version 2, J. Quant. Spectrosc. Ra., 112, 1551–1558, 2011.

Forget, F., Hourdin, F., Foumier, R., Hourdin, C., and Talagran, O.: Improved general circulation models of the Martian atmosphere from the surface to above 80 km, J. Geophys. Res., 104, 155–175, 1999.

Golubiatnikov, G. Y., Koshelev, M. A., and Krupnov, A. F.: Reinvestigation of pressure broadening parameters at 60-GHz band and single 118.75 GHz oxygen lines at room temperature, J. Mol. Spectrosc., 222, 191–197, 2003.

Hartogh, P., Lellouch, E., Crovisier, J., Banaszkiewicz, M., Bensch, F., Bergin, E., Billebaud, F., Biver, N., Blake, G., Błęcka, M., Blommaert, J., Bockelee-Morvan, D., Cavalié, T., Cernicharo, J., Courtin, R., Davis, G., Decin, L., Encrenaz, P., Encrenaz, T., González, A., de Graauw, T., Hutsemekers, D., Jarchow, C., Jehin, E., Kidger, M., Küppers, M., de Lange, A., Lara, L.-M., Lis, D., Lorente, R., Manfroid, J., Medvedev, A., Moreno, R., Naylor, D., Orton, G., Portyankina, G., Rengel, M., Sagawa, H., Sánchez-Portal, M., Schieder, R., Sidher, S., Stam, D., Swinyard, B., Szutowicz, S., Thomas, N., Thornhill, G., Vandenbussche, B., Verdugo, E., Waelkens, C., and Walker, H.: Water and related chemistry in the solar system. A guaranteed time key programme for Herschel, Planet. Space Sci., 57, 1596–1606, doi:10.1016/j.pss.2009.07.009, 2009.

Hartogh, P., Błęcka, M., Jarchow, C., Sagawa, H., Lellouch, E., de Val-Borro, M., Rengel, M., Medvedev, A. S., Swinyard, B., Moreno, R., Cavalié, T., Lis, D., Banaszkiewicz, M., Bockelée-Morvan, D., Crovisier, J., Encrenaz, T., Küppers, M., Lara, L.-M., Szutowicz, S., Vandenbussche, B., Bensch, F., Bergin, E. A., Billebaud, F., Biver, N., Blake, G., Blommaert, J., Cernicharo, J., Decin, L., Encrenaz, P., Feuchtgruber, H., Fulton, T., de Graauw, T., Jehin, E., Kidger, M., Lorente, R., Naylor, D., Portyankina, G., Sánchez-Portal, M., Schieder, R., Sidher, S., Thomas, N., Verdugo, E., Waelkens, C., Lorenzani, A., Tofani, G., Natale, E., Pearson, J., Klein, T., Leinz, C., Güsten, R., and Kramer, C.: First results on Martian carbon monoxide from Herschel/HIFI observations, Astron. Astrophys., 521, L48, doi:10.1051/0004-6361/201015159, 2010a.

Hartogh, P., Jarchow, C., Lellouch, E., Val-Borro, M., Rengel, M., Moreno, R., Medvedev, A., Sagawa, H., Swinyard, B., Cavalié, T., Lis, D., Błęcka, M., Banaszkiewicz, M., Bockelée-Morvan, D., Crovisier, J., Encrenaz, T., Küppers, M., Lara, L., Szutowicz, S., Vandenbussche, B., Bensch, F., Bergin, E. A., Billebaud, F., Biver, N., Blake, G., Blommaert, J., Cernicharo, J., Decin, L., Encrenaz, P., Feuchtgruber, H., Fulton, T., de Graauw, T., Jehin, E., Kidger, M., Lorente, R., Naylor, D., Portyankina, G., Sánchez-Portal, M., Schieder, R., Sidher, S., Thomas, N., Verdugo, E., Waelkens, C., Whyborn, N., Teyssier, D., Helmich, F., Roelfsema, P., Stutzki, J., LeDuc, H., and Stern, J.: Herschel/HIFI observations of Mars: first detection of $O_2$ at submillimetre wavelengths and upper limits on HCl and $H_2O_2$, Astron. Astrophys., 521, L49, doi:10.1051/0004-6361/201015160, 2010b.

Ho, W., Birnbaum, G., and Rosenberg, A.: Far-Infrared Collision-Induced Absorption in $CO_2$. I. Temperature Dependence, J. Chem. Phys., 55, 1028–1038, 1971.

Jakosky, B. M., Lin, R. P., Grebowsky, J. M., Luhmann, J. G., Mitchell, D. F., Beutelschies, G., Priser, T., Acuna, M., Andersson, L., Baird, D., Baker, D., Bartlett, R., Benna, M., Bougher, S., Brain, D., Carson, D., Cauffman, S., Chamberlin, P., Chaufray, J.-Y., Cheatom, O., Clarke, J., Connerney, J., Cravens, T., Curtis, D., Delory, G., Demcak, S., DeWolfe, A., Eparvier, F., Ergun, R., Eriksson, A., Espley, J., Fang, X., Folta, D., Fox, J., Gomez-Rosa, C., Habenicht, S., Halekas, J., Holsclaw, G., Houghton, M., Howard, R., Jarosz, M., Jedrich, N., Johnson, M., Kasprzak, W., Kelley, M., King, T., Lankton, M., Larson, D., Leblanc, F., Lefevre, F., Lillis, R., Mahaffy, P., Mazelle, C., McClintock, W., McFadden, J., Mitchell, D. L., Montmessin, F., Morrissey, J., Peterson, W., Possel, W., Sauvaud, J.-A., Schneider, N., Sidney, W., Sparacino, S., Stewart, A. I. F., Tolson, R., Toublanc, D., Waters, C., Woods, T., Yelle, R., and Zurek, R.: The Mars Atmosphere and Volatile Evolution (MAVEN) Mission, Space Sci. Rev., 195, 3–48, doi:10.1007/s11214-015-0139-x, 2015.

Kasai, Y., Sagawa, H., Kuroda, T., Manabe, T., Ochiai, S., Kikuchi, K., Nishibori, T., Baron, P., Mendrok, J., Hartogh, P., Murtagh, D., Urban, J., von Schéele, F., and Frisk, U.: Overview of the Martian atmospheric submillimetre sounder FIRE, Planet. Space Sci., 63–64, 62–82, 2012.

Larsson, R., Ramstad, R., Mendrok, J., Buehler, S. A., and Kasai, Y.: A Method for Remote Sensing of Weak Planetary Magnetic Fields: Simulated Application to Mars, Geophys. Res. Lett., 40, 5014–5018, 2013.

Larsson, R., Buehler, S. A., Eriksson, P., and Mendrok, J.: A treatment of the Zeeman effect using Stokes formalism and its implementation in the Atmospheric Radiative Transfer Simulator (ARTS), J. Quant. Spectrosc. Ra., 133, 445–453, 2014.

Larsson, R., Milz, M., Rayer, P., Saunders, R., Bell, W., Booton, A., Buehler, S. A., Eriksson, P., and John, V. O.: Modeling the Zeeman effect in high-altitude SSMIS channels for numerical weather prediction profiles: comparing a fast model and a line-by-line model, Atmos. Meas. Tech., 9, 841–857, doi:10.5194/amt-9-841-2016, 2016.

Mendrok, J. and Eriksson, P.: Microwave Propagation Toolbox for Planetary Atmospheres – Final Report (D13a), Tech. rep., ESTEC Contract No 4000104175/11/NL/AF, 2014.

Morschauser, A., Lesur, V., and Grott, M.: A spherical harmonic model of the lithospheric magnetic fieldof Mars, J. Geophys. Res., 119, 1162–1188, 2014.

Navas-Guzmán, F., Kämpfer, N., Murk, A., Larsson, R., Buehler, S. A., and Eriksson, P.: Zeeman effect in atmospheric $O_2$ measured by ground-based microwave radiometry, Atmos. Meas. Tech., 8, 1863–1874, doi:10.5194/amt-8-1863-2015, 2015.

Nimmo, F. and Tanaka, K.: Early Crustal Evolution of Mars, Annu. Rev. Earth Planet. Sc., 33, 133–161, 2005.

Rodgers, C. D.: Inverse methods for atmospheric sounding: Theory and practice, Vol. 2, World Scientific Publishing Co. Pte. Ltd., 2000.

Sandel, B., Gröller, H., Yelle, R., Koskinen, T., Lewis, N., Bertaux, J.-L., Montmessin, F., and Quémerais, E.: Altitude profiles of $O_2$ on Mars from SPICAM stellar occultations, Icarus, 252, 154–160, doi:10.1016/j.icarus.2015.01.004, 2015.

Sobis, P., Drakinskiy, V., Wadefalk, N., Karandikhar, Y., Hammar, A., Emrich, A., Zhao, H., Bryllert, T., Tang, A.-Y., Nilsson, P.-A., Schleeh, J., Kim, H., Jacob, K., Murk, A., Grahn, J., and Stake, J.: Low noise GaAs Schottky TMIC and InP Hemt MMIC based receivers for the ISMAR and SWI instruments, Tech. rep., ESA-ESTEC, Noordwijk, the Netherlands, 2014.

Sobis, P. J., Emrich, A., and Stake, J.: A Low VSWR 2SB Schottky Receiver, IEEE Transactions on Terahertz Science and Technology, 1, 403–411, doi:10.1109/TTHZ.2011.2166176, 2011.

Treuttel, J., Gatilova, L., Maestrini, A., Moro-Melgar, D., Yang, F., Tamazouzt, F., Vacelet, T., Jin, Y., Cavanna, A., Matéos, J., Féret, A., Chaumont, C., and Goldstein, C.: A 520–620-GHz Schottky Receiver Front-End for Planetary Science and Remote Sensing With 1070 K–1500 K DSB Noise Temperature at Room Temperature, IEEE Transactions on Terahertz Science and Technology, 6, 148–155, doi:10.1109/TTHZ.2015.2496421, 2016.

Zeeman, P.: On the Influence of Magnetism on the Nature of the Light Emitted by a Substance, Astrophys. J., 5, 332–347, doi:10.1086/140355, 1897.

# 9

# Noise in raw data from magnetic observatories

**Sergey Y. Khomutov**[1], **Oksana V. Mandrikova**[1], **Ekaterina A. Budilova**[1,2], **Kusumita Arora**[3], **and Lingala Manjula**[3]

[1]Institute of Cosmophysical Research and Radio Wave Propagation FEB RAS, Mirnaya str, 7, Paratunka 684034, Kamchatka, Russia
[2]Kamchatka State Technical University, Klyuchevskaya str, 35, Petropavlovsk-Kamchatsky 683003, Russia
[3]CSIR – National Geophysical Research Institute, Uppal Road, Hyderabad-500007, Telangana, India

*Correspondence to:* Sergey Y. Khomutov (khomutov@ikir.ru)

**Abstract.** In spite of significant progress in the development of new devices for magnetic measurements, mathematical and computational technologies for data processing and means of communication, the quality of magnetic data accessible through the data centres (for example, World Data Centres or INTERMAGNET) still largely depends on the actual conditions in which observation of the Earth's magnetic field is performed at observatories. Processing of raw data of magnetic measurements by observatory staff plays an important role. It includes effective identification of noise and elimination of its influence on final data. In this paper, on the basis of the experience gained during long-term magnetic monitoring carried out at the observatories of IKIR FEB RAS (Russia) and CSIR-NGRI (India), we present a review of methods commonly encountered in actual practice for noise identification and the possibility of reducing noise influence.

## 1  Introduction

Magnetic measurements at observatories are an important source of information for studying the processes in the Earth's interior and near-Earth environment that substantially supplements the data obtained from satellites during magnetic surveys or from temporary stations. Currently, about 130 magnetic observatories are integrated into the global observation network INTERMAGNET (www.intermagnet.org), which established the standards for measurements, processing and transmission of the data. INTERMAGNET standards define the requirements for technical parameters of variational and absolute observations, set the requirements for the accuracy of final data (INTERMAGNET Tech. Ref. Manual, 2012) and specify some noise characteristics of the raw data (Turbitt et al., 2014). However, the requirements for the detection and processing of some types of noise, such as spikes or jumps, are addressed at the level of individual observatories. Final data from INTERMAGNET observatories (quasi-definitive and definitive) undergo multistage control, but their validity and reliability greatly depend on the quality of the raw results of magnetic measurements.

There are a lot of reasons as to why the data quality decreases, for example methodological and hardware (technical) problems, organizational difficulties and noise caused by the environment. Methodological issues are solved in large part by standard requirements according to which measurements are carried out (for example, INTERMAGNET standards in INTERMAGNET Tech. Ref. Manual, 2012) or in guides for organization of magnetic measurements (Jankowski and Sucksdorff, 1996; Nechaev, 2006). Hardware problems are mainly solved by the developers at the level of production (development and production of magnetometers) and partly by a user at an observatory (calibration, comparisons, etc.). As a result of the influence of many external sources, noise manifests as signals in the magnetic field, which are recorded by magnetometers. Considerable parts of papers on noise in magnetic measurements are devoted to hardware noise or noise from uncontrolled sources. The first type is oriented to the developers, and in many cases developers are authors and co-authors (see, for example Denisov et al., 2006; Hegymegi et al., 2016; Khomutov et al., 2016). The second type, in turn, represents almost the whole of scientific research, with the results of an investigation of noise properties, physics of their sources, etc. (see Maule at al., 2009; Neska et al., 2013; Santarelli et al., 2014). At the same time, both the first and the second types of paper are not often given the practical recommendations to magnetologists – that is, specialists at observatories, who directly make magnetic measurements and process raw data.

Therefore, it seems necessary to make a review of manmade disturbances, which are most frequently encountered in the raw magnetic data and to show examples of possible methodological and software techniques that allow us to eliminate the noise with varying efficiency. As the initial data for the analysis, we considered the results of magnetic measurements, which are carried out at INTERMAGNET observatories: Paratunka, Magadan, Khabarovsk (Russia) and Hyderabad (India). In addition, data from the Cape Schmidt and Choutuppal observatories were used.

The topicality of the information is confirmed by the following:

1. A new generation of magnetologists ensures that all possible problems at the observatories can be solved by modern technologies. At the same time, the lack of experience of real work at magnetic observatories and the absence of full information about actual conditions of measurements can lead to serious negative consequences.

2. Due to the specific nature of the subject (mainly discussed by a narrow circle of specialists, who monitor the magnetic field at the observatories), the results and the conclusions of these discussions remain inaccessible through publications, and in the best case, they appear in the conference proceedings, i.e. are limited in distribution.

3. The amount of data with which scientists have to deal has grown significantly. It is almost impossible to perform a sufficiently correct estimation of the quality of these data, because we deal with the final results of measurements carried out at the observatories, and there is no information indicating the nature of potential problems in this data.

The term "noise" is relative and significantly depends on the specific problems to be solved, used equipment, requirements of accuracy and time resolution, and so on. For example, temporary signals in the magnetic field caused by the sources in the ionosphere–magnetosphere are considered as noise during ground magnetic survey and interpretation of its results. Concerning the magnetic measurements at INTERMAGNET observatories, the signals, which have sources closer than a few tens of kilometres, can be conventionally considered noise (see, for example, Santarelli et al., 2014). Of course, there are some powerful sources of manmade noise such as DC railways, which in the case of appropriate conductivity of the upper layer of the Earth's crust can produce a significant effect in the magnetic data at large distances. On the other hand, some natural phenomena can have

smaller spatial sizes than the limit value given above, for example tectonomagnetic and seismomagnetic effects with typical distances to the suspected source of tens of kilometres.

An indication of recognition of the signal as noise can provide the typical duration of this signal. We can be assume that this time is generally not significant, for example, about 1 h and shorter. At the same time, it is necessary to keep in mind that there are many natural signals with the considered durations. There are also many examples of noise with spectrums similar to natural signals, which can last from days up to months. However, even such random noise often represents a mixture (superposition) of many signals which have much shorter characteristic time of existence (seconds or minutes) and clear structure. For example, railway signals in magnetic measurements have a pulse or rectangular shape at a short distance from the railway (Neska et al., 2013). Therefore, the efficiency of the time criterion for an estimation of the noise is low.

In this work we will classify noise, which is encountered most frequently in actual magnetic measurements at the observatories of IKIR FEB RAS and CSIR-NGR, and illustrate them with some characteristic examples. Naturally, the description and the samples are quite limited, since the variety of noise is extremely large. We will consider only the noise, the man-made nature of which has already been proved or the structure of which allows us to interpret it unambiguously. Magnetic signals with features of noise, but with unknown sources, are not in the scope of this work. The data used in this paper were processed using the tools of a MATLAB mathematical software package (www.mathworks.com) and by application software applied to MATLAB and Octave (http://www.gnu.org/software/octave/) environments used at observatories.

## 2 Initial data, description of observatories

In this paper we apply the data collected during regular magnetic measurements at the observatories of IKIR FEB RAS (Russia) and CSIR-NGRI (India). The observatories are listed in Table 1, their location is shown in Fig. 1. Table 1 shows the name of each observatory, IAGA code, geographical coordinates, institute and status in INTERMAGNET network, and magnetometers.

All listed observatories, with the exception of the Hyderabad observatory (HYB), are located far enough from big cities, but in the vicinity of small settlements. There are no powerful sources of potential noise, such as factories and railways nearby (up to 10–30 km). HYB is located within the city and an above-ground subway line has been recently built a few hundreds of metres from the pavilions. Russian observatories CPS, MGD, PET and KHB were built in the 1960s according to the requirements for complex magnetic-ionospheric stations in the USSR, with a wide range of geophysical observations. As a result, there are other observa-

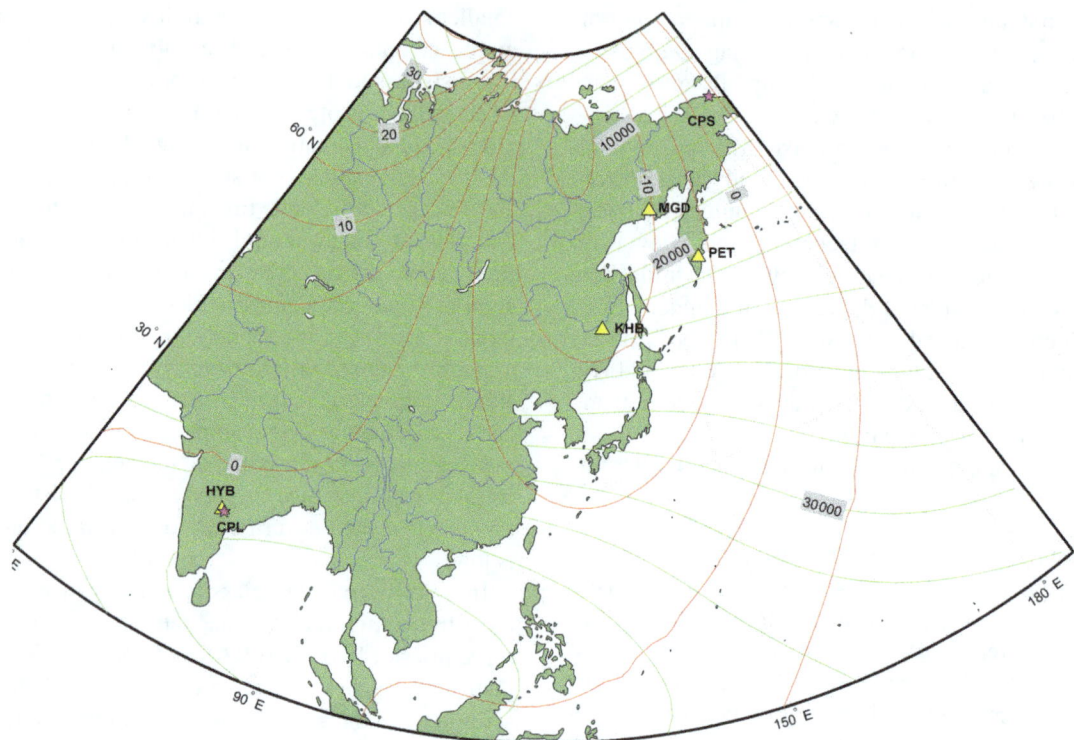

**Figure 1.** Location of magnetic observatories of IKIR FEB RAS (CPS, KHB, MGD and PET) and CSIR-NGRI (HYB and CPL). Isolines show horizontal component $H$ (green lines) and declination $D$ (red lines) according to IGRF12 model. Yellow triangles are the INTERMAGNET observatories, red stars are non-INTERMAGNET observatories.

**Table 1.** List of magnetic observatories, the data of which are used in this article. The magnetometers are marked by normal font for vector devices and italic font for scalar ones. IMO is INTERMAGNET magnetic observatory.

| Observatory | IAGA | Lat(N) | Lon(E) | Institute | Magnetometers |
|---|---|---|---|---|---|
| Cape Schmidt | CPS | 68.9 | 180.6 | IKIR | dIdD, Magdas, *POS-1* |
| Magadan | MGD | 60.1 | 150.7 | IKIR, IMO | FGE, FRG-601, Magdas, dIdD, *GSM-90, POS-1* |
| Paratunka | PET | 53.0 | 158.2 | IKIR, IMO | FGE, FRG-601, Magdas, dIdD, POS-4, *GSM-90, POS-1* |
| Khabarovsk | KHB | 47.6 | 134.7 | IKIR, IMO | dIdD, Quartz-06, *POS-1, GSM-19W* |
| Choutuppal | CPL | 17.3 | 78.9 | NGRI | FGE, GEOMAG-02M, GEOMAG-02MO, *GSM-90, GSM-19W* |
| Hyderabad | HYB | 17.4 | 78.6 | NGRI, IMO | FGE, GEOMAG-02, *GSM-90, GSM-19W* |

(1) GSM-90 (http://www.gemsys.ca/scalar-magnetometers/), GSM-19 (http://www.gemsys.ca/rugged-overhauser-magnetometer/) and POS-1 (http://magnetometer.ur.ru/content/view/15/30/lang,en/) are scalar Overhauser magnetometers. (2) dIdD GSM-19FD (http://www.gemsys.ca/vector-magnetometers/) and POS-4 (Sapunov et al., 2016) are vector magnetometers with Overhauser sensor in coil system. (3) MAGDAS (MAGDAS-A Installation Manual, 2005), FRG-601 (3-component fluxgate magnetometer FRG-601G, 2002), FGE (http://www.space.dtu.dk/english/research/instruments_systems_methods/3-axis_fluxgate_magnetometer_model_fgm-fge) and GEOMAG-02M (Nelapatla et al., 2017) are fluxgate magnetometers. (4) Quartz-06 is magnetometer with Bobrov's quartz sensors (IZMIRAN, Moscow).

tion systems in the immediate vicinity of magnetic pavilions, which are potential sources of interference, for example ionosondes for vertical sounding of the ionosphere. Moreover, due to the limited area, there are some facilities such as garages with heavy machinery, wells and electricity power equipment near the observatories, which may influence the magnetic measurements.

Extreme climate conditions at IKIR observatories are of great importance. For MGD and KHB observatories, sharply continental climate with seasonal changes of temperature from −40 to +30 °C is ordinary. The CPS observatory is located in the High Arctic zone with very hard climate conditions. The PET observatory is characterized by an abundance of precipitation and the level of snow in winter is up to 2 m. These conditions require special and expensive means with which to provide the required temperature conditions in magnetic pavilions. Special machinery is necessary to clear snow from the paths to the pavilions, which may also affect the quality of measurements. We should mention such features, as stability of an external power supply is difficult to

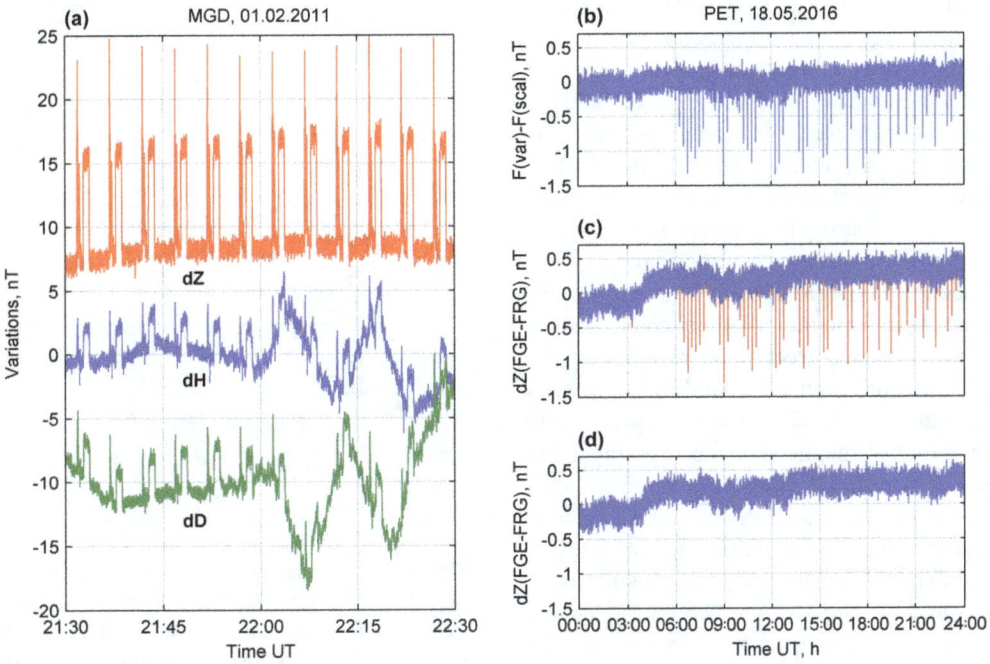

**Figure 2.** Example of regular noise in magnetic data arising during ionosonde operation. **(a)** At the Magadan observatory the ionosonde operates with a 5 min periodicity and makes noise signals in all components, recorded by FGE fluxgate magnetometer. **(b, c, d)** At the Paratunka observatory the ionosonde operates with 15 min periodicity. The effects appear in all records of fluxgate variometer FGE, mainly in the vertical component $Z$. These spikes are visible in the differences between total intensity $F(\mathrm{var})$, calculated from FGE data, and $F(\mathrm{scal})$ calculated from scalar magnetometer GSM-90 **(b)**. Subplot **(c)** shows the differences between $Z$ records of FGE (with noise from ionosonde) and Japanese fluxgate magnetometer FRG-601, which is not affected by ionosonde activity. Subplot **(d)** shows the result of clearing the ionosonde effects from FGE records.

provide at the observatories remote from densely populated areas, as well as ground connection quality due to the nature of the ground (at Cape Schmidt and Magadan). For Indian observatories climate conditions are also a significant problem: high temperatures throughout the year and a rainy season with high humidity.

## 3 Description and classification of noise in raw data

### 3.1 Regular and random noise

In order to develop a mechanism to deal with noise effectively, it is necessary to recognize its nature and to classify it as regular and random, frequent and rare, etc. Generally, regular noise is the result of technical problems with measuring equipment or interference from other devices operating in the vicinity of a magnetometer. An indicative example of regular noise is interference in the magnetic data that arises when ionosondes are operating. This is quite a common problem at integrated remote observatories, where it is physically and organizationally impossible to place magnetic and ionospheric measurements at a far enough distance.

Figure 2 (left panel) shows the hourly record from the Magadan observatory obtained by fluxgate magnetometer

FGE with Magdalog datalogger. It contains noise caused by ionosonde operation with regular sessions every 5 min. These spikes have quite large amplitude and duration and produce significant noise in the minute data obtained by averaging using the Gaussian filter (INTERMAGNET standard). A similar picture of interference caused by an ionosonde during vertical sounding every 15 min is also visible on the daily record obtained by a similar magnetometer at the Paratunka observatory; see Fig. 2 (right panel). Since the noise amplitude does not exceed 2 nT, in order to detect it, natural variations of the field were eliminated according to the data from other magnetometers installed in the same pavilion: $Z$ component variation was removed using the data of the fluxgate device FRG-601 (Fig. 2c), $F(\mathrm{scal})$ variation was removed using the data from scalar magnetometer GSM-90 and $F(\mathrm{var})$ was calculated from the FGE variations and the corresponding baseline values (Fig. 2b). The authors have also observed similar interferences in the raw data from Novosibirsk (NVS, fluxgate magnetometer LEMI-008) and Yakutsk (YAK, FGE magnetometer and Magdalog recorder) observatories. The distance to ionosondes did not exceed 200–300 m.

In all cases, the interference was recorded by variation fluxgate magnetometers. At the same time, in the records of the other fluxgate devices located in immediate proximity, interference from the ionosonde was not observed, for example in the results of the fluxgate magnetometer FRG-601, the data of which were used to remove natural geomagnetic variations (illustrated by the difference between the data obtained by FGE and FRG magnetometers shown in Fig. 2c and d). It should be noted that the interference is regular, because it is determined by sounding sessions, the beginning of which is synchronized with UTC. However, small shifts are observed, which are associated with a particular mode of sounding and the range of operating frequencies of the ionosonde.

Regularity and accurate synchronization of such interference with UTC makes it easier to identify them and to make a decision about the removal or correction. Certainly, this applies to relatively short noise signals, usually to spikes. A special software module was developed and integrated into the software package for raw magnetic data processing at the MGD and PET observatories. The module generates a temporary mask, a daily array of 0 and 1, where 0 corresponds to the measurements that include noise. Parameters of the 0–1 sequence are set in a special text file that contains the periodicity of sounding and the shifts of the beginning and the end of a removed interval with respect to the beginning of a sounding session for each day. During the processing of the raw data, the mask is applied to the original daily time series and the data marked by "0" are filled with special values NaN (Not-a-Number). An example of this clearing is shown in Fig. 2d.

Interference between closely spaced magnetometers may be a source of regular noise. It is a well known that proton magnetometers can cause interference either by the generation of additional external magnetic fields during polarization of the proton rich liquid or via direct influence of the DC current powering the proton (or Overhauser effect) magnetometers, which is modulated by the periodicity of measurements. For example, the effect up to 0.2 nT is expected at a distance of about 5 m from the proton magnetometer (Auster et al., 2007). Vector magnetometers using proton sensors in the coil systems (dIdD GSM-19FD, GEM System; POS-4, QMLab), in addition to the effects during polarization, produce significant additional magnetic fields affecting the measurements of devices located in the vicinity. Figure 3a and b show manifestation of the dIdD magnetometer operation at the Cape Schmidt observatory in the records of d$H$ and d$D$ variations from fluxgate magnetometer MAGDAS. Due to the polar specifics (necessity of heating and the absence of additional pavilions) both devices are installed on one pillar at a distance of about 2 m, almost on the same meridian. MAGDAS measurement frequency is 1 Hz. Oscillations in a range up to 2 nT with a period that is a multiple of dIdD measurement periodicity (2.5 s) are observed. There is also noise in the record of vertical component. The mechanism of influence of dIdD on MAGDAS measuring process is complex: it is associated with timer stability (noise amplitude "floats" over time) and practically it cannot be reliably corrected by software during the post-processing. Therefore, the only ef-

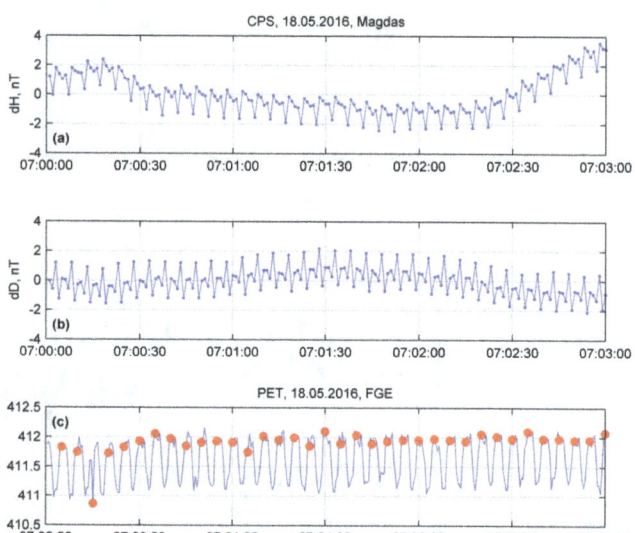

**Figure 3.** An example of influence of the magnetometers with Overhauser scalar sensors: **(a, b)** effect of vector dIdD GSM-19FD in the records of $H$ and $D$ components of MAGDAS fluxgate variometer at the Cape Schmidt observatory, **(c)** influence of the scalar magnetometer GSM-90 on the vertical component $Z$ of the FGE fluxgate variometer at the Paratunka observatory. Oscillations of d$Z$ have the period of GSM-90 measurements, undisturbed d$Z$ are found at upper part of oscillations; red dots show undisturbed values, which are used for the next processing step.

fective way to avoid this interference is to distance the magnetometers from each other.

Figure 3c shows the second example of the proton magnetometer influence on another device at the Paratunka observatory. The fluxgate variation magnetometer FGE signal (measuring frequency is 2 Hz) is modulated by the operation of a GSM-90 Overhauser magnetometer with a measuring rate of 5 s. The range of noise in the vertical component $Z$ is up to 1–2 nT (in other components it is less than 0.5 nT). Since the sensor GSM-90 is located at a distance of about 4 m from the FGE sensor, the impact through additional magnetic fields during polarization is hardly probable. Presumably, the interaction takes place at the hardware or communication level, because the devices are connected by a single datalogger Magdalog. The same noise is also observed on a similar set of magnetometers at the Magadan observatory. We cannot solve this problem technically. However, the synchronicity of measurements by two instruments (on one datalogger with a single timer) has allowed us to implement a software clearing. From the fluxgate magnetometer data we select only those which occurred during the frequency measurement of proton sensor precession, and fragments during polarization are removed. Unfortunately, due to a little synchronicity instability of measurements by two devices, only a few samples from the 5 s cycle of FGE can be reliably distinguished (for reliability actually only one sample is chosen,

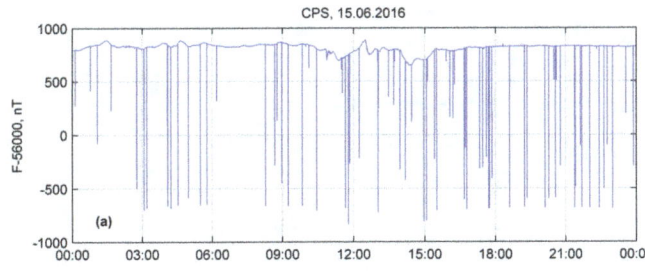

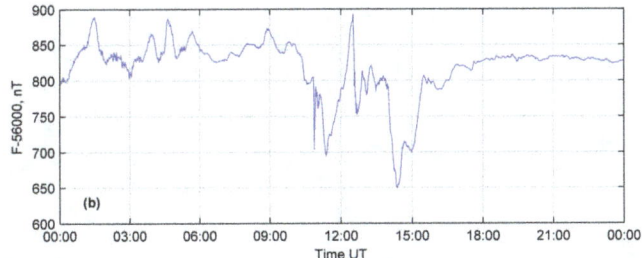

**Figure 4.** An example of spikes in the results of measurements by Overhauser magnetometer POS-1 at the Cape Schmidt observatory. **(a)** The total intensity $F$ raw record, **(b)** signal after noise removal.

which is shown in Fig. 3c by $\bigcirc$), i.e. forced 10-fold data loss takes place.

In minute values, calculated in accordance with the INTERMAGNET standard, normally distributed noise can (in special cases) be removed by filtering. However, the fact that noise is often asymmetrical creates at least two problems on minute intervals:

a. When averaging the noise with asymmetric signals, the obtained mean values are also biased. For the noise shown in Fig. 3c with magnitude up to $-1$ nT, the average estimate will be systematically biased downward by about 0.5 nT. This value is quite significant in relation to current requirements for long-term magnetic measurements.

b. If during the absolute observations a reading at the zero position of DIflux magnetometer fluxgate sensor coincides with noise in the variometer, then a difference between absolute and variation measurement arises; i.e. the accuracy of baseline values is decreased, and in the worst cases systematic errors in the final values of the total field vector arise.

## 3.2 Noise of different shapes

### 3.2.1 Spikes

A generalized view of possible shapes of structured noise is presented, for example in Lopez-de-Lacalle (2016, Fig. 1). Spikes are perhaps one of the most common types of signal in magnetic records, and with a sufficient probability are

not related to natural processes. Spike is interpreted as relatively short signal with a significant amplitude (duration is less than a few seconds or several measurements, if the magnetometer has a measurement frequency from ones to tenths of Hz), with well-defined sharp leading and back edges that are similar in amplitude. If one of the edges is weak or absent, then we can talk about the jump. All these properties of spikes can be used for their detection and removal. We can also note that spikes with a duration of one measurement cycle are most likely to be associated with hardware problems or interferences from nearby sources. The amplitude is also an important characteristic. Spikes with amplitudes of tens of nanotesla or more also have a low probability of being caused by natural sources.

Figure 4a shows an example of a daily record of the field total intensity $F$ recorded at the Cape Schmidt observatory by the Overhauser magnetometer, POS-1. The measurement periodicity is 3 s. For the CPS observatory, problems with the stability of the power supply and the quality of grounding are known, to which POS-1 is sensitive enough. On the record, outliers with the amplitude up to 1500 nT can be seen (over 90 events). The duration of these spikes does not exceed one measurement; i.e. they have sharp edges with almost equal magnitude of leading and back fronts. Therefore, it is not difficult to identify and to locate them. The results from using the simple method of spike detection are shown on Fig. 4b. It should be noted that in the Overhauser magnetometers POS-1, which are quite widely distributed at magnetic observatories, each record is accompanied by the estimation of measurement quality using a special parameter QMC (quality measurement criterion). The QMC value is related to the quality of the proton precession signal and gives qualitative estimates of measurement conditions such as signal-to-noise ratio, the duration of the precession signal and power supply voltage (POS-1 User manual, 2004; Denisov et al., 2006). Similarly, although a little less informative, estimations of signal quality are also performed for scalar magnetometers GSM (GSM-19 Instruction Manual, 2008, p. 54). For POS-1 these qualification parameters are used in the standard software at IKIR FEB RAS observatories to estimate the quality of measurements within the processing, which increases the efficiency of simple mathematical algorithms.

In many cases the visual control of the derivatives of magnetic field variations is provided by effective tools with which to detect the spikes in recorded raw data. It is clear that noise in the record shown in Fig. 4a does not represent a problem for programme processing, and in the case of smaller quantities they can be processed manually. However, difficulties arise in the case of more irregular shapes of noise, when they cannot be considered narrow isolated spikes. The examples are the noise described above, arising during ionosonde operation (Fig. 2a and b), which is often extended in time and can have a multimodel structure. The efficiency of their processing is provided by strict repeatability.

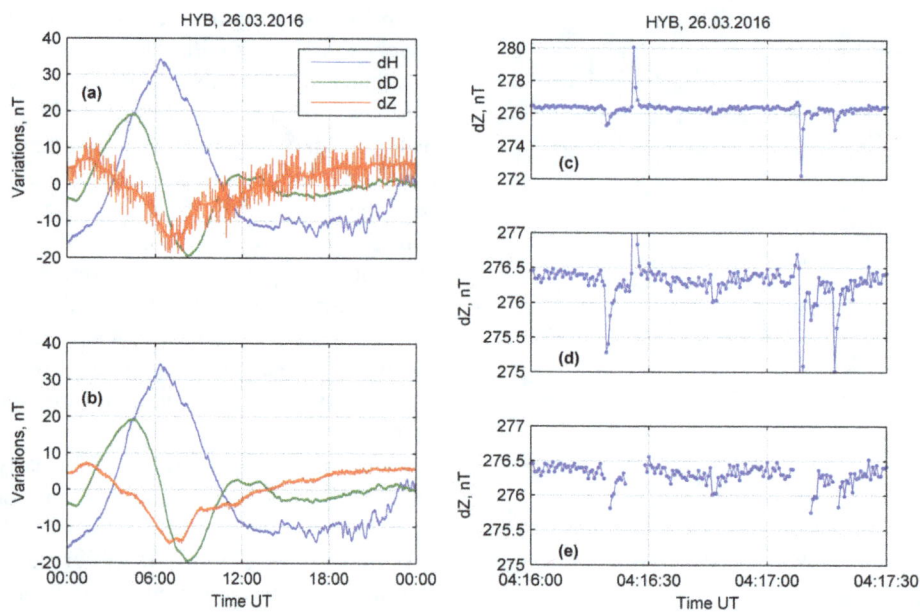

**Figure 5.** Example of frequent spikes at the Hyderabad observatory: **(a)** original daily records of variations d$H$, d$D$, d$Z$; **(b)** the same of **(a)**, but after noise identification and removal in vertical component; **(c)** 1.5 min fragment of daily d$Z$ record, showing the detailed structure of the spikes; **(d)** results of noise clearing using a simple algorithm; **(e)** result of clearing using an algorithm based on wavelet transform.

Figure 5 shows the daily record of d$H$, d$D$ and d$Z$ variations, obtained at the Hyderabad observatory using the fluxgate magnetometer FGE. The measurement frequency is 2 Hz. As can be clearly seen, there is irregular and frequent noise in the vertical component in the form of outliers (spikes) with amplitudes of more than 5 nT. The observatory is located on the territory of CSIR-NGRI Institute, within the city. The most probable reason for this noise is the metro line (above-ground) passing at a distance of 200–300 m, south of the observatory. Figure 5c shows a 1.5 min fragment for d$Z$, which shows that spikes have a sufficiently definite and stable structure, a sharp leading edge and exponentially falling back edge. The total spike duration is up to 3–6 samples, i.e. about 1–3 s.

The algorithm applied at the HYB observatory to detect and remove the noise caused by metro line is based on its structural stability and works as follows: when a change of d$Z$ between neighbouring measurements (in absolute value) exceeding a given threshold is found, three subsequent measurements are discarded. It is clear that this algorithm creates risks: (a) natural signals with sharp edges, for example, during magnetic disturbances, may be discarded; (b) noise may be removed not completely if its duration is longer than one; (c) noise can be missed if due to a small shift its leading edge takes two samples. However, to calculate the minute values of the magnetic field variations, that may be enough. Figure 5d shows the results of the algorithm application during real processing of raw HYB data (threshold value d$Z$ / d$t$ = 1 nT / 0.5 s was used).

An algorithm based on continuous wavelet transform showed higher efficiency in the detection of spikes. For the first time it was proposed in the paper Zhizhikina et al., 2016. This algorithm includes the following main operations: (1) wavelet decomposition of data on informative scale levels (determined during algorithm construction) is performed, and (2) spikes are detected on the basis of threshold functions (different thresholds for each scale level and for positive and negative values of the wavelet coefficients are used).

The algorithm efficiency is determined by the wavelet transform sensitivity to sharp changes of function values. Amplitudes of wavelet coefficients significantly increase in the areas containing local features in the form of sharp peaks (Daubechies, 1992). Figure 5b and e show the results of the algorithm.

The geomagnetic variations and noise are dependent on the location of the observatory. Therefore, preliminary tuning of the algorithm parameters is required for the selected observatory. Currently, the algorithm is adapted for the mid-latitude Paratunka observatory and for the equatorial Hyderabad observatory. The values of the parameters were defined for the criterion of the absence of false detections, using selected data. The effectiveness of the algorithm was estimated for quiet and disturbed magnetic fields. The spikes detected by the experienced magnetologist were considered as reference. Table 2 shows the results of the estimation of the algorithm effectiveness for the Hyderabad observatory.

The results show high reliability of the wavelet-based algorithm when detecting the main part of the spikes: about 99 % of spikes with high amplitude (> 3 nT) are isolated.

**Table 2.** The results of the estimation of the algorithm effectiveness for the Hyderabad observatory.

| Magnetic field conditions | Detected by magnetologist | Detected spikes, % | |
|---|---|---|---|
| | | Number of spikes: Wavelet-based algorithm | Simple algorithm |
| Quiet (local $K < 3$) | 1591 | 83.91 | 73.35 |
| Disturbed (local $K \geq 3$) | 1495 | 85.35 | 73.85 |

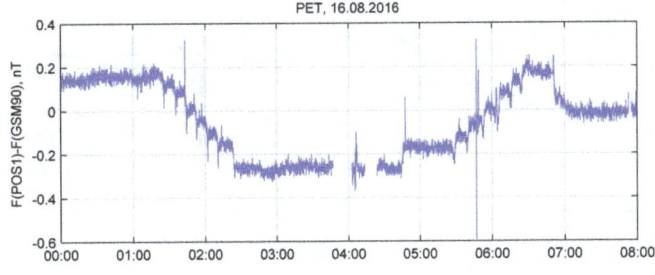

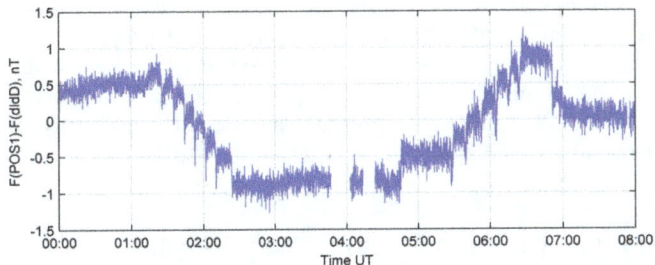

**Figure 6.** Example of magnetic field jumps at the Paratunka observatory during removal and lowering of casing steel pipe into a well of 80 m depth. Curves on the plots are differences of field total intensity $F$ measured by scalar magnetometers POS-1, GSM-90 and dIdD at various distances from the well. Each step corresponds to the operation with one pipe section.

However, spikes with small amplitude ($< 0.5$ nT), which are 25 % from all spikes, cannot be reliably detected by magnetologists (experts) and cannot be used for the estimation of the algorithm effectiveness. This also restricts the possibilities for the optimization of the algorithm for detecting small spikes.

### 3.2.2 Jumps

In a certain sense, the jumps in the results of measurements can be considered to be spikes as described above, but with a continuous interval of the record between the leading and the back edges or in the absence of a back edge. Because of the sharp edges and sufficient amplitude of a jump, it is not difficult to identify one. However, unlike spikes, in practice such jumps are quite rare in raw magnetic data (for example, Fig. 2). Jumps of magnetic record level with slow changed edges or with noisy edges are predominantly ob-

served. The reasons for such jumps are the technical operation with equipment, the changes of the magnetic field distribution in the pavilion or near it, the changes of the instrument parameters, etc.

An example of noise in the form of jumps caused by changes in the magnetic environment near the pavilions at the Paratunka observatory is shown in Fig. 6. The effect is visible in the field total intensity $F$, which occurred during removal and reinstallation of casing pipes in a well of 80 m in depth, located approximately at a distance of 100 m to the south from the magnetic pavilions. These operations were carried out within 6 h, and heavy machinery was used (truck crane, tractor). In general the effect does not exceed 1 nT, but it stands out well from the difference between the records of the two scalar magnetometers, located at different distances from the well (dIdD is the closest, POS-1 is the most remote, the distance between them is 30 m). Each removal of a pipe section from the well causes a jump of the magnetic field gradient between POS-1 and dIdD by about 0.1 nT. Lowering back into the well looks like a recovery process in $dF$. A much smaller effect is observed in the difference between records from the POS-1 and GSM-90 magnetometers, located at about the same distances from the well. This example represents a situation which is quite widespread at magnetic observatories and shows the following important points:

a. Identification of signals of a small amplitude by mathematical methods of pattern recognition, even if they are rather different from the field natural variations, can be practically implemented in very rare cases (only for typical noise and in the case of large samples of a priori data).

b. Such signals can be identified reliably only in difference data obtained by separated magnetometers.

c. Practically the only way to detect such signals is through an experienced, trained magnetologist, whose work is largely based on the additional information about measurement conditions.

In relation to the example given in Fig. 6, we can assert that only a magnetologist (expert) can recognize field variations as noise, by analysing the differences between measurement

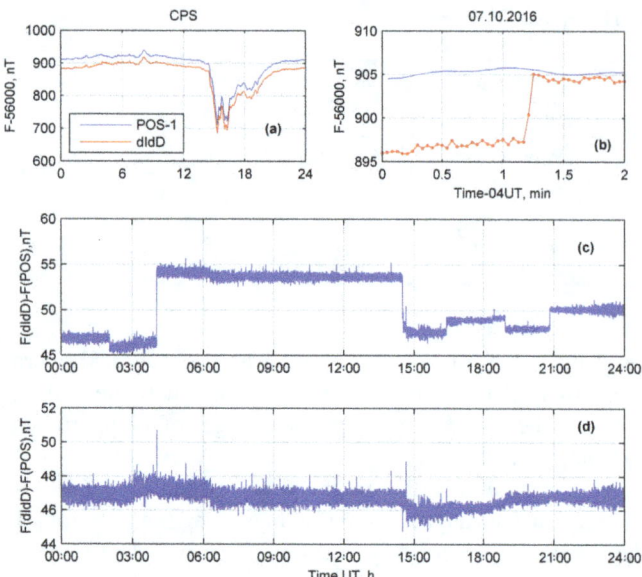

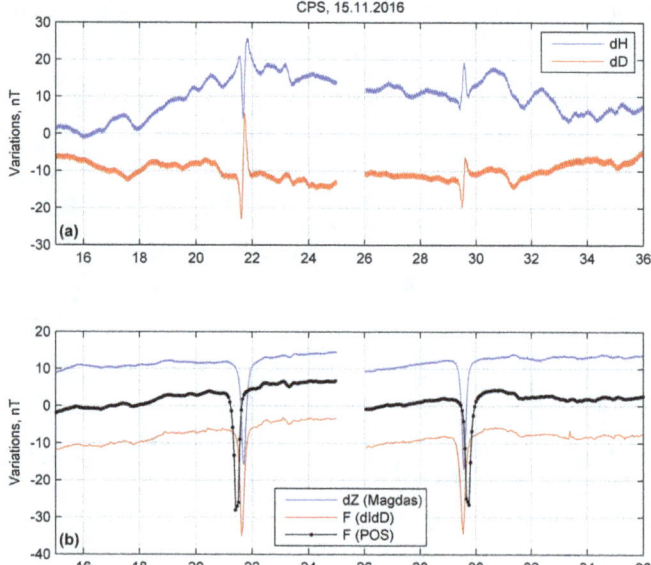

**Figure 7.** Example of noise in the form of jumps in the data from the dIdD GSM-19FD magnetometer at the Cape Schmidt observatory. **(a)** Daily records of $F$ of dIdD and scalar Overhauser magnetometer POS-1; **(b)** detailed fragment from **(a)** with a jump; **(c)** differences $dF = F(\text{dIdD}) - F(\text{POS})$ in which natural geomagnetic variations are removed; **(d)** the same as **(c)**, but jumps in $F(\text{dIdD})$ are removed.

**Figure 8.** Noise caused by a vehicle driving near the magnetic pavilions at the Cape Schmidt observatory (in one direction and back). **(a)** $dH$ (upper curve) and $dD$ (lower curve) variations, recorded by MAGDAS magnetometer. **(b)** $dZ$ (MAGDAS, upper curve), $dF$ (POS-1, middle curve with dots) and $dF$ (dIdD, lower curve). MAGDAS and dIdD magnetometers are located at a distance of about 30 m from the POS-1.

results obtained by the spatially separated devices, noting artificiality of jumps in the variations and knowing that works on the well were carried out at that time. Such anthropogenic disturbances are usually not corrected, and in most cases the record including the noise is removed.

Figure 7 shows an example of jumps in the daily record of the field total intensity $F$, obtained by dIdD GSM-19FD and POS-1 magnetometers at the Cape Schmidt observatory. It can be seen in Fig. 7c that the measurement difference between these two Overhauser magnetometers contains jumps with an amplitude of several nT. However, noise arises only in the dIdD record (Fig. 7b) and in most cases have sharp edges. The magnetometers are installed in different pavilions (dIdD is in a variational, POS-1 is in absolute), at a distance of about 30 m. The situation shown in Fig. 7c is rather characteristic of the observatory and occasionally it reoccurs. The source of the noise is not defined, but perhaps it is associated with interference of supply lines or communication cables in the variation pavilion or with the currents in moisture saturated soil near the dIdD.

In this example, our interest is not in the cause or mechanisms of noise, but in the possibility of its identification and correction. Since the amplitude is fairly significant, edges are sharp, and measurements are not burdened by this noise. The posed problem can probably be effectively solved by software tools. However, in this case, processing is performed by a magnetologist, who estimates the size and the location

of the jump using an $F(\text{dIdI}) - F(\text{POS})$ plot, followed by programme correction of $F(\text{dIdD})$ and necessary removal of unreliable data at the time of the jump if it has a significant duration. Figure 7d shows the results of the procedure described above.

It should be noted that the jumps, after which the record level is changed and retained for a long time (several days or longer, for example, after magnetometer reinstallation), appear in the baseline values of the variometers and are eliminated during calculation of the total field vector using the standard measurement technology at magnetic observatories (see also some remarks in Sect. 4).

### 3.2.3 Bay-like noise

Bay-like noise is a common type of noise at magnetic observatories. It is often the result of changes in the magnetic field near the magnetometer due to moving objects with magnetic effect, for example a car or a person with instruments. In the case of such noise, shapes of signals in the field components are defined and related. Nevertheless, this noise, if its amplitude does not reach extreme values, is hardly distinguishable from natural variations. The possible methods of identification are comparison with data obtained by other magnetometers (gradiometer principle) and analysis of the information about events at an observatory (logging of such events is the direct responsibility of the observatory and its staff).

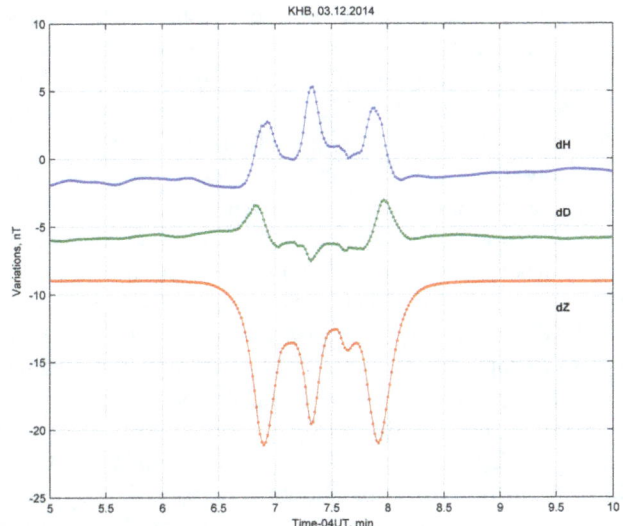

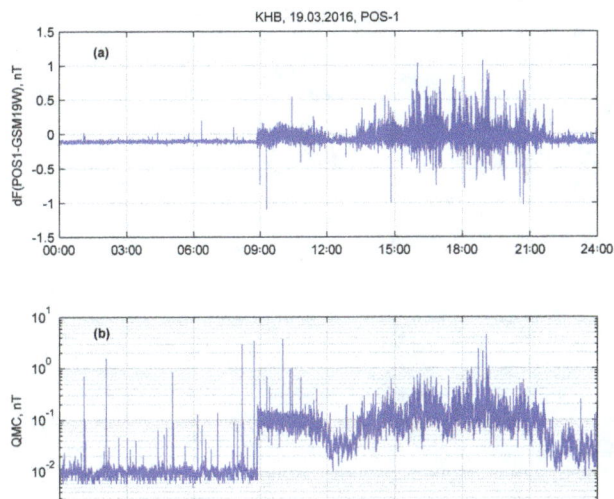

**Figure 9.** Noise caused by a snow plough clearing the path to the pavilions at the Khabarovsk observatory. Variations d$H$, d$D$ and d$Z$ were recorded by quartz magnetometer CAIS. Oscillations on the records are the result of the tractor work at a distance of about 30 m from the magnetometer.

**Figure 10.** Noise-like interference caused by the faulty power supply at the Khabarovsk observatory, manifested in the records of Overhauser magnetometer POS-1. **(a)** Daily record of $F$(POS-1) after elimination of natural geomagnetic variations by GSM-19W data; **(b)** behaviour of QMC (quality measurement criterion) instrumental parameter of POS-1.

Figure 8 shows bay-like noise occurring during off-road driving near the magnetic pavilions of the Cape Schmidt observatory and during the vehicle's return. Variations of d$H$, d$D$ and d$Z$ components are obtained with a MAGDAS magnetometer. Variations of field total intensity $F$ are obtained using dIdD and POS-1 magnetometers (Overhauser sensors). MAGDAS and dIdD are located in a variation pavilion, POS-1 is installed in an absolute pavilion. The distance between the pavilions is about 30 m. Noise duration is about 30–40 s, the amplitude is up to 20–30 nT, and the time shift between signals recorded by magnetometers in two remote pavilions is clearly defined. We may also note the dependence mentioned above between the form of a signal in different field components. The identification of a signal as noise was made by a magnetologist using the shape of a signal, time shift and by comparing it with similar signals that have been observed earlier (in this case there was no information about the source of the noise at the observatory). After localization, the noise was removed by a magnetologist during raw data processing.

One more example is shown in Fig. 9. A snow plough cleared the path to the magnetic pavilions at the Khabarovsk observatory. Field variations were recorded by the quartz magnetometer CAIS. Noise duration is about 1 min, amplitude is maximum in the vertical component (up to 10 nT). It would be difficult for the staff of the observatory to identify the signal in Fig. 9 as noise without information on the works conducted near the pavilions, because according the signal parameters, it is rather close to natural geomagnetic variations (an exception is possible only for $Z$ component, since its natural variations on this day did not exceed several

of nT). Just like in the previous case, the noise was removed by a magnetologist during raw data processing.

### 3.2.4   Random-like noise

Man-made disturbances, which are expressed as additional random noise at the background of original useful signal will be understood as random noise. This notion generalizes a very big class of noise which is often difficult to classify. In many cases this noise is not localized in frequency and/or time domain. They may have a hardware origin or may be connected with real field noisiness from external sources at the observatory. Almost always, the problems of this noise is solved either technically (fine adjustment of magnetometers, improvement of grounding, power supply, etc.) or organizationally (moving of a measurement point, replacement of a magnetometer by another one which is less sensitive to noise, etc.). It is hardly probable to find effective methodological and software approaches.

Identification of hardware noise in data is possible if we compare the results of measurements by different magnetometers or if we change the operating modes of a device if it is the only one at the observatory and if it is possible by the specification of a devices. The increased noise of the magnetic field at the observatory (as a result of total impact of many factors) can be identified, for example, if we make measurements by the same magnetometer at the observatory and in the place with obviously low noise. In general, assessment of the background noise is a labour-consuming task

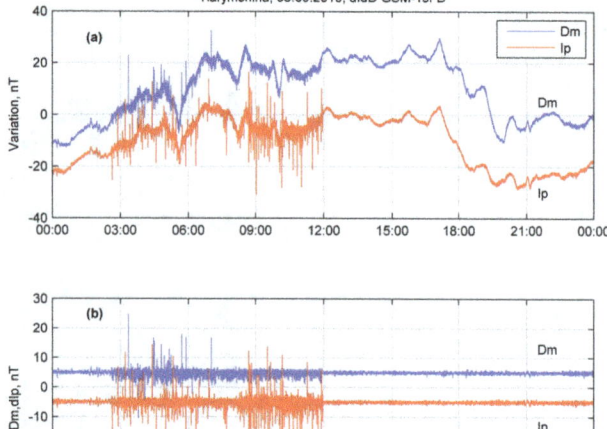

**Figure 11.** The noise in dIdD data which arose due to the failure of a power supply system at Karymshina station (15 km from the Paratunka observatory). **(a)** The total field intensity $F$ obtained in the case of additional fields with polarity west–east (Dm) and up–down (Ip) in the coil system (see dIdD Instruction Manual, 2010); **(b)** the same as in **(a)**, but slow variations are excluded (for clarity).

including a research aspect and very often it does not have effective practical results. The problem becomes more complicated due to the fact that in many cases it is almost impossible to distinguish the noise background from a natural signal which is a subject of scientific research, for example, seismomagnetic effects. Figures 10 and 11, as an example, show the random noise which arises in the results of magnetic measurements in the case of problems with power supply.

Figure 10 shows the results of daily measurements of the field total intensity $F$, made at the Khabarovsk observatory using scalar Overhauser magnetometer POS-1. There was a general power outage at the observatory and in the nearby settlements at about 08:50 UT. The measurements were continued using an autonomous power supply system at the observatory. At about 12:00 UT the external power supply was restored. On the record obtained by POS-1, noise with an amplitude up to 1 nT was recorded. It lasted for more than a day. For a descriptive graphical representation, natural geomagnetic variations were excluded using the data from another scalar magnetometer (GSM-19W). The noise caused by power failure did not manifest in these measurements (Fig. 10a). Figure 10b shows the behaviour of the QMC parameter, which estimates the quality of the precession signal of the POS-1 Overhauser sensor in nT units and was described in Sect. 3.2.1. In this example, the fact that the noise remained in the measurements of POS-1 after the restoration of external power is of interest; i.e. the presence of the noise could not be ascertained according to staff information on the situation at the observatory, it could be determined only

by the direct visual analysis of these measurements. And the second fact is that in this case an effective way to recognize the noise is to estimate the behaviour of the QMC parameter.

The second example (Fig. 11) shows an appearance of noise in the data from the dIdD GSM-19FD magnetometer, recorded at Karymshina station of IKIR FEB RAS. Karymshina station is located approximately 15 km from the Paratunka observatory, a place with a minimum of possible industrial sources of noise, including the absence of an external power supply by power lines. At about 02:45 UT a failure of a diesel generator occurred and an emergency scheme of power supply has been activated with an external battery package as a source of voltage and disconnection from all powerful devices, the standard power supply was restored at about 11:55 UT. It is shown in Fig. 11a that spikes with amplitude up to 10–20 nT and random noise up to 5 nT appeared during the operation of the emergency power system in two measurement channels of dIdD (with additional fields in coil systems D and I; see, for example dIdD Instruction Manual, 2010). For illustration purposes, Fig. 11b shows the record after eliminating the low-frequency variations. Since the amplitude of the noise is significant, it can be easily identified during the analysis.

In both examples given above, the only way to exclude noise is to remove all data fragments which contain noise. It is impossible to restore the original useful signal with an acceptable quality.

### 3.3 Critical and weak noise

This criterion is important due to the fact that it determines the extent of noise influence in magnetic data on the results of research that is carried out using these data. When a researcher uses the final data from the observatories, in most cases they have no information on what conditions the measurements were performed under, what procedures were applied during preliminary processing, etc. (except for the cases when these procedures are prescribed by standards). Thus, the responsibility for the quality of the data provided to the scientific community is very high and it lies entirely on the observatory.

Let us consider a classic case. An observatory of the INTERMAGNET network obtains primary 1 s data of d$H$ s, d$D$ s, d$Z$ s, $F$ s variations by direct measurements and then, using baseline values, calculates total $H$ s, $D$ s, $Z$ s, $F$ s, reduced to the main pillar and minute values $H$ m, $D$ m, $Z$ m, $F$ m using the procedure defined by INTERMAGNET standards (INTERMAGNET Tech. Ref. Manual, 2012). Filtering using a Gaussian filter is quite an effective method for suppressing random noise. Also, Gaussian filtering works acceptably with spikes of small amplitude, but it is not effective for jumps and bay-like noise. Thus, the residual effects from strong noise are included into the final minute data. At the same time they are smoothed; i.e. they are almost indistinguishable against the background of natural variations, but

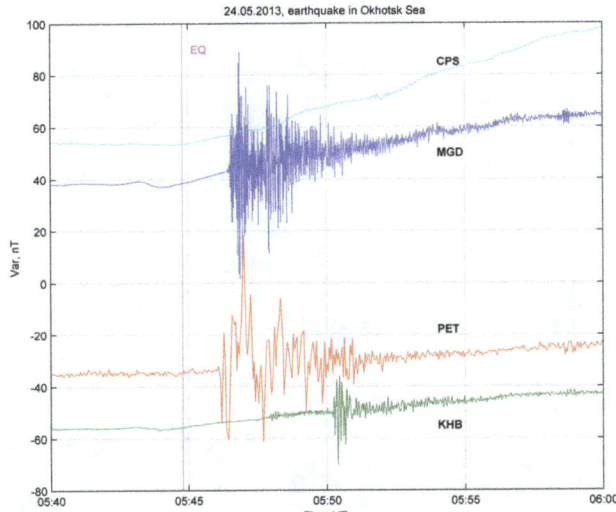

**Figure 12.** "Fictitious" variations of the magnetic field recorded by magnetometers with suspension system of measuring sensors: dIdD with scalar sensor in the coil system, FGE with fluxgate sensors and magnetometer Quartz-6 with Bobrov's quartz sensors. Records obtained at CPS, MGS, PET and KHB observatories during a strong earthquake in the Okhotsk Sea on 24 May 2013 are shown. Vertical line EQ shows the earthquake time.

they influence the results of further calculations performed with these data. That makes it necessary to identify such critical noise and to remove it during the primary data processing. It entails secondary problems, such as gaps in original data, which are used to calculate published minute values. In spite of the INTERMAGNET recommendation that mean values should be calculated in accordance with the 90 % availability rule: this rule is difficult to define and sometimes not applicable at all.

### 3.4  Noise with known and unknown sources

If it is known that a signal, which is suspected as noise, is the result of some sources which are not associated with natural variations of magnetic field, then there is a rather a powerful argument for removing the signal. At the same time, however, there are some fine points:

a. noise removal leads to the gaps in published data, therefore scientist in many cases before application of methods of analysis is forced to fill these gaps with dummy data, i.e. the data, calculated from available data set using some interpolation method. Errors of calculation results arising due to the filling may be comparable or even greater than the errors that occur due to the noise which was not deleted;

b. problems which are solved using the final data should be defined. Some signals can be as noise due to the criteria of its origin. However, at the same time these signals

can be the subject of other scientific research. For example, in practice of observatory measurements, a fictitious "seismomagnetic effect" is well known, when we observe oscillations in the data from magnetometers with a suspended system for compensation of sensor inclinations or in the records of induction magnetometers. These oscillations arise when a seismic wave from a near or strong earthquake is passing the place where the magnetometer is installed. As an example, Fig. 12 shows the records of magnetic field variations at IKIR FEB RAS observatories. On these fragments the effect of a strong earthquake with the magnitude of 8.3 which occurred on 24 May 2013 in the Sea of Okhotsk at the depth of about 600 km is illustrated. The following magnetometers were applied: dIdD GSM-19FD (Dm and Im modules with additional fields of coil system) at the CPS and PET, fluxgate FGE (H channel) at MGD, digital magnetometer with Bobrov's quartz sensors (H channel) at KHB. It is clear that the earthquake is manifested well at long distances, the magnetometer data based on different measurement principles. For the tasks of studying the variability of the magnetic field, the signals shown in Fig. 12 are noise and they should be removed. However, if, for example, seismomagnetic effects are investigated, then the recorded "fictitious" signals in Fig. 12 would be a good benchmark to estimate the passage of a seismic wave in the area where magnetometers are installed. Therefore, the researcher of seismomagnetic effects should understand well, what type of magnetometer is used to obtain magnetic data.

If a signal is suspected as a noise, but its structure does not allow us to recognize it as a noise, and there is no additional information about a possible source, then it is very complicated for a magnetologist to make a decision. In most cases, these signals are not removed, which creates risks of reducing the quality of observatory data.

## 4  Possible methods for noise removal from raw magnetic data

It is clear that identification of noise is the solution only of a part of the problem. The second part is to choose an effective method of further work with this noise. Unfortunately, the choice of possibilities is very small. Mainly a fragment is simply removed from a record that results in a gap, which can be filled by the most suitable "dummy" data, or if the structure of the noise is recognized, it is removed and the original useful signal remains.

Earlier, in the description of spikes (see Figs. 2, 4 and 5), possible techniques of dealing with these noise, using automatic removal of a record fragment after identification of noise, have already been shown. In other cases, noise can be removed manually by a magnetologist. There are different approaches to implement these procedures. If a file with

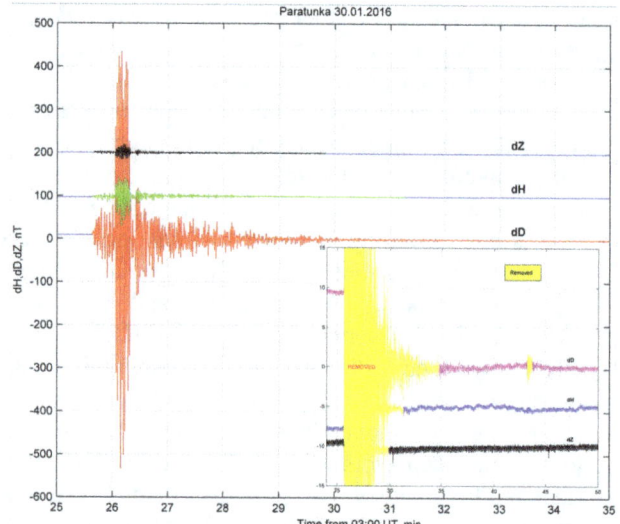

**Figure 13.** Example of manifestation of a nearby earthquake on 30 January 2016 (magnitude 5.7, distance is about 100 km) in the record of FGE magnetometer with a suspended system at the Paratunka observatory. The inset shows the results of data clearing performed by a magnetologist using a semi-automatic procedure (shown in yellow). In component $D$ the effect of an aftershock, which occurred in 17 min, is also removed. Jumps of average level of the records, reaching 10 nT in $D$ (about 1.5′) are clearly visible in the inset.

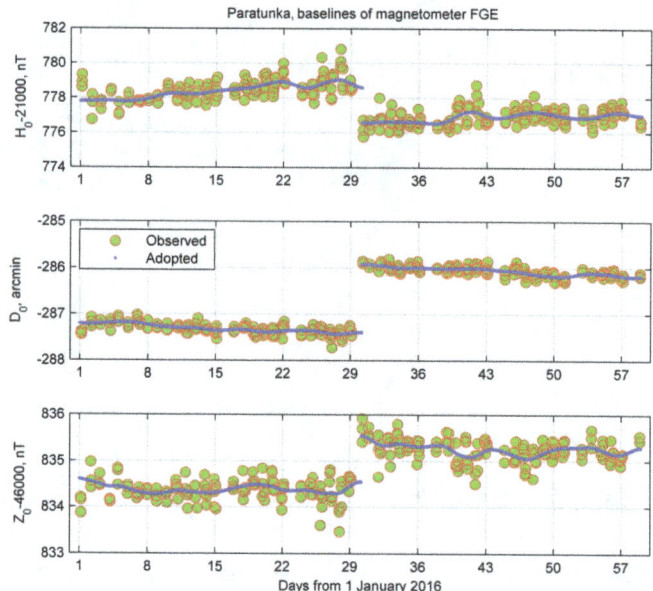

**Figure 14.** Baseline values of FGE variation magnetometer, Paratunka observatory, from January to February 2016. Marker ○ shows individual baseline values (observed); continuous curve shows the values adopted for each minute (adopted). The jumps in baselines due to the earthquake on 30 January 2016 are clearly visible.

raw data has text format (for example, POS-1 User manual, 2004 and dIdD Instruction Manual, 2010), it is possible to remove the unreliable records manually. In this case, during further work with the corrected file, there will be gaps not only in the measured magnetic values, but also in a timestamp, which is not always convenient if we need a uniform time grid. Another possible approach is to create an intermediate file, usually in text format, in which noisy data are manually or semi-automatically replaced by the values that indicate unreliable measurements, for example 99 999. This option is particularly useful if the initial measurements are recorded in files in binary format, e.g. MAGDAS-A Installation Manual (2005). In this case special converter software is usually used, including those which allow an operator to encode the required data as unreliable.

In the software for magnetologists developed in IKIR FEB RAS, which is based on MATLAB and Octave mathematical packages (Khomutov, 2016), the third approach is used. It is based on the following methodical principles: (a) original files are always used in processing of raw (original) magnetic data, and (b) the intermediate files are not formed; i.e. though stream processing is performed from the raw data to the required final result, all the results of intermediate computations remain only in the computer's random access memory. Information about unreliable data during a time interval is stored in special text files, with a date, start time and end time and special features, for example for the choice

of the magnetic components and for comments. An example of such a record for the FGE magnetometer at the Paratunka observatory is shown below.

| Date | UT1 | UT2 | HDZ |
|------|------|------|------|
| 30 Jan 2016 | 03.4272 | 03.5211 | 100 % earthquake |
| 30 Jan 2016 | 03.4269 | 03.5795 | 010 % earthquake |
| 30 Jan 2016 | 03.4274 | 03.4973 | 001 % earthquake |
| 30 Jan 2016 | 03.7199 | 03.7283 | 010 % aftershock |

A maximally simplified text format is used, which reduces the probability of errors during manual typing and speeds up the file reading. The example shows the information, used to remove the noise caused by a nearby earthquake on 30 January 2016 in the data from the FGE magnetometer. In the processing of measurement results from 30 January 2016, $D$ values from 03:25:37($=$03.4269) to 03:34:46($=$03.5795) UT, which are noisy due to the mechanical influence of a seismic wave on the suspended sensor, will be filled with a special symbol, NaN, and will be excluded from further processing. The boundaries of intervals with noise are defined by a magnetologist in interactive mode by record plots for the corresponding components and with required scaling that is provided by convenient interface with graphics in MATLAB and Octave. Noise in other components is removed similarly. The effect of the described earthquake and the results of applying the clearing procedure are shown in Fig. 13.

Considerably a more difficult situation arises in the attempt to restore an original signal, i.e. removal only of noise from the measurement results. It is impossible in most cases in the actual working practice of the observatory due to the unknown structure of a useful signal and the unknown structure of noise. Analysis of a concrete situation is specific, and become a scientific research question in itself. However, in some simple cases the problem can be solved. These situations include, for example, noise, leading to jumps of record level, sometimes followed by restoration which is also in the form of a jump. It is clear that in this case the noise is a constant addition to the useful signal, and after its subtraction, the initial undisturbed signal will be restored. The main problem is the correct estimation of the value of the noise (magnitude of a jump) and the need to be sure that noise has a constant value in the analysed interval of the record.

In practice, identification of a jump and estimation of its parameters (time and magnitude) is performed by a magnetologist, usually in an interactive mode of work with magnetic record graphs. If the jump was very rapid and the time of a transitional process was comparable with the duration of the interval between samples and generally faster than possible natural variations in the magnetic field, then the jump parameters can be estimated visually with sufficient reliability. If the transitional process is long enough, then natural variations in the field can significantly affect the accuracy of the estimation. In this case, it may be useful to compare the analysed record with measurement results obtained by another magnetometer, because natural geomagnetic variation will be excluded from the difference between two records, and the jump will be manifested as two permanent levels of the difference. However, in this case, there are also a lot of limitations, for example the nature of the cause of the jump. If the jump is caused by a change in the magnetic field, the source of this change is fairly close to the place of measurements, and the compared magnetometers are located closely (usually in the same or in neighbouring pavilions), then noise effect will be manifested in the data from both devices and it will be difficult to make any reliable estimations. If the reason of the jump is of technical origin or its impact on the supporting magnetometer is small, it is possible to estimate the required parameters.

The situation described above is partly similar to that which occurs in the case of elimination of long-term changes of the level of magnetic record from variation magnetometer, using absolute observations. Figure 14 shows an example of compensation of jumps that arose in the records of the FGE magnetometer at the Paratunka observatory after the earthquake on 30 January 2016 (see Fig. 13). Regular absolute magnetic observations and dense series of obtained baseline values allow us to effectively eliminate the effect of such jumps in the series of full values of the magnetic field components; for example, the total values of the declination $D = D0 + dD$ since 31 January 2016 will have been free of the jump effects in the case of an earthquake.

## 5 Conclusions

The review of noise in raw magnetic data and some methods of its identification and removal given above represent only a part of the real situation which a magnetologist deals with at an observatory when processing the measurement results. However, even with such a volume it is possible to draw the following important conclusions:

1. In most cases, the correct identification of noise can be performed only according to the estimates made by an expert, a magnetologist of the observatory, who uses raw (original) results of measurements, and all available additional information on the measurement conditions.

2. The most important source of information about noise is the comparison of measurement results obtained by different magnetometers, including those using different measuring principles, as well as careful monitoring of the environment at the observatory.

3. Automatic identification and correction of noise by computer programs are of auxiliary nature and are principally an interactive tool used to help a magnetologist with data processing.

All this indicates that the data processing should be carried out by a qualified magnetologists directly at a workplace, i.e. at the observatory, and should include the whole set of requirements (Jankowski and Sucksdorff, 1996; Nechaev, 2006; INTERMAGNET Tech. Ref. Manual, 2012). A similar opinion is presented by Linthe et al. (2012). Unfortunately, at the present time, many observatories, especially those that are newly created and located in remote areas, have problems with staff and their qualifications. In these cases it seems reasonable to create centres for collecting raw magnetic data, where a full cycle of data processing would be performed. Examples of such centres are BGS, which has a centre in Edinburgh (http://www.geomag.bgs.ac.uk/data_service/space_weather/current_conditions.html) or GC RAS (http://geomag.gcras.ru/) which collects and processes the data from the Russian magnetic observatories. At the same time, some of the problems, including those connected with incorrect noise processing, can be more difficult to resolve.

*Competing interests.* Kusumita Aurora is a member of the editorial board of the special issue of the journal.

*Special issue statement.* This article is part of the special issue "The Earth's magnetic field: measurements, data, and applications from ground observations (ANGEO/GI inter-journal SI)". It is a result of the XVIIth IAGA Workshop on Geomagnetic Observatory Instruments, Data Acquisition and Processing, Dourbes, Belgium, 4–10 September 2016.

*Acknowledgements.* The authors are grateful to the staff at the observatories of IKIR FEB RAS for providing qualitative magnetic measurements. Stanislav Nechaev, main magnetologist of Patrony observatory (Irkutsk, IRT), Pavel Borodin, magnetologist of Arti observatory (Arti, Ekaterinburg, ARS), Zinaida Dumbrava, Head of Khabarovsk observatory (IKIR FEB RAS, KHB) and Vladimir Sapunov, Head of Quantum Magnetic Laboratory (Ural Federal University, Ekaterinburg) are acknowledged for helpful discussions over many years, which improved our understanding of noise in magnetic observations. Phani Chandrasekhar and K. Chandashakhar Rao are acknowledged for their contributions to the observatories in Hyderabad and Choutuppal. The director of CSIR-NGRI is acknowledged for his permission to publish this work. The DST-RFBR collaboration is acknowledged for funding the joint studies (Grant of DST no. INT/RUS/RFBR/P-234, dated 28-9-2016 and Grant of RFBR no. 16-55-45007). Magnetometers POS-1 and POS-4 for the observatories of IKIR FEB RAS were purchased with grant funds of the Russian Science Foundation, project number 14-11-00194. The authors are very grateful to Anna Larionova for her help with English corrections, to the referees and the editorial advisor for the comments and corrections made in the manuscript. They greatly improved the quality of presentation.

Edited by: Alexandre Gonsette

# References

3 component fluxgate magnetometer FRG-601G: Ver1.1. Operation manual, Tierra Tecnica Ltd, p. 9, 2002.

Auster, H. U., Mandea, M., Hemshorn, A., Korte, M., and Pulz, E.: GAUSS: Geomagnetic Automated System, Proceedings of the XII IAGA Workshop on Geomagnetic Observatory Instruments, Data Acquisition and Processing, Belsk, 19–24 June 2006, Publication of the Institute of Geophysics Pol, edited by: Reda, J., Acad. Sci., 398, 51–61, 2007.

Daubechies, I.: Ten Lectures on Wavelets, SIAM, 377 pp., 1992.

Denisov, A. Y., Denisova, O. V., Sapunov, V. A., and Khomutov, S. Y.: Measurement quality estimation of proton-precession magnetometers, Earth Planets Space, 58, 707–710, 2006.

dIdD v.7.0 Insrtuction Manual: GEM Systems, Canada, 50 pp., 2010.

GSM-19 v7.0 Instruction Manual: GEM Systems, Canada, 149 pp., 2008.

Hegymegi, L., Csontos, A., and Merenyi, L.: Monitoring of long term mechanical stability of a suspended dIdD sensor applying optical observation, J. Ind. Geophys. Union, 2, 19–23, 2016.

INTERMAGNET technical reference manual: Version 4.6, edited by: St Louis, B., 92 pp., available at: http://intermagnet.org/publications/intermag_4-6.pdf (last access: 7 February 2017), 2012.

Jankowski, J. and Sucksdorff, C.: Guide for magnetic measurements and observatory practice, Warsaw, 235 pp., 1996.

Khomutov, S. Y.: Methodological and software approaches to processing of magnetic measurements at observatories of IKIR FEB RAS, Russia, J. Ind. Geophys. Union, 2, 54–61, 2016.

Khomutov, S., Sapunov, V., Denisov, A., Savelyev, D., and Babakhanov, I.: Overhauser vector magnetometer POS-4: Results of continuous measurements during 2015–2016 at geophysical observatory "Paratunka" of IKIR FEB RAS, Kamchatka, Russia, E3S Web Conf., 11, 1–5, https://doi.org/10.1051/e3sconf/20161100007, 2016.

Linthe, H.-J., Reda, J., Isac, A., Matzka, J., and Turbitt, C. W.: Observatory Data Quality Control – the instrument to ensure valuable research, in: XVth IAGA Workshop on Geomagnetic Observatory Instruments and Data Processing, Cadiz, Spain, 4–14 June 2012, 173–177, 2012.

Lopez-de-Lacalle, J.: tsoutliers R Package for Detection of Outliers in Time Series, available at: https://jalobe.com/doc/tsoutliers.pdf (last access: 20 January 2017), 2016.

MAGDAS-A Installation Manual, SERC, Kyushu Univ., edited by: Maeda, G. N., 35 pp., 2005.

Maule, C. F., Thejll, P., Neska, A., Matzka, J., Pedersen, L., and Nilsson, A.: Analyzing and correcting for contaminating magnetic fields at the Brorfelde geomagnetic observatory due to high voltage DC power lines, Earth Planets Space, 61, 1233–1241, 2009.

Nechaev, S. A.: Manual for stationary geomagnetic observations, Publ. Institute of Geography Siberian Branch of RAS, Irkutsk, 140 pp., 2006 (in Russian).

Nelapatla, P. C., Kumar Potharaju, S. V., Arora, K., Kasuba, C. S. R., Rakhlin, L., Tymoshyn, S., Merenyi, L., Chilukuri, A., Bulusu, J., and Khomutov, S.: One second vector and scalar magnetic measurements at low latitude observatory, CPL, Geosci. Instrum. Method. Data Syst. Discuss., https://doi.org/10.5194/gi-2017-16, in review, 2017.

Neska, A., Reda, J., Neska, M., and Sumaruk, Y.: On the influence of DC railway noise on variation data from Belsk and Lviv geomagnetic observatories, Acta Geophys., 61, 385–403, https://doi.org/10.2478/s11600-012-0058-0, 2013.

POS-1: Processor Overhauser Sensor – Magnetometer, User Manual, Ural State Technical University, 21 pp., 2004.

Santarelli, L., Palangio, P., and De Lauretis, M.: Electromagnetic background noise at L'Aquila Geomagnetic Observatory, Ann. Geophys.-Italy, 57, G0211, https://doi.org/10.4401/ag-6299, 2014.

Sapunov, V. A., Denisov, A. Y., Saveliev, D. V., Soloviev, A. A., Khomutov, S. Y., Borodin, P. B., Narkhov, E. D., Sergeev, A. V., and Shirokov, A. N.: New vector/scalar Overhauser DNP magnetometers POS-4 for magnetic observatories and directional oil drilling support Magnetic Resonance in Solids, Electron. J., 18, 16209, 2016.

Turbitt, C., St-Louis, B., Rasson, J., Matzka, J., Stewart, D., Lalanne, X., Schwarz, G., and Shanahan, T. INTERMAGNET Definitive One-second Data Standard, INTERMAGNET Technical Note, ver.1.0, TN6, 7 pp., 2014.

Zhizhikina, E. A., Mandrikova, O. V., and Khomutov, S. Y.: Algorithm for detection of artificial disturbances in geomagnetic data, Bulletin of Kamchatka State Technical University, 35, 21–26, 2016 (in Russian).

# 10

# Mass spectrometry of planetary exospheres at high relative velocity: direct comparison of open- and closed-source measurements

**Stefan Meyer**[1,*]**, Marek Tulej**[1]**, and Peter Wurz**[1]

[1]Physics Institute, Space Research and Planetary Sciences, University of Bern, Sidlerstrasse 5, 3012 Bern, Switzerland
* *Invited contribution by Stefan Meyer, recipient of the EGU Outstanding Student Poster and PICO Award 2016.*

*Correspondence to:* Stefan Meyer (stefan.meyer@space.unibe.ch)

**Abstract.** The exploration of habitable environments on or inside icy moons around the gas giants in the solar system is of major interest in upcoming planetary missions. Exactly this theme is addressed by the JUpiter ICy moons Explorer (JUICE) mission of ESA, which will characterise Ganymede, Europa and Callisto as planetary objects and potential habitats.

We developed a prototype of the Neutral Gas and Ion Mass spectrometer (NIM) of the Particle Environment Package (PEP) for the JUICE mission intended for composition measurements of neutral gas and thermal plasma. NIM/PEP will be used to measure the chemical composition of the exospheres of the icy Jovian moons. Besides direct ion measurement, the NIM instrument is able to measure the inflowing neutral gas in two different modes: in neutral mode, where the gas enters directly the ion source (open source), and in thermal mode, where the gas gets thermally accommodated to the wall temperature by several collisions inside an equilibrium sphere, called antechamber, before entering the ion source (closed source).

We performed measurements with the prototype NIM using a neutral gas beam of 1 up to $4.5\,\mathrm{km\,s^{-1}}$ velocity in the neutral and thermal mode. The current trajectory of JUICE foresees a flyby velocity of $4\,\mathrm{km\,s^{-1}}$ at Europa; other flybys are in the range of 1 up to $7\,\mathrm{km\,s^{-1}}$ and orbital velocity in Ganymede orbits is around $2\,\mathrm{km\,s^{-1}}$. Different species are used for the gas beam, such as noble gases Ne, Ar, Kr as well as molecules like $H_2$, methane, ethane, propane and more complex ones.

The NIM prototype was successfully tested under realistic JUICE mission conditions. In addition, we find that the antechamber (closed source) behaves as expected with predictable density enhancement over the specified mass range and within the JUICE mission phase velocities. Furthermore, with the open source and the closed source we measure almost the same composition for noble gases, as well as for molecules, indicating no additional fragmentation of the species recorded with the antechamber for the investigated parameter range.

## 1 Introduction

JUpiter ICy moons Explorer (JUICE) is an L-class mission of ESA, which will investigate and characterise Ganymede, Europa and Callisto as planetary objects and potential habitats (JUICE Team, 2012; Grasset et al., 2013). The current trajectory of the JUICE spacecraft foresees a flyby velocity of $4\,\mathrm{km\,s^{-1}}$ at Europa, other flybys in the range of 1 up to $7\,\mathrm{km\,s^{-1}}$ and orbital velocity in Ganymede orbits around $2\,\mathrm{km\,s^{-1}}$.

The Particle Environment Package (PEP) carried by JUICE combines remote global imaging with in situ measurements to study the atmospheres, plasma environments, and magnetospheric interactions and to determine global surface composition and chemistry, especially as related to habitability (Barabash et al., 2013).

NIM, the Neutral Gas and Ion Mass spectrometer, is part of the PEP suite and will be used to measure the chemical composition of the regular atmosphere produced by sublimation, energetic particle bombardment and photon interaction with the surface of the icy Jovian moons (e.g. Wurz et al., 2014; Vorburger et al., 2015). The NIM measurements include volatile species, contributions from non-ice material

**Figure 1.** Photograph of CASYMIR calibration facility in clean room area (class 100) with attached UHV chamber for the NIM prototype in the rear.

on the surface and the isotopic composition of major species. In addition, the ion composition of the ionospheres will be measured by direct ion measurement of NIM (ion mode).

We developed a prototype of the NIM instrument, part of PEP, for the JUICE mission and performed measurements with the prototype instrument using a neutral gas beam of 1 up to $4.5\,\mathrm{km\,s^{-1}}$ velocity, containing different species in the neutral (open-source) and thermal (closed-source) mode. The reason for inventing the thermal mode is the large field of view for the neutral gas, which covers basically a half-sphere opening angle. This allows measurement not only near closest approach of a flyby but also at farther distances where the ram direction is outside the narrow field of view of the neutral mode. Furthermore, thermalised neutrals possess low energy spread, which results in best possible performance in terms of mass resolution and transmission. In contrast, the operation in thermal mode may cause complex interactions of the incoming species with the inner wall material of the sphere resulting in fragmentation and chemical alteration. Therefore, the results of these measurements with respect to fragmentation and density enhancements in the closed-source mode are presented here. Furthermore, we give a direct comparison between open- and closed-source mode measurements.

## 2 Experimental methods

### 2.1 Calibration facility

To obtain a neutral gas beam at reasonable velocities, the CASYMIR (calibration system for the mass spectrometer instrument ROSINA) calibration facility has been used in combination with the UHV chamber, housing the NIM prototype.

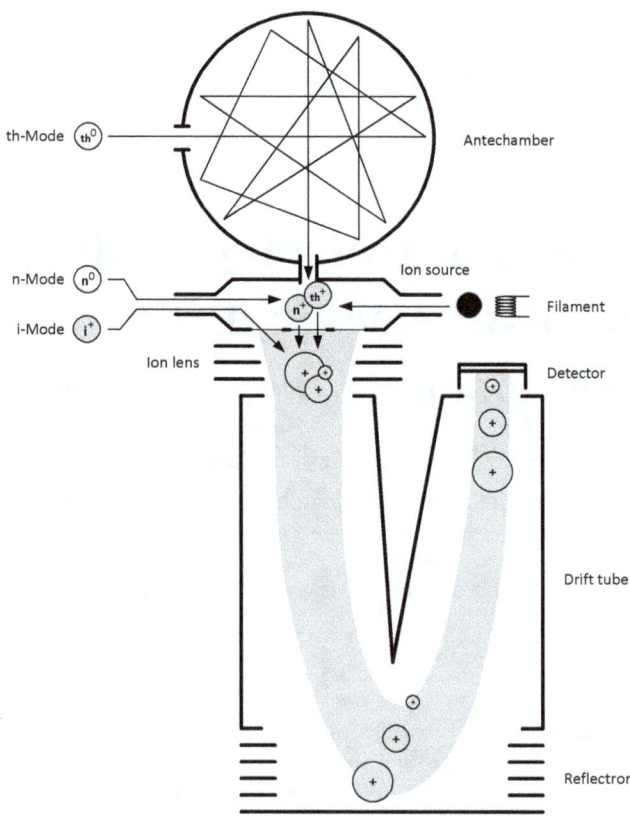

**Figure 2.** Scheme of NIM prototype illustrating its three operational modes.

The scope of the CASYMIR project was the development of a calibration chamber with the aim of testing and calibrating the two ROSINA mass spectrometers for the Rosetta mission (Graf et al., 2004). CASYMIR consists of a vacuum system with several pumping stages leading to an ultra-high vacuum of a few $10^{-10}\,\mathrm{mbar}$, where the instrument is attached. A supersonic molecular beam is created by thermal expansion from a heated nozzle up to a maximum speed of $4.5\,\mathrm{km\,s^{-1}}$, which is reached with $H_2$, the lightest gas. The gas mixture for the beam is supplied from a special designed gas mixing unit (GMU) that allows mixtures of different gases. A photograph of the CASYMIR calibration facility with attached UHV chamber in clean room is displayed in Fig. 1. All tests in neutral mode and thermal mode were performed with the NIM prototype installed in the UHV chamber attached to the CASYMIR facility.

### 2.2 NIM prototype

NIM is a time-of-flight instrument with heritage from the RTOF sensor of the ROSINA instrument on the Rosetta mission (Scherer et al., 2006; Balsiger et al., 2007) and the P-BACE instrument (Abplanalp et al., 2009). The NIM is designed to operate in three different modes:

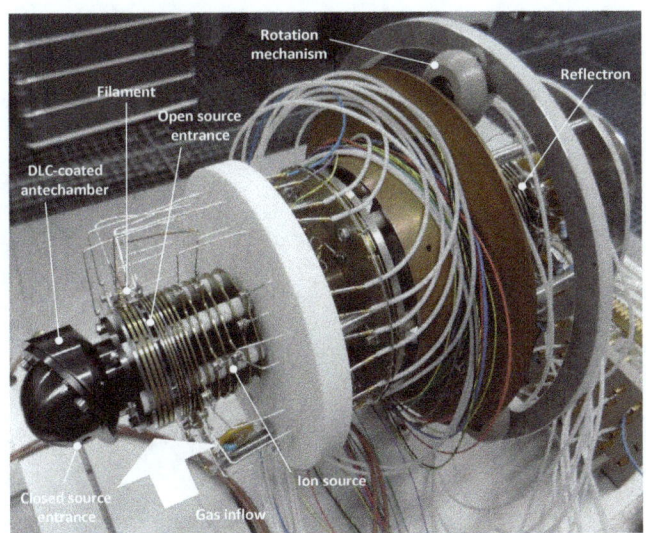

**Figure 3.** Photograph of NIM prototype with cabling and rotation mechanism.

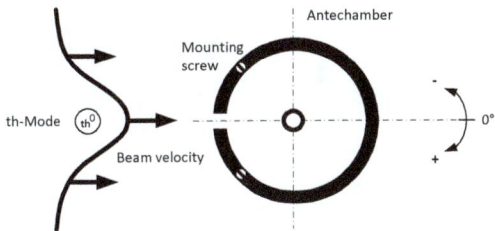

**Figure 4.** Scheme of neutral gas beam entry in th-mode at CASYMIR measurements. View from top; see Fig. 2.

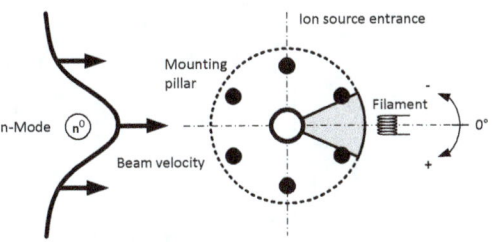

**Figure 5.** Scheme of neutral gas beam entry in n-mode at CASYMIR measurements. View from top; see Fig. 2.

- In the thermal mode (th-mode), the neutral gas is decelerated from spacecraft velocity down to thermal energies by an equilibrium sphere, called antechamber, and passed on into the ion source in thermal state (th$^0$ species in Fig. 2), to be ionised by an electron beam and stored until the subsequent guidance through the ion optics to the detector. This mode (also referred to as closed source in the literature) is used for neutral gas measurements at any mission phase, mainly during Europa torus crossing and all other flybys.

- In the neutral mode (n-mode), the neutral gas (n$^0$ species in Fig. 2) enters directly the ion source with the speed relative to the spacecraft, is ionised by an electron beam and subsequently guided through ion optics to the detector. This mode (also referred to as open source in the literature) is used for neutral gas measurements close to the moon, mainly during closest approach at flyby or in orbit phase.

- In the ion mode (i-mode), thermal ions from the ambient plasma (i$^+$ species in Fig. 2) enter the ion source with the speed relative to the spacecraft and are directly guided through ion optics to the detector. This mode (also referred to as open source in the literature) is used for thermal ion measurements of ionospheric ions close to the moon, mainly in orbit phase or near closest approach during flyby.

These three different modes are illustrated in Fig. 2 together with the entire ion-optical scheme of the NIM prototype instrument. In ion mode (i-mode) the ions from the ambient plasma enter directly into the ion source with spacecraft velocity in ram direction. In neutral mode (n-mode) the neutral gas is also entering directly into the ion source with spacecraft velocity in ram direction, which is a so-called open source. The neutral gas is then ionised around the central axis of the ion source by an electron beam produced by a filament and the generated ions are deflected by ion optics towards ion source exit. In thermal mode (th-mode) the neutral gas is first entering a sphere and then decelerated by hitting the interior wall many times, thus establishing thermal equilibrium of the gas with the walls of the antechamber. This is exactly the principle of an antechamber, which is a so-called closed source. The thermalised neutral gas from the antechamber enters the ion source, where it is ionised by the electron beam and the generated ions are again deflected by ion optics towards ion source exit.

The advantage of operation in th-mode is the large field of view for the neutral gas entering a small hole in the antechamber, which covers basically a half-sphere opening angle. This allows measurement not only near closest approach of a flyby but also at farther distances where the ram direction is outside the narrow field of view of the neutral mode. Furthermore, thermalised neutrals do not possess large energy spread (constant energy instead of constant velocity), which results in best possible performance in terms of mass resolution and transmission. The disadvantage of operation in th-mode is the complex interactions of the incoming species with the inner wall material of the sphere resulting in fragmentation and chemical alteration. However, the measurements obtained in th-mode can then be compared with those obtained by the n-mode at closest approach of a flyby or in orbit phase. Furthermore, inter-calibrations between th-mode and n-mode can be done in the laboratory.

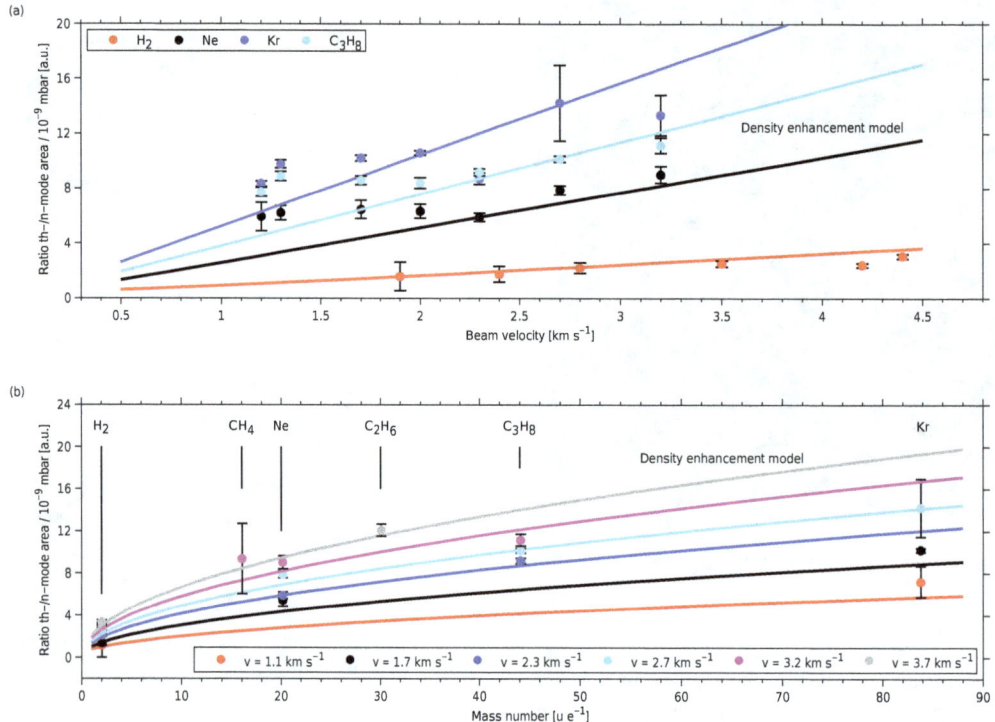

**Figure 6.** Ratio of signal in thermal mode to signal in neutral mode as function of beam velocity in **(a)** and as function of mass in **(b)**.

Figure 3 shows a photograph of the NIM prototype with cabling and rotation mechanism, which covers a measurement range of 180° (±90° with respect to ram direction and antechamber entrance hole). On the left-hand side of Fig. 3, the tested antechamber with DLC coating can be seen. DLC stands for diamond-like carbon, which is an amorphous carbon material that displays some of the typical properties of diamond. A gold-coated antechamber, which is not depicted here, was tested as well. As has already been discussed above, the antechamber is used in th-mode as closed source to equilibrate the incoming neutral gas beam to thermal energies. The field of view covers basically a half-sphere, in which the effective particle number density is modulated by a cosine of the entrance angle with respect to the normal of the opening hole.

Because of the thermalisation of the incoming gas, the antechamber produces a density enhancement compared to the open source, which can be calculated as follows (Wurz et al., 2007):

$$\frac{n_{cs}}{n_{os}} = \sqrt{\frac{T_a}{T_s}} \cdot \frac{F(S) \cdot k \cdot \sin^2\left(\frac{\omega}{2}\right) \cdot \cos^2\left(\frac{\omega}{2}\right)}{1 - k \cdot \cos^2\left(\frac{\omega}{2}\right)} \cdot \frac{d_i^2}{d_i^2 + d_s^2}$$

$$F(S) = e^{-S^2} + \pi^{\frac{1}{2}} \cdot S \cdot (1 + \text{erf}(S))$$

$$S = v_{sc} \cdot \cos\chi \cdot \sqrt{\frac{m}{2k_B T_a}}, \qquad (1)$$

where $n_{cs}$ is the closed-source number density, $n_{os}$ is the open-source number density, $T_a$ the ambient gas temperature,

$T_s$ the antechamber temperature, $\omega$ is the cone half-angle of the open source, $v_{sc}$ the spacecraft velocity and $\chi$ its angle with respect to the surface normal of the entrance aperture, $m$ the mass of the gas, $k$ the probability of a molecule being re-emitted after colliding with the interior surface, $d_i$ the opening hole diameter and $d_s$ the diameter of the exit hole to the ion source.

## 2.3 Measurement method

Measurements with the NIM prototype, built into the UHV chamber, using a neutral gas beam of realistic velocities in the n-mode (open source) and th-mode (closed source) have been performed. To obtain a neutral gas beam at realistic velocities, the CASYMIR facility has been used (see Sect. 2.1). Different species and gas mixtures are used, such as noble gases Ne, Ar, Kr as well as molecules like $H_2$, $CH_4$, $C_2H_6$, $C_3H_8$ and more complex ones.

The used measurement characteristics are discussed in the following. On the one hand, Fig. 4 shows the scheme of neutral gas beam entry in th-mode at the CASYMIR measurements, where the neutral gas beam with Gaussian beam profile impinges on the antechamber opening hole with 4 mm diameter, which can be rotated ±90°. The antechamber measures 40 mm in diameter and is DLC-coated inside (gold-coated antechamber is also tested) with mounting screws at ±45° from 0° position.

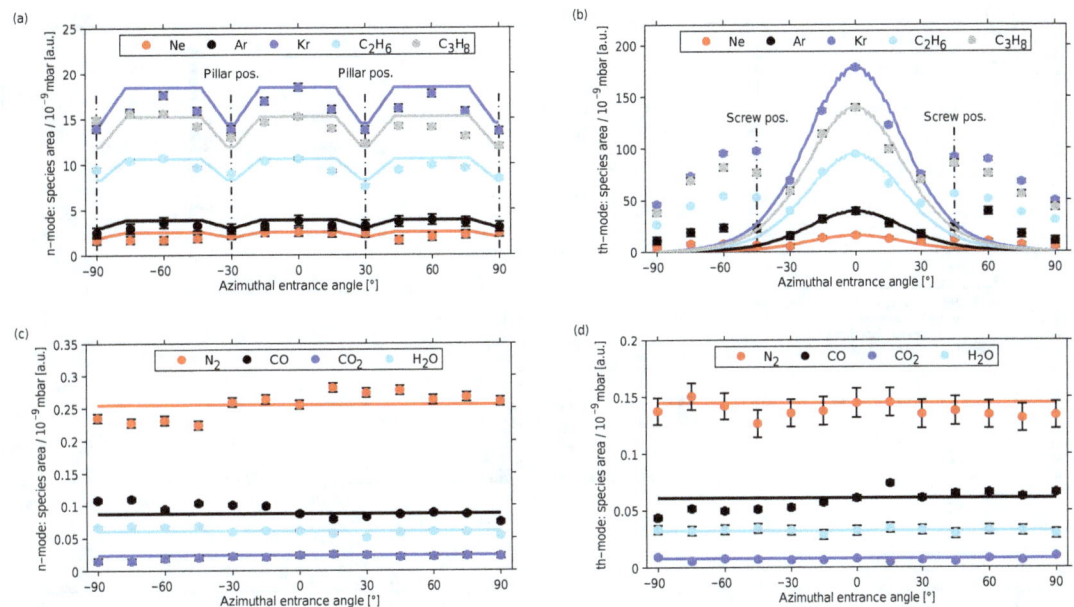

**Figure 7.** The results of azimuthal rotation campaign conducted in CASYMIR facility. Panels (**a, b**) are for species in the neutral beam; panels (**c, d**) are for species of the residual gas. Panels (**a, c**) are for n-mode measurements; panels (**b, d**) are for th-mode measurements.

On the other hand, Fig. 5 shows the scheme of neutral gas beam entry in n-mode for the CASYMIR measurements, where the neutral gas beam with Gaussian profile enters the open (line-of-sight) entrance rings, which can be rotated $\pm 90°$. The opening angle with respect to elevation is $\sim 11°$, which is given by the line-of-sight geometry for the entrance with dimensions $2\,\mathrm{mm} \times 6\,\mathrm{mm}$ (thickness × width) and additional mounting pillars of $4\,\mathrm{mm}$ every $30°$.

Generally, the Gaussian gas beam, with width of about $11\,\mathrm{mm}$ at instrument entrance, has a flux of about $2 \times 10^{12}\,\mathrm{cm}^{-2}\,\mathrm{s}^{-1}$, which corresponds to a pressure of $10^{-9}\,\mathrm{mbar}$ of a selected gas mixture at a pressure of about $10^{-10}\,\mathrm{mbar}$ of residual gas in the UHV chamber, mainly consisting of $N_2$, $H_2O$, CO and $CO_2$ gas. Furthermore, the ionising electron beam was kept at $100\,\mu\mathrm{A}$ for all measurements, but can be operated between 10 and $1000\,\mu\mathrm{A}$. In fact, all analyses were made based on a final mass spectrum consisting of $100\,000$ accumulated single spectra, which is called one measurement in our case.

## 3 Results and discussion

### 3.1 Beam velocity results

For the beam velocity campaign, measurements with all stated gas mixtures at different velocities (1 up to $4.5\,\mathrm{km\,s}^{-1}$) for both modes, n-mode and th-mode, at $0°$ position each, have been performed. The result summary of this campaign is presented in Fig. 6, where the ratio of the measured peak area normalised to $10^{-9}\,\mathrm{mbar}$ between th-mode and n-mode is shown. In the upper panel a of Fig. 6, the ratio of peak

areas of the th-mode to the n-mode, the c–o ratio, is plotted versus the beam velocity for the different species in the measured gas mixtures. Additionally, the density enhancement model, described above (see Eq. 1), is given as solid lines for the corresponding molecular masses of the analysed species. In the lower panel b of Fig. 6, the c–o ratio is plotted versus the species mass for different beam velocities. Likewise, the density enhancement model, described above (see Eq. 1), is given as solid lines for the corresponding beam velocities.

The measured c–o ratio is in good agreement with the calculated density enhancement for noble gases like Ne and Kr, as well as molecules like $CH_4$, $C_2H_6$ and $C_3H_8$. Because of the good agreement, it can be extrapolated for all JUICE mission phases. Therefore, the behaviour of the antechamber (closed source) together with the NIM prototype could successfully be verified for gas beam velocities of 1 up to $4.5\,\mathrm{km\,s}^{-1}$, which covers the majority of JUICE mission phases, including Europa flyby and Ganymede orbit. The verified mass range in n-mode and th-mode goes from 2 ($H_2$) up to $84\,\mathrm{u}$ (Kr), which is most relevant for the JUICE mission. However, NIM supports a mass range up to $1300\,\mathrm{u\,e}^{-1}$ for a $25\,\mu\mathrm{s}$ measurement.

### 3.2 Azimuthal rotation results

For the azimuthal rotation campaign, measurements of all stated gas mixtures at different azimuthal entrance angles ($\pm 90°$ from $0°$ position) for both modes, n-mode and th-mode, at the same velocity for each gas mixture (2 up to $3\,\mathrm{km\,s}^{-1}$) have been conducted. The result summary of this campaign is presented in Fig. 7, where the measured species

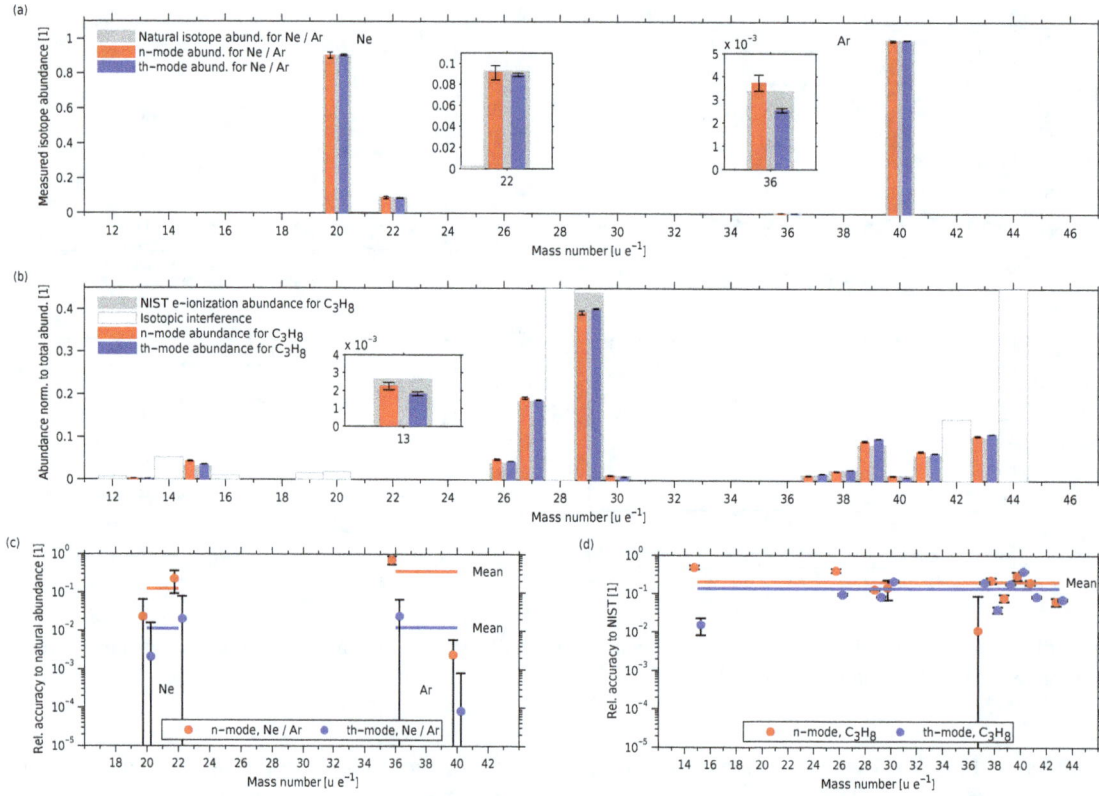

**Figure 8.** Isotope analysis results of mass spectra from CASYMIR measurements, obtained by summing up 100 000 individual spectra.

area normalised to $10^{-9}$ mbar versus azimuthal rotation angle is shown. In the upper left panel a of Fig. 7, the n-mode area is plotted for different species in the measured gas mixtures. The solid lines give the expected behaviour, which is independent (constant) with respect to the azimuthal entrance angle, except for a drop of the signal, where the mounting pillars of the ion source are in line of sight (see Fig. 5). This could be verified with the open-source measurements ranging from $-90$ to $+90°$ for different species in the measured gas mixtures. The beam velocity for this campaign was kept between 2 and 3 km s$^{-1}$ depending on the gas mixture.

In the upper right panel b of Fig. 7, the th-mode area is plotted for different species in the measured gas mixtures. The solid lines give again the expected behaviour, which follows a cosine of the entrance angle modulated by the Gaussian beam profile. In space, the curves are expected to be purely cosine of the angle between entrance aperture of the antechamber and the spacecraft velocity, since the incoming gas is equally distributed in space (at least locally) and has in principle a flat profile. However, the th-mode measurements are in good agreement with the expected profile around $\pm30°$. Outside of $\pm30°$, unexpected side lobes occur with maxima at $\pm45°$, which we attribute to scattering of the gas beam at the mounting screws of the antechamber (see Fig. 4).

The measured species from the residual gas in the vacuum chamber are expected to be independent of the azimuthal rotation angle for both the n-mode and the th-mode, as is shown by the measurement of the residual gas species $N_2$, CO, $CO_2$ and $H_2O$ in n-mode, in the lower left panel c of Fig. 7, and in th-mode, in the lower right panel d of Fig. 7.

### 3.3 Isotope analysis results

For the isotope analysis, measurements with all stated gas mixtures at velocities between 2 and 3 km s$^{-1}$ for both modes, n-mode and th-mode, at 0° position each, have been selected. The isotope analysis results for both th-mode and n-mode are shown in Fig. 8. In the top panel a of Fig. 8, the measured isotope abundance is plotted for elements (noble gases like Ne and Ar) in comparison with the natural table of isotopic abundances (CIAAW, 2016). In the bottom left panel c of Fig. 8, its relative accuracy to natural isotopic abundances is given. Here, the measured relative isotopic accuracy to natural abundance reaches from 100 ppm for $^{40}$Ar in th-mode up to almost 1 for $^{36}$Ar in n-mode, depending on the signal-to-noise ratio (SNR) up to the detection limit. Two additional insets in Fig. 8a show a zoom on the $^{22}$Ne isotope and $^{36}$Ar isotope, whereas $^{21}$Ne and $^{38}$Ar are not visible and below the detection limit at this beam flux and NIM operation conditions. The n-mode and th-mode mean accuracies

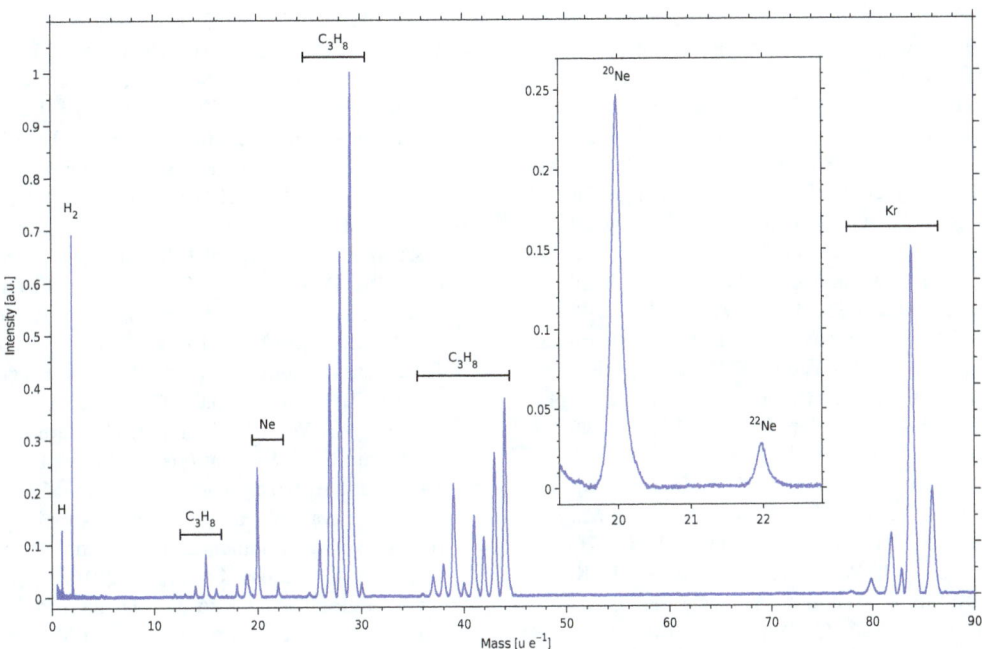

**Figure 9.** Typical mass spectrum of a th-mode measurement of a gas mixture of $H_2$, Ne, Kr and $C_3H_8$. The spectrum is a background corrected sum of 100 000 individual spectra.

are different due to the density enhancement, which also results in a SNR enhancement and therefore a better accuracy for th-mode. In addition, NIM was not optimised for isotope measurements during these campaigns.

In the middle panel b of Fig. 8, the fragmentation pattern normalised to the total abundance is plotted for the molecule $C_3H_8$ (propane) in comparison with the NIST electron ionisation abundance from NIST Chemistry Webbook (Web NIST, 2016). Mass lines with isotopic interference are marked and omitted for the analysis. An additional inset in Fig. 8b shows a zoom on the mass line of $13\,\mathrm{u\,e^{-1}}$. The measured mean relative isotopic accuracy to NIST electron ionisation abundance, shown in the bottom right panel d of Fig. 8, is between 10 and 20 % for both n-mode and th-mode, which is about the NIST measurement accuracy and suggests, together with the noble gas results, a much higher accuracy for the NIM prototype instrument. Moreover, the direct comparison of the n-mode (open source) and th-mode (closed source) shows the same fragmentation pattern; i.e. there is no evidence for any fractionation effects inside the antechamber, since both modes measure almost the same composition, at least for the measured molecules $H_2$, $CH_4$, $C_2H_6$ and $C_3H_8$, as well as methanol vapour and propanol vapour.

A typical mass spectrum, measured in th-mode with a gas mixture of $H_2$, Ne, Kr and $C_3H_8$, is illustrated in Fig. 9. An additional inset in Fig. 9 shows a zoom on the $^{20}$Ne and $^{22}$Ne isotopes, whereas $^{21}$Ne is not visible and below the detection limit at this beam flux and NIM operation conditions.

## 4   Conclusion

The NIM prototype has been successfully tested under realistic JUICE mission conditions. In addition, the antechamber (closed source) behaves as expected with predictable density enhancement over the specified mass range and within the JUICE mission phase velocities. Moreover, n-mode (open source) and th-mode (closed source) measure almost the same composition for noble gases, as well as for molecules, indicating no additional fragmentation of the species inside the antechamber. This holds for both versions of the antechamber, the DLC-coated one and the gold-coated one, respectively, which were both successfully tested.

*Acknowledgements.* The authors would like to acknowledge the contribution of a number of people helping in technical preparation of the NIM prototype for the presented investigations, including Stefan Brüngger, Philippe Németh and mechanical workshop, as well as Matthias Lüthi and electronics workshop. The financial support from Swiss National Science Foundation and the PRODEX programme of the Swiss Space Office is acknowledged.

Edited by: G. Kargl

## References

Abplanalp, D., Wurz, P., Huber, L., Leya, I., Kopp, E., Rohner, U., Wieser, M., Kalla, L., and Barabash, S.: A neutral gas mass spectrometer to measure the chemical composition of the stratosphere, Adv. Space Res., 44, 870–878, doi:10.1016/j.asr.2009.06.016, 2009.

Balsiger, H., Altwegg, K., Bochsler, P., Eberhardt, P., Fischer, J., Graf, S., Jäckel, A., Kopp, E., Langer, U., Mildner, M., Müller, J., Riesen, T., Rubin, M., Scherer, S., Wurz, P., Wüthrich, S., Arijs, E., Delanoye, S., De Keyser, J., Neefs, E., Nevejans, D., Rème, H., Aoustin, C., Mazelle, C., Médale, J.-L., Sauvaud, J. A., Berthelier, J.-J., Bertaux, J.-L., Duvet, L., Illiano, J.-M., Fuselier, S. A., Ghielmetti, A. G., Magoncelli, T., Shelley, E. G., Korth, A., Heerlein, K., Lauche, H., Livi, S., Loose, A., Mall, U., Wilken, B., Gliem, F., Fiethe, B., Gombosi, T. I., Block, B., Carignan, G. R., Fisk, L. A., Waite, J. H., Young, D. T., and Wollnik, H.: ROSINA – Rosetta Orbiter Spectrometer for Ion and Neutral Analysis, Space Sci. Rev., 128, 745–801, doi:10.1007/s11214-006-8335-3, 2007.

Barabash, S., Wurz, P., Brandt, P., Wieser, M., Holmström, M., Futaana, Y., Stenberg, G., Nilsson, H., Eriksson, A., Tulej, M., Vorburger, A., Thomas, N., Paranicas, C., Mitchell, D. G., Ho, G., Mauk, B. H., Haggerty, D., Westlake, J. H., Fränz, M., Krupp, N., Roussos, E., Kallio, E., Schmidt, W., Szego, K., Szalai, S., Khurana, K., Xianzhe, J., Paty, C., Wimmer-Schweingruber, R. F., Heber, B., Asamura, K., Grande, M., Lammer, H., Zhang, T., McKenna-Lawlor, S., Krimigis, S. M., Sarris, T., and Grodent, D.: Particle Environment Package (PEP), European Planetary Science Congress 2013, 8–13 September, London, UK, Vol. 8, EPSC2013, 2013.

CIAAW: available at: http://www.ciaaw.org/isotopic-abundances.htm, Commission on Isotopic Abundances and Atomic Weights, last access: 2 February 2016.

Graf, S., Altwegg, K., Balsiger, H., Jäckel, A., Kopp, E., Langer, U., Luithardt, W., Westermann, C., and Wurz, P.: A cometary neutral gas simulator for gas dynamic sensor and mass spectrometer calibration, J. Geophys. Res., 109, E07S08, doi:10.1029/2003JE002188, 2004.

Grasset, O., Dougherty, M. K., Coustenis, A., Bunce, E. J., Erd, C., Titov, D., Blanc, M., Coates, A., Drossart, P., Fletcher, L. N., Hussmann, H., Jaumann, R., Krupp, N., Lebreton, J.P., Prieto-Ballesteros, O., Tortora, P., Tosi, F., and Van Hoolst, T.: JUpiter Icy moons Explorer (JUICE): An ESA mission to orbit Ganymede and to characterise the Jupiter system, Planet. Space Sci., 78, 1–21, doi:10.1016/j.pss.2012.12.002, 2013.

JUICE Team: JUICE assessment study report (Yellow Book), ESA/SRE(2011)18, JUICE Science Study Team, ESA, 2012.

Scherer, S., Altwegg, K., Balsiger, H., Fischer, J., Jäckel, A., Korth, A., Mildner, M., Piazza, D., Rème, H., and Wurz, P.: A novel principle for an ion mirror design in time-of-flight mass spectrometry, Int. J. Mass Spectrom., 251, 73–81, doi:10.1016/j.ijms.2006.01.025, 2006.

Vorburger, A., Wurz, P., Lammer, H., Barabash, S., and Mousis, O.: Monte-Carlo Simulation of Callisto's Exosphere, Icarus, 262, 14–29, doi:10.1016/j.icarus.2015.07.035, 2015.

Web NIST: available at: http://webbook.nist.gov/chemistry, National Institute of Standards and Technology, Standard Reference Database, last access: 1 February 2016.

Wurz, P., Balogh, A., Coffey, V., Dichter, B. K., Kasprzak, W. T., Lazarus, A. J., Lennartsson, W., and McFadden, J. P.: Calibration Techniques, Calibration of Particle Instruments, in: Space Physics, edited by: Wüest, M., Evans, D. S., and von Steiger, R., International Space Science Institute, 2007.

Wurz, P., Vorburger, A., Galli, A., Tulej, M., Thomas, N., Alibert, Y., Barabash, S., Wieser, M., and Lammer, H.: Measurement of the Atmospheres of Europa, Ganymede, and Callisto, European Planetary Science Congress 2014, EPSC Abstracts, Vol. 9, ID EPSC2014-504, 2014.

# 11

# Optimal site selection for sitting a solar park using multi-criteria decision analysis and geographical information systems

**Andreas Georgiou and Dimitrios Skarlatos**

Civil Engineering & Geomatics Dept., Cyprus University of Technology, 30 Archbishop Kyprianou Str., 3036 Limassol, Cyprus

*Correspondence to:* Andreas Georgiou (angeocy@gmail.com)

**Abstract.** Among the renewable power sources, solar power is rapidly becoming popular because it is inexhaustible, clean, and dependable. It has also become more efficient since the power conversion efficiency of photovoltaic solar cells has increased. Following these trends, solar power will become more affordable in years to come and considerable investments are to be expected. Despite the size of solar plants, the sitting procedure is a crucial factor for their efficiency and financial viability. Many aspects influence such a decision: legal, environmental, technical, and financial to name a few. This paper describes a general integrated framework to evaluate land suitability for the optimal placement of photovoltaic solar power plants, which is based on a combination of a geographic information system (GIS), remote sensing techniques, and multi-criteria decision-making methods.

An application of the proposed framework for the Limassol district in Cyprus is further illustrated. The combination of a GIS and multi-criteria methods produces an excellent analysis tool that creates an extensive database of spatial and non-spatial data, which will be used to simplify problems as well as solve and promote the use of multiple criteria. A set of environmental, economic, social, and technical constrains, based on recent Cypriot legislation, European's Union policies, and expert advice, identifies the potential sites for solar park installation. The pairwise comparison method in the context of the analytic hierarchy process (AHP) is applied to estimate the criteria weights in order to establish their relative importance in site evaluation. In addition, four different methods to combine information layers and check their sensitivity were used. The first considered all the criteria as being equally important and assigned them equal weight, whereas the others grouped the criteria and graded them according to their objective perceived importance. The overall suitability of the study region for sitting solar parks is appraised through the summation rule.

Strict application of the framework depicts 3.0 % of the study region scoring a best-suitability index for solar resource exploitation, hence minimizing the risk in a potential investment. However, using different weighting schemes for criteria, suitable areas may reach up to 83 % of the study region. The suggested methodological framework applied can be easily utilized by potential investors and renewable energy developers, through a front end web-based application with proper GUI for personalized weighting schemes.

## 1 Introduction

Energy is an essential part of modern life as almost all human activities are strongly connected with it. The availability and secure supply of energy are considered important prerequisites of economic and social development of a country. Although in the current economic situation, the rational use of the available resources and the need to overcome the negative environmental impacts and other problems associated with fossil fuels have forced many countries to enquired into and change to more environmentally friendly alternatives, which are renewable in order to sustain the increasing energy demand (Sanchez-Lozano et al., 2013; Bahadori and Nwaoha, 2013).

Among the renewable power sources, solar has grown exponentially worldwide during the last decade. This is not surprising as the sun can provide more than 2500 terawatts (TW)

of technically accessible energy over large areas of Earth's surface and solar energy technologies are no longer cost prohibitive (Hernandez et al., 2014). However, currently it only covers a minor portion of global energy demands (0.05 % of the total primary energy supply) as photovoltaic (PV) power generates less than 1 % of total electricity supply (Solangi et al., 2011); nevertheless, solar energy has great future potential.

Solar energy is obviously environmentally advantageous relative to any other non-renewable energy source and the linchpin of any sustainable development program. It can be exploited through the solar thermal and PV routes for various applications. The main direct or indirectly derived advantages of solar energy are no emission of greenhouse or toxic gasses, reclamation of degraded land, reduction of transmission lines from electricity grids, and increase of regional/national energy independence. In addition, it can provide diversification and security of the energy supply, acceleration of rural electrification in developing countries, job opportunities, improvement of life quality in developing countries, and investment security for park development as solar panels are resistant to extreme climate conditions with a life expectancy greater than 35 years (Solangi et al., 2011; Tsoutsos et al., 2005; Torres-Sibille et al., 2009; Hernandez et al., 2014). However, conflicts can also arise between renewable energy and nature conservation policy. The environmental impacts from photovoltaic power generation include general effects on visual impact, land use intensity, wildlife impacts, reflection effects, depletion of natural resources, and waste management (Torres-Sibille et al., 2009; Tsoutsos et al., 2005; Turney and Ftenakis, 2011). Although the number of direct animal deaths at solar parks, is thought to be negligible (Katzner et al., 2013). The worst impacts of ground-mounted solar installations occur when all natural habitat in the vicinity is cleared, stripping vegetation and compacting soil. This can reduce the carbon content of the soil compared to undisturbed areas and, in arid regions, allows for the transport of dust, which can reduce the efficiency of solar panels (Hernandez et al., 2014). Other risks to wildlife from solar park operation include chemicals such as dust suppressants and rust inhibitors (Hernandez et al., 2014). Water is also used to clean the panels, which may pressurize scarce resources in dry regions (Cameron et al., 2012). It is also important to take into account the life-cycle assessment: processes involved in obtaining rare materials used for making solar panels may lead to biodiversity impacts elsewhere, e.g., at the source of extraction (European Commission, 2014).

The sitting of photovoltaic power facilities is important in order to maximize the potential of the PV technology implementation in reality. Any site selection and assessment procedure must address the technical, economic, social, and environmental aspects of the project to determine whether it is suitable for solar energy development. As a result, energy and electricity industry professionals and policy groups have developed a variety approaches to mitigate sitting of so-

lar parks. A geographic information system (GIS) is a popular and effective decision-making tool for the selection of optimal sites for different types of activities and installations (Carrion et al., 2008; Tegou et al., 2010; Kontos et al., 2005). Applications of GISs and renewable energy source planning include wind farm sitting, photovoltaic electrification, biomass evaluation, visual impact assessment of wind farm, etc. (Tegou et al., 2010; Georgiou et al., 2012; Masera et al., 2006; Ramachandra and Shruthi, 2007). One of the most common GIS-based strategies that have been designed to facilitate decision making in site evaluation and land suitability is multi-criteria analysis (MCA) (Torres-Sibille et al., 2009). The analytic hierarchy process (AHP) method that introduced by Saaty (1980) is a flexible and easily implemented MCA technique and its use has been largely explored in the literature with many examples in locating facilities and land suitability analysis (Tegou et al., 2010; Kontos et al., 2005; Georgiou et al., 2012; Masera et al., 2006).

The scope of this paper is to develop and present an integrated framework to quantify and evaluate land suitability for the optimal photovoltaic solar power plant placement with an application to the Limassol district in Cyprus. This should be considered as a tool, which different users can change its respective weights in order to produce a custom made map for their own "most suitable" areas for solar park investment. The proposed framework comprises of a combination of already established methods and tools for solar resource assessment, remote sensing techniques, spatial analysis, and multi-criteria decision-making methods. The AHP has been chosen as a means of weighting the suitability criteria, the simple additive weighting (SAW) method has been used as an aggregation algorithm, and a GIS as an integrated platform of analysis and presentation. Innovative aspects comprise of a unique and balanced approach among practice, law, and theory of solar park siting. In order to do so, a real application area was selected for implementation. A novel methodology was adopted, which takes into consideration several constraints and many criteria that have been pre-quantified. In addition, a straightforward integration was developed using seamless existing tools for analysis, modeling and representation in a single GIS environment, allowing for a flexible tool that encourages several "what if" scenarios to be easily implemented. The tool, currently implemented within a local GIS, has a prospect for future web automation.

## 2   Material and methods

The methodological framework considers that each potential site that may host a solar park should satisfy a number of functional parameters and assesses their comparative importance. To do so, a combination of MCA with a GIS were used, with the AHP method as additional tool to assign weight of relative importance to each evaluation criterion. An

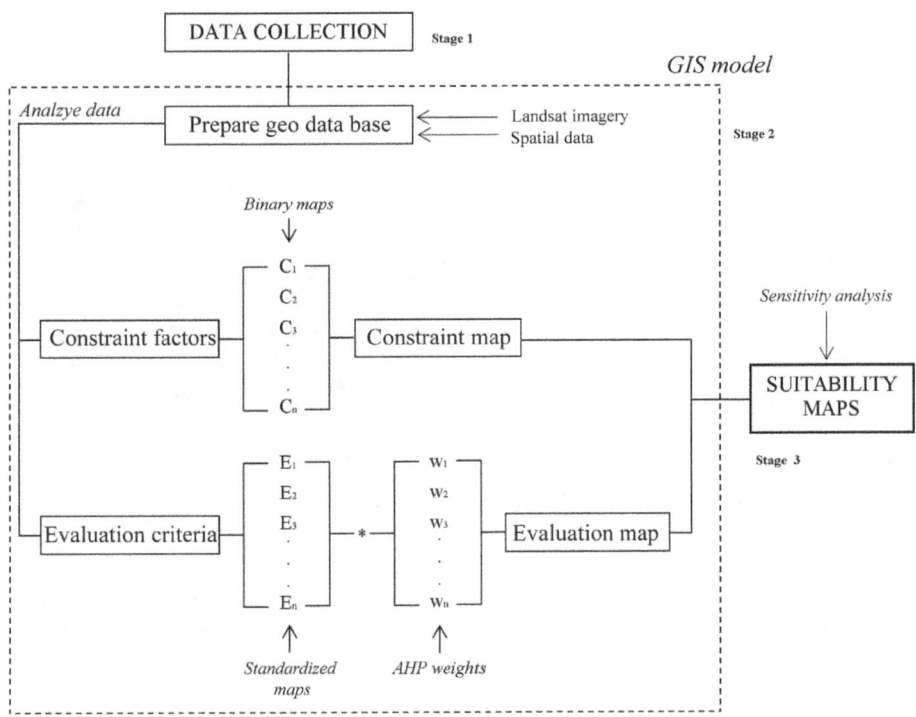

**Figure 1.** Flow chart of proposed methodology framework.

overall suitability index (SI) is then calculated for each potential cell in the map using the weighted overlay technique.

The presented methodological framework involves several stages as presented in Fig. 1 and were grouped as (i) collection of data, (ii) setup of the GIS model, (iii) sensitivity analysis, and (iv) extraction of the suitability maps. More specific, the first step is to define and gather all appropriate data layers needed for the analysis in order to set up the digital geo-database. The next step is to establish the constraint factors that will determine unsuitable areas and will be in the form of a binary map, where "0" refers to unsuitable areas and "1" to areas suitable for further examination for solar exploitation. At the exclusion areas, local and EU legislation was used to define criteria in addition to GIS and remote sensing techniques. The next step is to establish the cost functions for all available criteria and estimation of weights of the evaluation criteria according the AHP algorithm. These weights are based on subjective criteria that can be changed according to the needs of researchers. The final step consists of the formulation and calculation of the final suitability index map using the SAW method and the presentation of the results in thematic maps.

The definition of both bounding constraints and evaluation criteria depended on standing legislation and on the characteristics of the study area. All factors were selected in accordance with the Cypriot legislation for renewable energy sources (RES) sitting (Law 29(I), 2005) and in some cases, under the advice of the experts of the Ministry of Agriculture Natural Resources & Environment. In addition, European's

Union policy (European Commission, 2014) and previous similar research in the renewable energy systems field (European Commission, 2014; Carrion et al., 2008; Katsaprakakis, 2012; Mari et al., 2011) are used to configure the list of parameters that are used.

## 2.1 The AHP Method

The AHP is a multi-criteria decision-making approach that can be used for solving complex and unstructured problems. It helps to capture both qualitative and quantitative aspects of a decision problem and provides a powerful yet simple way of weighting the decision criteria, thus reducing bias in decision-making (Saaty, 1987; Georgiou et al., 2012). The AHP is based on pairwise comparisons and used to derive normalized absolute scales of numbers, whose elements are then used as priorities. By comparing pairs of criteria one at a time and using integer numbers from the 1 to 9 scale of the AHP, decision-makers can quantify their judgment about the relative importance of criteria. Then a pairwise comparison matrix is formed where the relative importance weight of each criterion is computed as the normalized geometric mean of each row of the matrix. A consistency index (CI) that measures the inconsistencies of pairwise comparison calculated as follows (Eq. 1), where $\lambda$max is the largest eigenvalue and $n$ the number of rows or columns:

$$CI = \frac{\lambda max - n}{n-1}. \tag{1}$$

A measure of coherence of the pairwise comparisons is calculates in the form of consistency ratio (CR) where RI is the average CI of the randomly generated comparisons (Pilavachi et al., 2009):

$$CR = \frac{CI}{RI}. \qquad (2)$$

CR value of 10 % or less is considered as acceptable; otherwise, one has to revise his judgments.

## 2.2 Simple additive weighting method

The SAW method is the simplest way for aggregating the used criteria in order to compute a SI for each cell in the study area. More specific, each evaluation criterion is multiplied by the respective weight and then all criteria are summed in order to provide a total performance score for each cell. The SI lies between 0 and 100, corresponding to the "worst" and "best" sites, respectively. The applied formulation is (Georgiou et al., 2012)

$$SI_i = \sum_{j=1}^{n} W_j \cdot V_{ij}, \qquad (3)$$

where $SI_i$ is the overall suitability index for cell $i$, $W_j$ the relative importance weight of criterion $j$, $V_{ij}$ the score of cell $i$ under criterion $j$, and $n$ the total number of criteria.

## 3 Case study

### 3.1 Study site

Located in the southern part of the island of Cyprus (Fig. 2), the study area of Limassol district covers an area of about 1370 km$^2$. The island of Cyprus is located in the northeastern part of the Mediterranean Sea and therefore, has a typical eastern Mediterranean climate with a long hot dry summer, mild winter, and more than 3000 h of sunshine annually. One of the most important aspects of the Cypriot budget is energy, as it is characterized by high dependence on imported energy sources, the intense use of oil in the energy balance, isolation from European energy networks, and a low degree of exploitation of renewable energy sources. Regarding primary energy, 90 % is oil based, 6 % is coal based and the remaining 4 % is based in solar energy and basically in solar thermal energy (Pilavachi et al., 2009; Maxoulis and Kalogirou, 2008). For those reasons, as well as the fact that Cyprus is an island, it must be as energy independent as possible.

### 3.2 Preparation of the geo-database

This study aims to develop a framework model using a GIS system, supporting satellite imagery and both rasters and vectors as input data. Spatial data sets of archaeological sites, road network, electricity grid, solar radiation, digital elevation model (DEM), NATURA 2000 areas, rivers, land use,

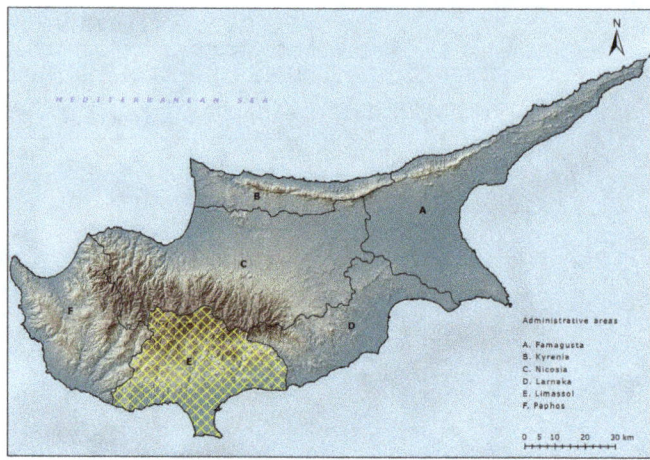

**Figure 2.** Study area of Limassol district.

built-up areas, surface waters, airport area, slope, and aspect are all part of the geo-database. This geo-database can be easily expanded with more layers of information, once they are available.

The land use, built-up areas, and surface waters were produced from the analysis of a Landsat-8 OLI/TIRS image as further illustrated. The Landsat-8 OLI/TIRS image has been chosen, as the spatial resolution provided is suitable for the analysis and it is the same as DEM. The image was acquired on 26 September 2015 and contains 11 bands. Vector data such as archaeological sites, road network, rivers, and an airport were digitized by 1 : 50 000 maps of Cyprus while NATURA 2000 areas and electricity grid were produced by the Ministry of Agriculture Natural Resources & Environment and Electricity Authority of Cyprus (EAC), respectively. Finally, the DEM with spatial resolution of 30 m was produced by the Cyprus Geological Survey Institute using three parameters. The main parameter for DEM construction was the contours derived from 1 : 50 000 maps of Cyprus, to this was then added several topographical points derived from aerial photography and photogrammetry, and, finally, rivers were used for parameter identification and geographic correction.

### 3.3 Landsat-8 OLI/TIRS data pre-processing

Landsat-8 satellite images are available through the U.S. Geological Survey (Zanter, 2016) Earth Resources Observation and Science (EROS) center. Landsat-8 carries two instruments: the Operational Land Imager (OLI) sensor includes refined heritage bands, along with three new bands: a deep blue band for coastal/aerosol studies, a shortwave infrared band for cirrus detection, and a Quality Assessment band (30 m resolution). The Thermal Infrared Sensor (TIRS) provides two thermal bands (100 m resolution). These sensors both provide improved signal-to-noise radiometric (SNR) performance quantized over a 12-bit dynamic range (Zanter, 2016). The satellite collects images of the Earth with a 16-day repeat cycle with the approximate scene size to be at 170 km north–south by 183 km east–west (Zanter, 2016).

Prior to deriving the spectral indices necessary for the analysis, the Landsat-8 OLI/TIRS data had to undergo radiometric calibration and atmospheric correction. The digital number (DN) values of the multispectral and thermal bands had to be converted into top-of-atmosphere (TOA) reflectance and be corrected with sun angle.

The TOA spectral radiances of the multispectral and thermal bands of the Landsat-8 OLI/TIRS imagery can be calculated using Eq. (4).

$$L_{\lambda'}(\text{Landsat}-8) = M_L Q_{\text{cal}} + A_L, \qquad (4)$$

where $M_L$ and $A_L$ are, respectively, the band-specific multiplicative and additive rescaling factors from the metadata; and $Q_{\text{cal}}$ is the quantized and calibrated standard product pixel values (DN) (Zanter, 2016).

In the correction of the reflectance with the sun angle, we used the TOA planetary reflectance without the sun correction ($L_{\lambda'}$) and the local sun elevation angle ($\theta$SE) using Eq. (5). The scene center sun elevation angle in degrees is provided in the metadata. Source: (Zanter, 2016).

$$L_\lambda = \frac{L_{\lambda'}}{\cos\theta\text{SE}} \qquad (5)$$

### 3.4 Classification of main area categories

In order to exclude certain areas from selection, the area of study (AoS) was grouped into three generalized categories, i.e., vegetation, open water, and built-up land. Based on these three elements, three indices, NDVI, MNDWI, and NDBI, were selected in this study to be used for extraction of those three major land-use classes, respectively.

### 3.4.1 NDVI – derived vegetation image

There are various vegetation indices to enhance vegetation information in remote sensing imagery usually by ratioing a near-infrared (NIR) band to a red band. This takes advantage of the high vegetation reflectance in NIR spectral range and high pigment absorption of the red light (Hangiu, 2007). Normalized Difference Vegetation Index (NDVI) is the best indicating factor for plant growth status and the spatial distribution of vegetation, which has a linear relationship with the density of vegetation distribution (Haoxu et al., 2011); the formula is shown as Eq. (6):

$$\text{NDVI} = \frac{\text{NIR} - \text{RED}}{\text{NIR} + \text{RED}}, \qquad (6)$$

where NIR presents near-infrared wavelength and RED represents red wavelength. They belong to bands of the Landsat-8 OLI/TIRS and respectively represented the fifth and fourth band.

Once the NDVI was finalized, a threshold of 0.45 was selected as most appropriate for the extraction of high vegetation locations.

### 3.4.2 MNDWI – derived water image

As the study area is crossed by several rivers and distributed with some reservoirs and small lakes, in order to extract surface water, the modified normalized difference water index (MNDWI) was adopted (Haoxu et al., 2011). The formula is as follow:

$$\text{MNDWI} = \frac{\text{Green} - \text{MIR}}{\text{Green} + \text{MIR}}, \qquad (7)$$

where Green represents the green wavelength, MIR represents the middle-infrared wavelength, and they belong to bands of the Landsat-8 OLI/TIRS and respectively represented the third and sixth band.

Based on the ground survey data and hence the information about the known eater body location, a threshold of 0.2 was selected as most appropriate for the extraction of surface water.

### 3.4.3 NDBI – derived built-up image

The built-up land image was produced using the normalized difference building index (NDBI), which takes advantage of the unique spectral response of the built-up lands that have a higher reflectance in MIR wavelength range than in NIR wavelength range (Zha et al., 2003); the formula is shown as Eq. (5):

$$\text{NDBI} = \frac{\text{MIR} - \text{NIR}}{\text{MIR} + \text{NIR}}, \qquad (8)$$

where MIR represents the middle-infrared wavelength, NIR the near-infrared wavelength, and they belong to bands of the Landsat-8 OLI/TIRS and respectively represented the fifth and sixth band. However, the resulted index map found that many vegetated areas have positive NDBI values and in some circumstances, water bodies can also reflect MIR stronger than NIR. Consequently, the contrast of the NDBI images is not as good as NDVI and MNDWI images, because many pixels of vegetation and water areas having positive NDBI values show medium gray tones and are presented as noise mixed with built-up features. Some studies address similar problems (Hangiu, 2007; Zha et al., 2003) with low accuracy in the final extraction of NDBI. These suggest that the urban built-up land features could not be extracted merely based on a NDBI image. In this study, a combination of NDBI with NDVI and MNDWI is used to extract urban built-up land features. This combination can remove the vegetation and water noise, and hence improve the extraction accuracy.

The method that used to extract built-up land features based in an "if-the-else" logic calculation through a band spectral signature analysis (Hangiu, 2007). A new image data

**Table 1.** Constraint factors of the case study.

|  | Constraint factors | Type |
|---|---|---|
|  | *The solar park must not be within:* |  |
| $C_1$ | 50 m from primary and secondary roads | Social impact |
| $C_2$ | High vegetation | Environmental/technical |
| $C_3$ | 200 m from NATURA 2000 areas | Environmental |
| $C_4$ | 200 m from national forest | Environmental |
| $C_5$ | 200 m from urban zones | Social |
| $C_6$ | 100 m from surface waters | Environmental |
| $C_7$ | 2000 m from airport | Safety |
| $C_8$ | 200 m from archaeological sites | Social impact |
| $C_9$ | 200 m from shoreline | Social impact |
| $C_{10}$ | Areas with aspect: east ǀ west ǀ north ǀ northeast ǀ northwest | Technical |

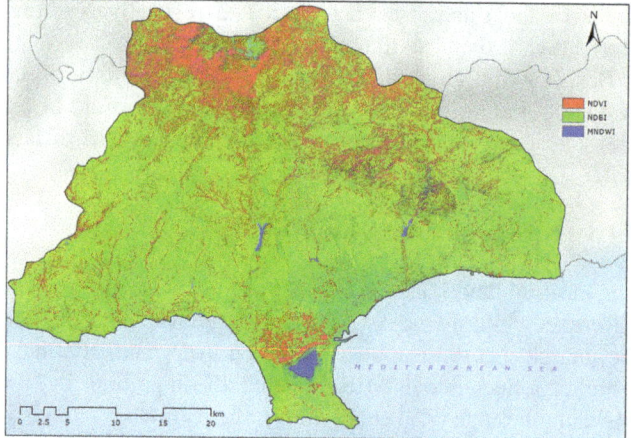

**Figure 3.** Results from Landsat 8 OLI/TIRS image classification in NDVI, NDBI, and MNDWI.

set was created, which used NDVI (Band1 – RED), NDBI (Band2 – GREEN) and MNDWI (Band3 – BLUE) images as three bands and the new classification image is presented in Fig. 3. A simple rule-based logic tree is used to segment urban built-up lands from non-urban built-up features. Examining the signatures of the three new bands found that there are no major differences between means of NDVI and NDBI that might cause confusion between built-up land and vegetation classes. Therefore, the logic calculation that is used to assist in the extraction is as follows:

If BAND 1 <0.15 and BAND 2 >BAND 3 then 1 Else 0

The maximum of built-up land class in Band 1 (NDVI) is 0.15, whereas the minimum of vegetation class in that band is 0.45. Therefore, using 0.15 as a threshold value can help avoid the confusion between vegetation and built-up land classes and greatly increase the extraction accuracy.

To compare the extraction accuracy, the extracted data of built-up areas, high vegetation, and surface waters is checked by a reference map. A GeoEye Ikonos with finer spatial res-

olution provided as base map in ArcGIS™ was used as a reference data set from which the extraction results were compared. A random sampling method was used to visually check the classification results against the higher-resolution satellite image.

### 3.5 Establishment of constraints factors

The constraint factors that were used are presented in Table 1 and comprise of environmental, safety, social (in terms of pressure in society), and technical parameters. A binary GIS mask is created for each constraint, with cells falling within a constrained area assigned "0" and the rest of them assigned "1".

The constraints $C_3$, $C_4$, $C_5$, $C_8$, and $C_9$ are according the national legislation, while the $C_{10}$ is set by experts to exploit the best performance of a solar panel that derives from areas with aspects south, southeast, and southwest. The $C_1$ and $C_7$ are set in way to avoid any reflections from the solar park in these directions and, finally, $C_2$ and $C_6$ are set by researchers under environmental and technical concern, respectively.

The constraint factors exclude 17 % (227 km$^2$) of the district area.

### 3.6 Establishment of evaluation criteria and normalization

The evaluation criteria that score the potential sites are based mainly on financial parameters as presented in Table 2. After the evaluation criteria were determined and assessed, they were normalized through cost functions in a scale from 0 to 100 in order to allow for direct comparability, with 100 representing the most desired value (low cost) and 0 representing the most undesired value (high cost). Some standardize evaluation layers were calculated using an inverse distance cost function, i.e., main roads and distance from them. Figure 4 presents each standardized evaluation layer. This research focuses on developing a workbench GIS model for sitting solar parks and as such does not focus in detail on the cost functions themselves. It should be noted that once the

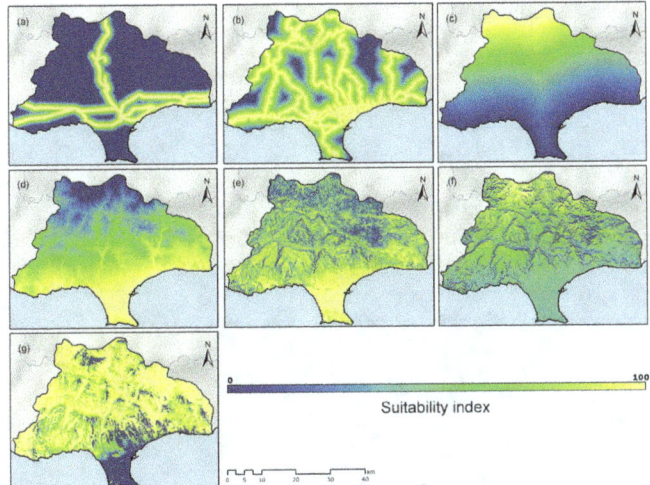

**Figure 4.** Standardized evaluation layers **(a)** electricity grid, **(b)** road network, **(c)** land value, **(d)** elevation, **(e)** slope, **(f)** solar energy, and **(g)** viewshed from primary roads.

**Table 2.** Evaluation criteria of the case study.

| | Evaluation criteria | Type |
|---|---|---|
| $E_1$ | Elevation | Technical |
| $E_2$ | Slope < 45° | Technical |
| $E_3$ | Viewshed from primary roads | Social |
| $E_4$ | Land value | Financial |
| $E_5$ | Distance from road network < 2500 m | Financial/technical |
| $E_6$ | Distance from electricity grid < 2000 m | Financial/technical |
| $E_7$ | Solar radiation > mean radiation of the area | Technical |

GIS model has been established the cost functions and the weighting schemes can be easily adapted to support a more precise and detailed cost-function scheme. It should be noted here, that if the suggested method is to be used for financial investment analysis, then each of these cost functions can be further adjusted and updated to reflect local and contemporary financial practice. Further analysis of the cost functions will not be further analyzed here, as this is not the scope of the paper.

In technical terms, very steep slopes of land are not suitable for solar park installation. For that reason, land slopes greater than 45° were excluded while the remaining got grading values of 0 to 100. In addition, high altitude areas have higher transportation cost and are not preferable. Finally, solar radiation values greater than the mean value of the study area were taken into consideration, getting grading values from 0 to 100.

On the other hand, in financial terms, the distance from road network and electricity grid increase the investment cost since additional infrastructure is necessary. In that way, areas farther than 2500 m from the road network and 2000 m from the electricity grid are considered as not economically viable and are assigned the value of "0". Finally, the land value is strongly correlated with the distance from the shoreline, as seaside areas cost more and are therefore not affordable for such installations. Finally, in social terms, the visibility of potential sites from primary roads was taken into consideration with grading from 0 to 100, where "0" presents high observation frequency and "100" zero visibility.

### 3.7 Rationale for weights in AHP

The AHP method is used to assign weight to the criteria as not all of them are equally important. The pairwise weight matrix for the calculation of the overall weights of the evaluation criteria is created (Table 3), and the priority weights estimated (Table 3). The AHP parameters are also shown, indicating that the original judgments are consistent.

The rationale behind the particular criteria weighting, is highlighted in the following.

- The solar radiation is considered to be the most important criterion since it determines the output of the solar park.

- The distance from electricity grid (EAC) and from roads follow, as they determine the final cost of installation.

- The slope and elevation pose are technical criteria that might increase the investment.

- The land value is thought to be less significant as it has only to do with the cost of the land that will host the solar park.

- Finally, the viewshed from primary roads is placed last, as the social concern is considered less significant compared to the other criteria.

### 4 Results

#### 4.1 Suitable areas suggestions

The suitability index map (Fig. 5c) is derived from the multiplication of the binary constraint map (Fig. 5a) with the evaluation map (Fig. 5b), hence totally removing the restricted areas from the evaluation map. The most appropriate areas for solar park installation are those shown in light yellow, with a suitability index of 70–80. Nevertheless, there are no best-ranked sites (with score 100) in the study area, showing that there are no sites with best grades in all criteria. It is also noticeable that most of the study area (83 %) is restricted from solar park installation, while only a considerably small percentage (1 %) of the area achieved a suitability index of 80, even though solar energy is favorable in more areas. Finally, the distribution of the suitability index pixels presented in Fig. 5d, shows that most of the pixels have values around 60 and few of them with SI > 75.

**Table 3.** Case 2: pair-comparison matrix and relative importance weights in the last column.

| | Viewshed | Land value | EAC | Slope | Solar | Elevation | Roads | Weight |
|---|---|---|---|---|---|---|---|---|
| Viewshed | 1.000 | 1.000 | 0.333 | 0.500 | 0.111 | 0.500 | 1.000 | 0.037 |
| Land Value | 1.000 | 1.000 | 0.500 | 2.000 | 0.143 | 2.000 | 0.500 | 0.078 |
| EAC | 3.000 | 2.000 | 1.000 | 3.000 | 0.143 | 3.000 | 2.000 | 0.133 |
| Slope | 2.000 | 0.500 | 0.333 | 1.000 | 0.111 | 1.000 | 0.333 | 0.051 |
| Solar | 9.000 | 7.000 | 7.000 | 9.000 | 1.000 | 9.000 | 9.000 | 0.545 |
| Elevation | 2.000 | 0.500 | 0.333 | 1.000 | 0.111 | 1.000 | 0.500 | 0.052 |
| Roads | 1.000 | 2.000 | 0.500 | 3.000 | 0.111 | 2.000 | 1.000 | 0.105 |

$CR = 0.071 < 0.1$

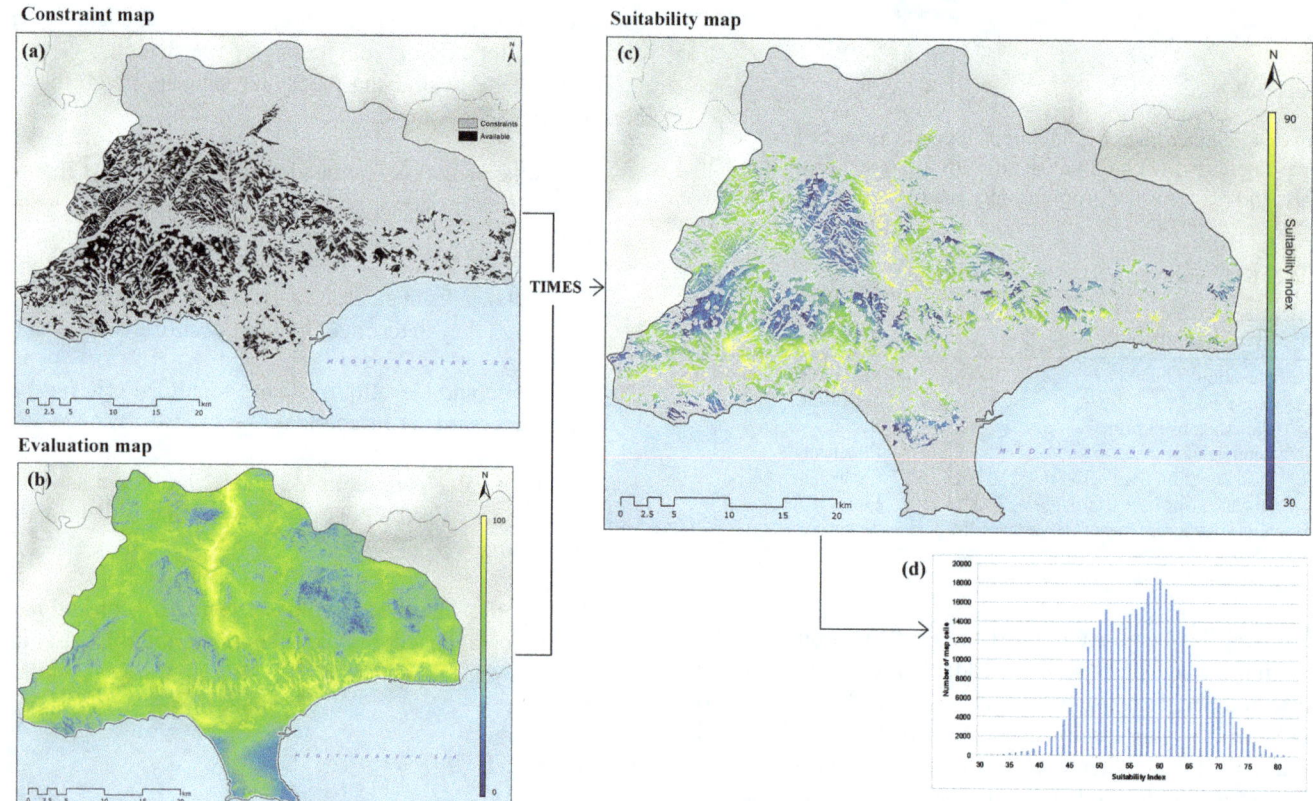

**Figure 5. (a)** Constraint map, **(b)** evaluation map, **(c)** the final suitability index map of study area as derived from the merging of constraint map and evaluation map, and **(d)** distribution of suitability index pixels over the final map.

## 4.2 Sensitivity analysis

In a multi-criteria analysis a "what if", sensitivity analysis is recommended as a means of checking the stability of the results against the subjectivity of the expert judgments. The most common method is to modify the weighting obtain from the experts, while the assumption of equal weighting is also used (Cameron et al., 2012). In this project, the sensitivity analysis performed considers the effect of changes of criteria weights upon the overall suitability index. To that aim, the following four cases were examined.

– Case 1: all criteria have the same weights.

– Case 2: the weight of the criterion "solar energy" has the biggest score and the rest are equally distributed.

– Case 3: the weight of the criteria "solar energy" and "land value" have the biggest score while the rest are distributed equally.

– Case 4: all economical criteria (road network, electricity grid, and land value) have weights equal to zero (0).

The results and statistics information of the four cases are illustrated in Fig. 6. As observed, the present framework is

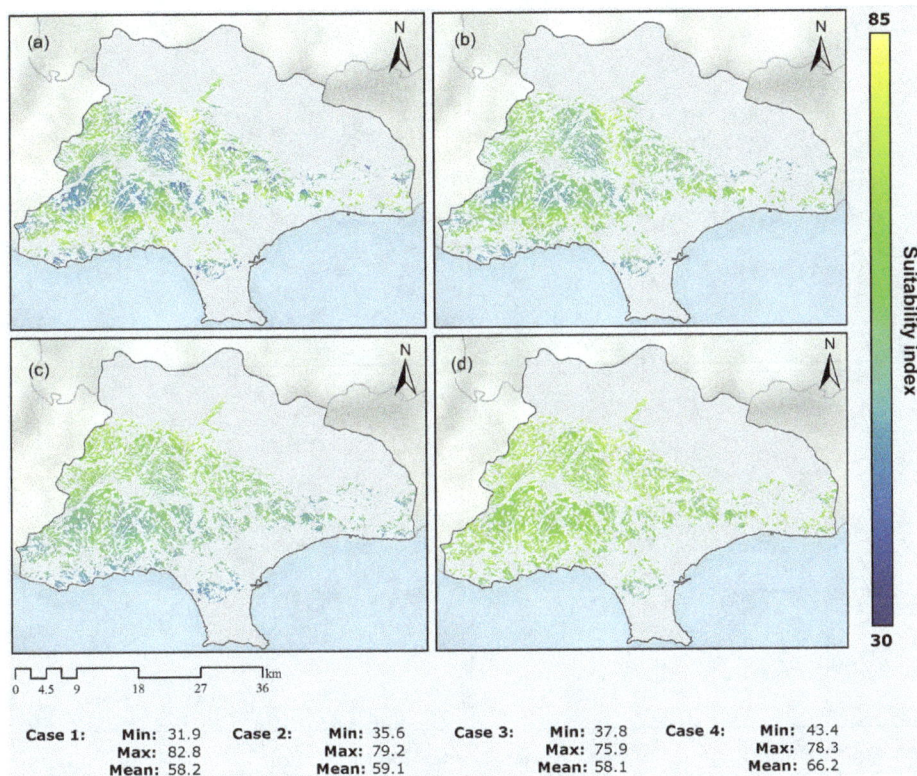

**Figure 6.** Sensitivity analysis results and comparison among different scenarios, which are not differentiating results: (**a**) case 1, (**b**) case 2, (**c**) case 3, and (**d**) case 4.

sensitive to the criteria weights. This was expected since the evaluation criteria are selected with respect to the specific characteristics of the study area. The change of the final suitability map that is derived from the changing of criteria weights implies that each selected criterion is influential in the evaluation of the study area.

Figure 7 presents the results of the sensitivity analysis as classified into five classes in order to understand the variation of suitability index. It is obvious that depending the parameters that the user changes for each scenario, the respective suitability pattern varies over the study area.

It is noticeable that, although the resulting maps for the four cases of the sensitivity analysis show considerable modification in the suitability index, Fig. 8 shows that the number of the most suitable areas (SI > 75) for solar park sitting remains low and in some cases null. In case 4, where no economic criteria are taken into consideration, a noteworthy variation is observed; i.e., the majority of potential sites are classed from 45 to 60 with a few high scoring potential sites. In addition, a noticeable lack of potential sites with SI > 75 observed in case 2 and case 3 with most of the pixels to be concentrated in SI $\sim$ 54 and SI $\sim$ 50, respectively. Finally, only in case 1 are pixels with high values presented, with SI values evenly distributed.

## 5  Discussion

The proposed method has been implemented in Limassol district. The SAW method was used to quantify several features that rule the decision of citing a solar park and create suitability index maps for every feature. The AHP method was used to establish weighting among the different features and merge the suitability index maps into a single evaluation map. Remote sensing classification was used to detect urban areas and water bodies and create a mask of exclusion areas. Additional masks of exclusion areas were created based on legal restrictions. The evaluation map and the exclusion masks were merged to create the final combined index map. Although the design of the proposed method seems complicated, it is straight forward and takes into consideration all possible aspects of a solar park siting. Any aspects that are overlooked can be easily added, and the potential user, may easily vary the weights according to his own priorities and local specifics.

In order to check the method's versatility, sensitivity, and adaptability, different scenarios were utilized to access the different results. As demonstrated (Figs. 6 and 7), the method produces different results upon different inputs; hence, one may adapt it to one's personal requirements.

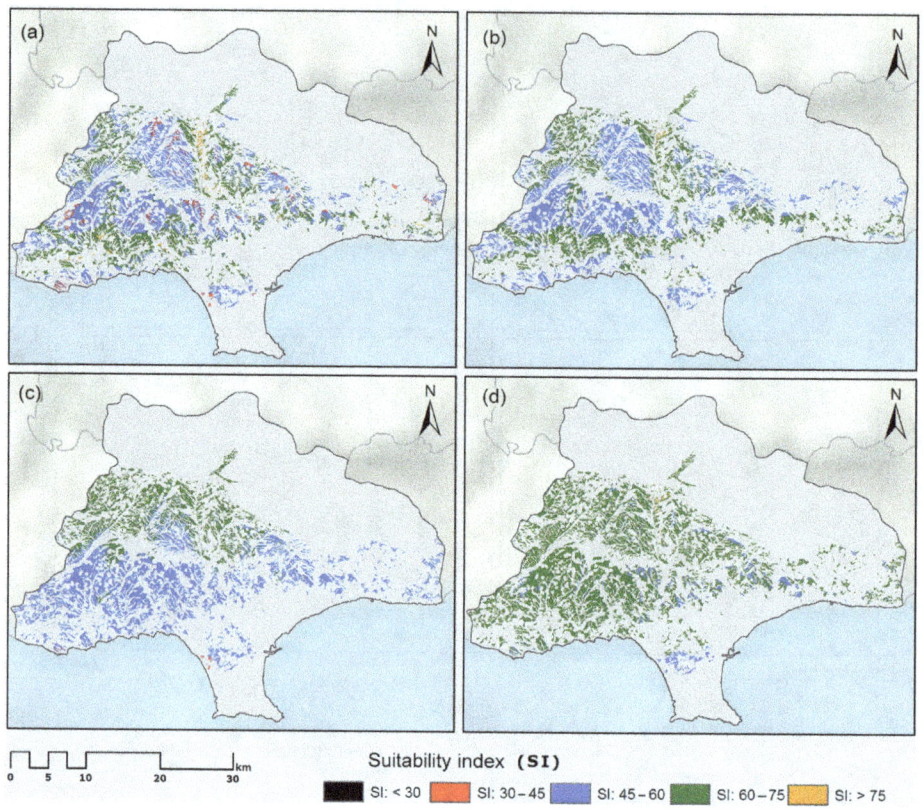

Suitability index **(SI)**

SI: < 30    SI: 30–45    SI: 45–60    SI: 60–75    SI: > 75

**Figure 7.** Classified results of sensitivity analysis and comparison among different scenarios: **(a)** case 1, **(b)** case 2, **(c)** case 3, and **(d)** case 4.

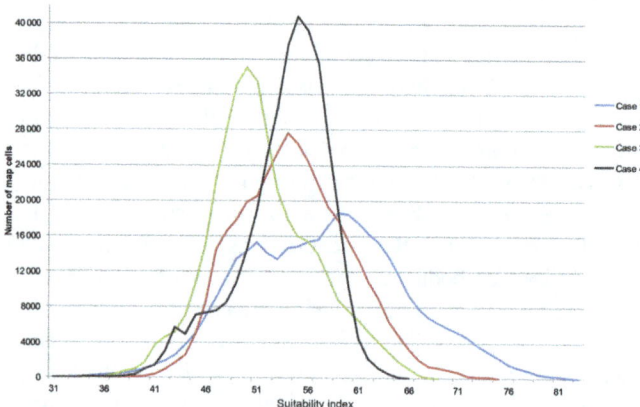

**Figure 8.** Number of map cells for each case of sensitivity analysis.

## 6   Conclusions

This study presents a model, which can be easily used to evaluate large areas for optimal site selection for a solar park. Such a model can be very helpful for potential investors to locate potential sites for solar energy exploitation, before carry out a detailed field survey.

Although there are several implementations of siting models within the GIS framework (Carrion et al., 2008; Tegou et al., 2010; Kontos et al., 2005; Georgiou et al., 2012; Maser et al., 2006; Ramachandra et al., 2007), the combination of MCA, AHP, and a GIS in an integrated platform is not common. In this article, a decision analysis methodological framework for solar energy exploitation and site evaluation is developed and applied in the Limassol district in Cyprus. The framework it is a combination of already existing tools such as multi-criteria analysis, AHP, and integrated site evaluation in a straightforward way. It also combines GIS and remote sensing techniques for spatial analysis, modeling, and visualization. The objective of the paper is to propose a method for solar park installation suitability analysis, taking into account a number of financial, social, environmental, and technical criteria. The pairwise comparison method in the context of the AHP was utilized to assign relative weights to the evaluation criteria, whereas the SAW was method used as a way for aggregating the used criteria, in order to compute the SI for each cell in the study area. A GIS established the spatial dimension of constrains and evaluation criteria and elaborated them for the production of the overall suitability map. A sensitivity analysis on the weights of the evaluation criteria was also performed, showing that each criterion is influential in the evaluation of the suitability of a site.

The results identified promising sites for electricity generation from solar energy, excluding over 80 % of the whole study region. The best score areas (SI > 75) cover only 3.0 % (40.3 km$^2$) of the study area. However, the proposed methodology allows the analyst to consider even less suitable sites, by reducing the acceptable threshold of suitability index. This would result in the identification of more areas as appropriate for solar park development in combination with field inspection. Thus, future work could include the individual assessment of the optimal locations in conjunction with field inspection in order to make the final selection of sites.

The innovative aspect of this work derives from the proposed tool as currently implemented within a local GIS, which provides a versatile platform of analysis and semi-automation of the operations, which might also extended into full automation and has a prospect for future web automation platform. This work derives a holistic approach from criteria selection and evaluation, data gathering and multi-criteria analysis. That makes the tool flexible that encourages several "what if" scenarios to be easily implemented. In addition, an innovator dimension gives the balanced approach among practice, the way that evaluation criteria were used in conjunction with the legislative boundary constrains under a unified multi-criteria decision aiding. Finally, it provides accuracy and precision in less evaluation time, allowing for checking the robustness and stability of the results obtained. For these reasons, it may well be helpful for potential investors in solar park investments and also in other kinds of project sitting, due to the generic nature of the framework. In addition, the proposed GIS model may be further developed with contributions from EAC's experts, in order to become a valuable tool for sitting small, medium, or large solar parks, through adaptation of the basic model presented here.

*Acknowledgements.* The authors would like to thank the public services of Cyprus (Ministry of Agriculture Natural Resources & Environment, Electricity Authority of Cyprus and Geological Survey Department) for offering their data, valuable help, and remarks for this research.

Edited by: L. Eppelbaum

# References

Bahadori, A. and Nwaoha, C.: A review on solar energy utilization in Australia, Renew. Sust. Energ. Rev., 18, 1–5, 2013.

Cameron, D. R., Cohen, B. S., and Morrison, S. A.: An approach to enhance the conservation-compatibility of solar energy development, PloS One, 7, e38437, doi:10.1371/journal.pone.0038437, 2012.

Carrion, A. J., Estrella, E. A., Dols, A. F., Toro, Z. M., Rodriquez, M., and Ridao, R. A.: Environmental decision-support systems for evaluating the carrying capacity of land areas: Optimal site selection for grid-connected photovoltaic power plants, Renew. Sust. Energ. Rev., 12, 2358–2380, 2008.

European Commission: Science for Environment Policy. Future Brief: Wind & Solar energy and nature conservation, Issue 9, December 2014.

Georgiou, A., Polatidis, H., and Haralambopoulos D.: Wind energy resource assessment and development: Decision analysis for site evaluation and application, Energy Sources, Part A: Recovery, Utilization, and Environmental Effects, 34, 1759–1767, 2012.

Hangiu, X.: Extraction of Urban Built-up Land Features from Landsat Imagery Using a Thematic-oriented Index Combination Technique, Photogramm. Eng. Rem. S., 73, 1381–1391, 2007.

Haoxu, L., Yaowen, X., Lin, Y., and Lijun, W.: A Study on the land cover classification of the arid region based on Multi-temporal TM images, ESIAT, 2011.

Hernandez, R. R., Easter, S. B., Murphy-Mariscal, M. L., Maestre, F. T., Tavassoli, M., Allen, E. B., Barrows, C. W., Belnap, J., Ochoa-Hueso, R., Ravi, S., and Allen, M. F.: Environmental impacts of utility-scale solar energy, Renew. Sust. Energ. Rev., 29, 766–779, 2014.

Katsaprakakis, D. A.: A review of the environmental and human impacts from wind parks. A case study for the Prefecture of Lasithi, Crete, Renew. Sust. Energ. Rev., 16, 2850–2863, 2012.

Katzner, T., Johnson, J. A., Evans, D. M., Garner, T. W. J., Gompper, M. E., Altwegg, R., Branch, T. A., Gordon, I. J., and Pettorelli, N.: Challenges and opportunities for animal conservation from renewable energy development, Anim. Conserv., 16, 367–369, 2013.

Kontos, Th., Komilis, D., and Halvadakis, K.: Sitting MSW landfills with a spatial multiple criteria analysis methodology, Waste Manage., 25, 818–832, 2005.

Law 29(I)/2005: Town and Country planning, Cypriot House of Representatives, 2005.

Mari, R., Bottai, L., Busillo, C., Calastrini, F., Gozzini, B., and Gualtieri, G.: A GIS-based interactive web decision support system for planning wind farms in Tuscany (Italy), Renew. Energ., 36, 754–763, 2011.

Masera, O., Ghilardi, A., Drigo, R., and Trossero, M. A.: WISDOM: A GIS-based supply demand mapping tool for woodfuel management, Biomass and Bio-energy, 30, 618–637, 2006.

Maxoulis, N. C. and Kalogirou, A. S.: Cyprus energy policy: The road to the 2006 world renewable energy congress trophy, Renewable Energ., 33, 355–365, 2008.

Pilavachi, P. A., Kalamalikas, N. G., Kakouris, M. K., Kakaras, E., and Giannakopoulos, D.: The energy policy of the Republic of Cyprus, Energy, 34, 547–554, 2009.

Ramachandra, T. V. and Shruthi, B. V.: Spatial mapping of renewable energy potential, Renew. Sust. Energ. Rev., 11, 1460–1480, 2007.

Saaty, T.: The Analytic Hierarchy Process, McGraw – Hill, New York, 1980.

Saaty, W. R.: The Analytic Hierarchy Process – What it is and how it is used, Math Modelling, 9, 161–176, 1987.

Sanchez-Lozano, M. J., Teruel-Solano, J., Soto-Elvira, L. P., and Garcia-Cascales, S. M.: Geographical Information Systems (GIS) and Multi-Criteria Decision Making (MCDM) methods for the evaluation of solar farms locations: Case study in southeastern Spain, Renew. Sust. Energ. Rev., 24, 544–556, 2013.

Solangi, K. H., Islam, M. R., Saidur, R., Rahim, N. A., and Fayaz, H.: A review on global solar energy policy, Renew. Sust. Energ. Rev., 15, 2149–2163, 2011.

Tegou, L. I., Polatidis, H., and Haralambopoulos, D.: Environmental management framework for wind farm sitting: Methodology and case study, J. Environ. Manage., 91, 2134–2147, 2010.

Torres-Sibille, A. C., Cloquell-Ballester, V. A., Cloquell-Ballester, V. A., and Ramirez, M. A. A.: Aesthetic impact assessment of solar power plants: An objective and a subjective approach, Renew. Sust. Energ. Rev., 13, 986–999, 2009.

Tsoutsos, T., Frantzeskaki, N., and Gekas, V.: Environmental impacts from the solar energy technologies, Energ. Policy, 33, 289–296, 2005.

Turney, M. and Ftenakis V.: Environmental impacts from the installation and operation of large-scale solar power plants, Renew. Sust. Energ. Rev., 15, 3261–3270, 2011.

Zanter, K.: LANDSAT 8 (L8) Data users handbook, USGS, 17–65, 2016.

Zha, Y., Gao, J., and Ni, S.: Use the normalized difference built-up index in automatically mapping urban areas from TM imagery, Int. J. Remote Sens., 24, 583–594, 2003.

# 12

# In situ vector calibration of magnetic observatories

**Alexandre Gonsette, Jean Rasson, and François Humbled**

Centre de Physique du Globe, Royal Meteorological Institute, 5670 Dourbes, Belgium

*Correspondence to:* Alexandre Gonsette (agonsett@meteo.be)

**Abstract.** The goal of magnetic observatories is to measure and provide a vector magnetic field in a geodetic coordinate system. For that purpose, instrument set-up and calibration are crucial. In particular, the scale factor and orientation of a vector magnetometer may affect the magnetic field measurement. Here, we highlight the baseline concept and demonstrate that it is essential for data quality control. We show how the baselines can highlight a possible calibration error. We also provide a calibration method based on high-frequency "absolute measurements". This method determines a transformation matrix for correcting variometer data suffering from scale factor and orientation errors. We finally present a practical case where recovered data have been successfully compared to those coming from a reference magnetometer.

## 1 Introduction

Most magnetic observatories are built according to a standardized or universally adopted scheme (Jankowski and Sucksdorff, 1996) including at least a set of three major instruments: a variometer, an absolute scalar magnetometer, and a declination and inclination flux instrument (DI-flux instrument). The different data streams are combined to build a unique vector of magnetic field data. The variometer is a vector magnetometer, which records variations of the magnetic field components at a regular interval (e.g. at 1 Hz). However, this is not an absolute instrument. In particular, reference directions, the vertical and geographical north, are not available. They usually work as near-zero sensors, so that an offset must be added to the relative value of each component in order to adjust it and therefore determine the complete vector. Those offsets or baselines should be as constant as possible but may drift more or less depending on the en-

vironment stability and device quality. For instance, thermal variations may affect the pillar stability. A baseline can also suffer from sudden variation due to an instrumental effect after a (unwanted) motion like a shock due to maintenance staff or a change in the surrounding environment (Fig. 1). A regular determination of the baselines is thus necessary to take their change into account. This is the main goal of the well-known "absolute measurements" that are carried out by the two other instruments.

First, a scalar magnetometer records the intensity of the field $|\boldsymbol{B}|$. Most of the time, a proton precession or an Overhauser magnetometer is used for this task. Overhauser magnetometer exploits the fact that protons perform precession at a frequency proportional to the magnetic field according to

$$\omega_{\text{precession}} = \gamma \, \|\boldsymbol{B}\|, \tag{1}$$

where $\gamma$, the gyromagnetic ratio, is a fundamental physical constant (Mohr et al., 2016). Therefore, this magnetometer can be considered an absolute instrument.

The last instrument serves to determine the magnetic field orientation according to reference direction. Magnetic declination is the angle between true north and the magnetic field in a horizontal plane, and the inclination is the angle between the horizontal plane and the field. In a conventional observatory, a DI-flux instrument (non-magnetic theodolite-embedding single-axis magnetic sensor) is manipulated by an observer according to a particular procedure (Kerridge, 1988) taking about 15 min per measurement. This instrument is also considered absolute because angles are measured according to geodetic reference directions. Due to this man-power dependency, the frequency of absolute measurements does not exceed once per day (St Louis, 2012). However, new automatic devices such as AutoDIF (automatic DI-flux instrument; Gonsette et al., 2012) close the loop by automa-

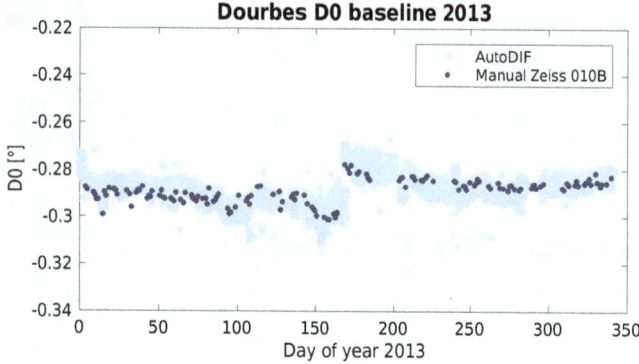

**Figure 1.** Baseline example computed from conventional manual measurements (dark blue) and the automatic system (light blue). In mid-2013, a baseline jump corresponding to an instrumental effect occurred, proving that regular absolute measurement are crucial.

tising the DI-flux measurements procedure. Moreover, AutoDIF is able to increase the frequency of baseline determination by performing several measurements per day.

After collecting synchronised data from the three instruments, baselines are computed by using the relation for the Cartesian coordinate system:

$$\begin{bmatrix} X_0(t) \\ Y_0(t) \\ Z_0(t) \end{bmatrix} = \begin{bmatrix} X(t) \\ Y(t) \\ Z(t) \end{bmatrix} - \begin{bmatrix} \delta X(t) \\ \delta Y(t) \\ \delta Z(t) \end{bmatrix}, \tag{2}$$

where $X(t)$, $Y(t)$ and $Z(t)$ are, at the time $t$, the tree conventional components of the field, pointing to the geographic north, eastward and downward respectively. The "0" index refers to the baseline spot measurements, while $\delta$ refers to the variometer data. The full baseline measurement protocol including a set of four absolute declinations and four absolute inclinations (even if only two are required for determining all the unknowns) can be found in the literature; this can also be found for spherical and cylindrical configurations (Rasson, 2005). The need for eight (at least six) measurements is justified by the DI-flux sensor offset and misalignment. A baseline function is then applied on these measurements using various methods such as a least-squares polynomial or spline approximation. Finally, the vector field is constructed by adding the variometer values to the adopted baselines.

Equation (2) assumes a variometer properly set up with the $Z$ axis vertical and the $X$ axis pointing toward the geographic north. The scale factor of each component is also assumed to be perfect.

A correct orientation is usually ensured by paying attention during the set-up step, but its stability in time is not always evident. Permafrost areas are examples of drifting regions (Eckstaller et al., 2007) where variometer orientation is not guaranteed. If the orthogonality errors are neglected,

the problem of calibration can be expressed as follows:

$$\begin{bmatrix} X \\ Y \\ Z \end{bmatrix} = \mathbf{R}_z(\gamma)\,\mathbf{R}_y(\beta)\mathbf{R}_x(\alpha) \begin{bmatrix} k_1 & 0 & 0 \\ 0 & k_2 & 0 \\ 0 & 0 & k_3 \end{bmatrix} \begin{bmatrix} \delta U \\ \delta V \\ \delta W \end{bmatrix}$$
$$+ \begin{bmatrix} X_0 \\ Y_0 \\ Z_0 \end{bmatrix}, \tag{3}$$

where the $\mathbf{R}_{x,y,z}$ are an elementary rotation matrix and the $k_i$ variables are the scale factors for each component. $U$, $V$ and $W$ are the three variometers output into the sensors reference frame. Calibration procedures can be divided into two categories. On one hand, the scalar calibration compares scalar values computed from the vector magnetometer to absolute scalar values. This technique is exploited by satellites because the vector reference field is not available. Nevertheless, instruments are orbiting around the Earth (Olsen et al., 2003). The different scalar measurements from the scalar instrument can therefore be compared to the scalar values computed from the vector instrument. On the other hand, the vector calibration directly compares vector magnetometer measurements to the reference vector value. Marusenkov et al. (2011) used a second variometer already calibrated as the reference. Previously, Jankowski and Sucksdorff (1996) proposed a comparison between the variometer data and the absolute measurements performed during disturbed days in order to calibrate the observatory. The development was made for small angle errors (no more than 1–2°), but Jankowski and Sucksdorff (1996) suggested that the method could remain valid for any angle. Jankowski and Sucksdorff (1996) also pointed out the difficulty in getting sufficiently strong magnetic activity at low latitudes. The method presented in this paper is relatively close to the latter, except for the fact that the automatic DI-flux instrument can generate a large number of absolute measurements within a short time (e.g. 48 absolute measurements every 24 h), leading to a fast automatic calibration process also at low latitude or during significantly magnetic periods.

The method presented in this document is related to a variometer in XYZ configuration. However, other configurations may also be considered. For instance, many observatories set up their magnetometers in an HDZ configuration, where $H$ is the direction of the magnetic north, $D$ the declination and $Z$ the vertical component. Working directly with the $D$ component would lead to non-linear equations. Nevertheless, most modern variometers are based on fluxgate sensor technology. Thus, the recorded signal is the orthogonal projection of the field along the fluxgate sensitive axis. The residue ($\delta E$) expressed in nT (nanotesla) can be used like any geographic component and converted afterward into a declination value according to

$$\delta D = \frac{180}{\pi}\,\mathrm{asin}\left(\frac{\delta E}{H}\right). \tag{4}$$

The same approach can be used for a DFI magnetometer. However, the reader should keep in mind that not all variometer axes might have a compensating coil allowing them to work in the entire field. Indeed, recording the $D$ and $I$ variations is similar to a DI-flux process. The sensor is quasi-perpendicular to the field so that the residues are close to zero. The recorded signal could rapidly saturate.

## 2  Calibration error detection

Before solving the calibration problem, it could be useful to give some clues for detecting required adjustments. Indeed, it is difficult, when only examining definitive data, to detect a few nanotesla errors in daily amplitude. Direct comparison with other observatories requires them to be close enough while many observatories cannot afford to buy an auxiliary variometer. Fortunately, baselines are useful tools for checking data. As described below, they are affected by calibration errors, and, if they are measured with a sufficiently high frequency, particular errors can be highlighted.

### 2.1  Scale factor error

Let us consider an observatory working with a variometer, such as a LEMI-025, in a Cartesian coordinate system. Each sensor converts a real magnetic signal expressed in nanotesla into a more suitable format (usually a voltage). This converted signal passes through an ADC providing, in turn, a digital representation of the initial signal. A scale factor is then used to convert the true signal into a digitised signal. Consider the $X$ component:

$$\delta X_{\text{voltage}} = k_1 \, \delta X_{\text{real}}, \tag{5}$$

$$\delta X_{\text{digital}} = k_2 \delta X_{\text{voltage}} = k \, \delta X_{\text{real}}, \tag{6}$$

where $\delta X_{\text{real}}$ is the real magnetic variation in nT toward the $X$ direction, $k_1$ is a scale factor in volt/nT converting the magnetic field signal into an electrical signal, $\delta X_{\text{voltage}}$ is the image of the field signal expressed in volts, $k_2$ is a scale factor in nT/volt converting the electric signal into a digital value, $k = k_2 k_1$ is the dimensionless scale factor converting the real magnetic signal into its digital representation. $k$ should be as close to 1 as possible.

Supposing now a difference between the digital and real variation of a component resulting from a badly calibrated scale factor, the baseline measurement will be affected by this error:

$$\begin{bmatrix} X_0^*(t) \\ Y_0^*(t) \\ Z_0^*(t) \end{bmatrix} = \begin{bmatrix} X(t) \\ Y(t) \\ Z(t) \end{bmatrix} - \begin{bmatrix} k_x & 0 & 0 \\ 0 & k_y & 0 \\ 0 & 0 & k_z \end{bmatrix} \begin{bmatrix} \delta X(t) \\ \delta Y(t) \\ \delta Z(t) \end{bmatrix}, \tag{7}$$

$$\begin{bmatrix} X_0^*(t) \\ Y_0^*(t) \\ Z_0^*(t) \end{bmatrix} = \begin{bmatrix} X_0(t) \\ Y_0(t) \\ Z_0(t) \end{bmatrix} + \begin{bmatrix} (1-k_X) & 0 & 0 \\ 0 & (1-k_Y) & 0 \\ 0 & 0 & (1-k_Z) \end{bmatrix}$$
$$\begin{bmatrix} \delta X(t) \\ \delta Y(t) \\ \delta Z(t) \end{bmatrix}. \tag{8}$$

The (*) symbol denotes the erroneous baseline affected by a scale factor error. The baseline then varies with respect to its corresponding variometer component value, meaning that a correlation exists between both.

The scale factor is usually factory calibrated and should be stable over time. It is certainly true but there are many situations for which the scale factor is not known exactly (e.g. a homemade instrument) or differs from its factory value (e.g. a repair after a lightning strike may affect the instrument parameters). The impact of a scale factor error also depends on the magnitude of the magnetic activity. A 1 % error for the $H$ component scale factor at mid-latitude would lead to no more than 0.5 nT during quiet days. On the other hand, the same percent error at high latitude during a stormy day may affect the data by several nT.

### 2.2  Orientation error

Now, let us consider once again the same XYZ variometer but this time presenting an orientation error. That could be due, for instance, to a levelling error caused by a bad set-up or an unstable basement and/or an $X$ axis pointing to any other direction than the conventional one. The given components are affected by this orientation error and do not correspond to the expected ones. This is the reason why instruments such as ASMO (Alldregde, 1960) or any other three-axis magnetometers will never be considered as a full magnetic observatory.

Rasson (2005) treated the simplified case of a rotation $\theta$ around the $Z$ axis. The orthogonality between components was assumed to be perfect. In that particular case, the relative real values at time $t$ are given by

$$\begin{bmatrix} \delta X(t) \\ \delta Y(t) \\ \delta Z(t) \end{bmatrix} = \begin{bmatrix} \cos(\theta) & -\sin(\theta) & 0 \\ \sin(\theta) & \cos(\theta) & 0 \\ 0 & 0 & 1 \end{bmatrix} \begin{bmatrix} \delta U(t) \\ \delta V(t) \\ \delta W(t) \end{bmatrix}. \tag{9}$$

The $X_0$ baseline, for instance, should be computed as

$$X_0 = X(t) - \cos(\theta)\delta U(t) + \sin(\theta)\delta V(t). \tag{10}$$

If no correction is applied, the observed baseline gives the following form:

$$X_0^* = X_0 - (1 - \cos(\theta))\delta U(t) - \sin(\theta)\delta V(t). \tag{11}$$

In this case, a correlation exists between the baseline and another relative component. Figure 2 shows an example of a variometer rotated around its vertical axis by 1.7°. The high-resolution baseline (blue) computed by means of an automatic DI-flux instrument presents the same trend as the $\delta Y$ (red) component. The peak–peak amplitude is more than 2 nT.

The general case is much more complex in particular if the orientation error is combined with a significant scale factor error. Indeed, the term $(1 - \cos(\theta))$ in Eq. (11) may be interpreted either as a scale factor error or as an orientation error.

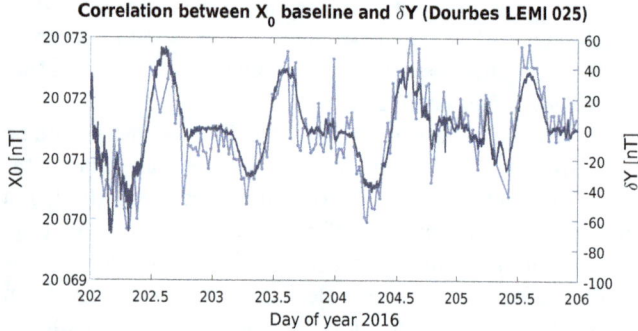

**Figure 2.** Light blue: $X_0$ baseline computed from high-frequency absolute measurements. Dark blue: variometer $Y$ component from the LEMI-025. Because the variometer is not properly oriented, a strong correlation appears between $X_0$ and $Y$.

## 3  Calibration process

Absolute measurements, before giving baselines, provide absolute or spot values of the magnetic field. When performed with a sufficiently high frequency (e.g. once per hour), the generated magnetogram can be compared to the variometer value. Therefore, a vector calibration can be done as if a reference variometer was available.

A DI-flux instrument, either a manual system such as a Zeiss 010-B or an automatic system like the AutoDIF, is affected by the sensor offset and misalignments errors. A single spot measurement is therefore computed from a set of four declination (index 1 to 4) and four inclination (index 5 to 8) records. The eight synchronised variometer values as well as the eight scalar measurements are averaged. Thus, each spot value and corresponding variometer value is computed as follows:

$$X_m = \frac{\sum F_i}{8} \cos\left(\frac{I_5 + I_6 + I_7 + I_8}{4}\right)$$
$$\cos\left(\frac{D_1 + D_2 + D_3 + D_4}{4}\right), \tag{12}$$

$$Y_m = \frac{\sum F_i}{8} \cos\left(\frac{I_5 + I_6 + I_7 + I_8}{4}\right)$$
$$\sin\left(\frac{D_1 + D_2 + D_3 + D_4}{4}\right), \tag{13}$$

$$Z_m = \frac{\sum F_i}{8} \sin\left(\frac{I_5 + I_6 + I_7 + I_8}{4}\right), \tag{14}$$

$$\begin{bmatrix} \delta U_m \\ \delta V_m \\ \delta W_m \end{bmatrix} = \frac{1}{8} \begin{bmatrix} \sum \delta U_i \\ \sum \delta V_i \\ \sum \delta W_i \end{bmatrix}, \tag{15}$$

where, the "$i$" index refers to the records 1 to 8 synchronised with the four declinations and the four inclinations.

Let us consider a series of $n$ samples built from Eqs. (12)–(15). The general case, including orthogonality errors, can be

expressed by rewriting Eq. (3) as follows:

$$\begin{bmatrix} X_m \\ Y_m \\ Z_m \end{bmatrix}^{\mathbf{T}} = \begin{bmatrix} a & b & c \\ d & e & f \\ g & h & i \end{bmatrix} \begin{bmatrix} \delta U_m \\ \delta V_m \\ \delta W_m \end{bmatrix}^{\mathbf{T}} + \begin{bmatrix} X_0 \\ Y_0 \\ Z_0 \end{bmatrix}, \tag{16}$$

where $X_m = [X_{m1}, \ldots, X_{mn}]^T$, $Y = [Y_{m1}, \ldots, Y_{mn}]^T$, $Z = [Z_{m1}, \ldots, Z_{mn}]^T$ are the time series of $X$, $Y$ and $Z$ spot values recorded by means of the absolute instruments and $\delta U = [\delta U_{m1}, \ldots, \delta U_{mn}]^T$, $\delta V = [\delta V_{m1}, \ldots, \delta V_{mn}]^T$, $\delta W = [\delta W_{m1}, \ldots, \delta W_{mn}]^T$ are the three component time series of the variometer. Because the period of acquisition is relatively small (a few days is enough), the baseline values $X_0$, $Y_0$, $Z_0$ are assumed to be constant. For each component $X$, $Y$ and $Z$, the problem consists of solving a linear system, where a time series of spot values and the quasi-synchronised three variometer components are the input. Assuming the system to be overdetermined, the latter is solved in the least-squares sense. Equation (17) gives the coefficients corresponding to the $X$ component (others are similar):

$$\begin{bmatrix} a \\ b \\ c \\ X_0 \end{bmatrix} = \left(\mathbf{A^T A}\right)^{-1} \mathbf{A^T X}, \tag{17}$$

where $\mathbf{A} = \begin{bmatrix} \delta \mathbf{U} & \delta \mathbf{V} & \delta \mathbf{W} & \mathbf{1} \end{bmatrix}$. Once the whole coefficients matrix is determined, the variometer data are redressed:

$$\begin{bmatrix} \delta X \\ \delta Y \\ \delta Z \end{bmatrix}^T = \begin{bmatrix} a & b & c \\ d & e & f \\ g & h & i \end{bmatrix} \begin{bmatrix} \delta U \\ \delta V \\ \delta W \end{bmatrix}^T. \tag{18}$$

Equation (18) refers to all variometer data and not only the averaged data obtained from Eq. (15). For each set of absolute measurements, the three corrected baselines can be processed in a conventional way. Considering an XYZ variometer, the $Z_0$ baseline is first computed and then $X_0$ and $Y_0$ are computed:

$$Z_0 = \frac{F_5 + F_6 + F_7 + F_8}{4} \sin\left(\frac{I_5 + I_6 + I_7 + I_8}{4}\right)$$
$$- \frac{\delta Z_5 + \delta Z_6 + \delta Z_7 + \delta Z_8}{4}, \tag{19}$$

$$H_i = \sqrt{F_i^2 - (Z_0 + \delta Z_i)^2}, \tag{20}$$

$$X_0 = \frac{H_1 + H_2 + H_3 + H_4}{4} \cos\left(\frac{D_1 + D_2 + D_3 + D_4}{4}\right)$$
$$- \frac{\delta X_1 + \delta X_2 + \delta X_3 + \delta X_4}{4}, \tag{21}$$

$$Y_0 = \frac{H_1 + H_2 + H_3 + H_4}{4} \sin\left(\frac{D_1 + D_2 + D_3 + D_4}{4}\right)$$
$$- \frac{\delta Y_1 + \delta Y_2 + \delta Y_3 + \delta Y_4}{4}. \tag{22}$$

A function (polynomial, cubic-spline, etc.) is then fitted on them. Finally, the magnetic vector is built according to Eq. (2).

**Figure 3.** LEMI-025 installed in the Dourbes magnetic observatory. The red arrow indicates the true north direction. The orange arrows highlight the bubble-levels saturation.

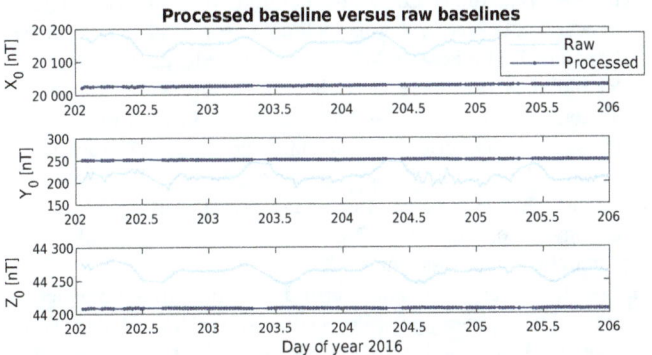

**Figure 4.** LEMi-025 baselines. Light blue: before processing. Dark blue: after processing.

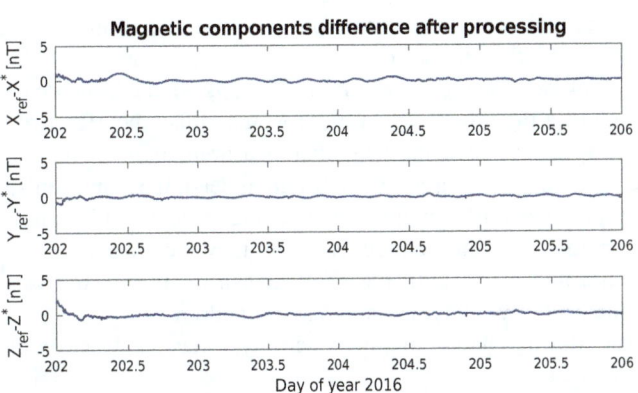

**Figure 5.** Variometer difference between a reference variometer and the case study variometer. The value are clearly within 1 nT.

## 4 Case study

A LEMI-025 variometer has been installed in the Dourbes magnetic observatory. The device has deliberately been set up in a non-conventional orientation as shown in Fig. 3. The levelling and orientation error have been strongly exaggerated compared to those encountered in conventional observatories, but, if we consider a possible future automatic deployment using systems such as a GyroDIF (Gonsette et al., 2017), the orientation could be completely random. An AutoDIF installed in the Dourbes absolute house has been used for performing absolute declination and inclination measurements because of its high-frequency measurement capability. An Overhauser magnetometer recorded the magnetic field intensity at the same time. One measurement every 30 min has been made during 4 days from 20 to 24 July 2016. The mean Kp index over this period is 2 while the maximum is 5 (only three periods of 3 h reached level 5 of the Kp index).

Before processing, the baseline computation clearly highlights the set-up error as shown in Fig. 4. Actually, such big variations do not meet the international standards (St Louis, 2012) and could discard the concerned magnetic observatory. Indeed, most observatories perform absolute $D$ and $I$ measurements no more than once a day, introducing an aliasing in the baseline computations. The amplitude of the baseline variations in Fig. 4 is such that the 5 nT tolerated errors are

not met anymore. However, after solving the system for each components and applying the transformation matrix to the variometer data, the baseline computation gives more correct data. In this case, a cubic-spline function has been used for fitting to the baseline measurements.

A second LEMI-025 is installed in the variometer house of the Dourbes observatory. This one is correctly set up, so it could be used for a posteriori comparison. Figure 5 shows the difference between vector components built from the case study variometer and the reference variometer. Notice that, even if both are separated by as much as 10 m, the observatory environment should ensure minimal difference. If we exclude the borders for which the cubic-spline baselines are badly defined, the three curves meet the INTERMAGNET 1 s standards requiring an absolute accuracy not worse than $\pm 2.5$ nT. The $Y$ and $Z$ curves remain within $\pm 0.44$ nT. The $X$ component is slightly more noisy, with the upper and lower borders being $+1.11$ and $-0.38$ nT respectively. The mean differences are 0.06, 0.009 and 0.002 nT for $X$, $Y$ and $Z$ curves respectively, and the corresponding standard deviations ($1\sigma$) are 0.26, 0.15 and 0.23 nT respectively.

## 5 Discussion

In this paper, the measurement errors have not been taken into account. In particular, absolute measurements were performed sequentially so that the magnetic field could be changed between the first and the last measurement. Equations (12)–(14) do not take the variations between the mean declination time and the mean inclination time into account. Indeed, using Eqs. (19)–(22) with the badly set up variometer for compensating the magnetic activity would lead to a nonlinear system. Nevertheless, AutoDIF achieves a complete protocol of absolute measurement within less than 5 min including the geographic north measurement at the beginning. Because of the high number of measurements during a few days, the error due to this delay can be considered as random. Assuming that the measurement errors are a random noise, their effects are therefore cancelled according to the Gauss–Markov theorem.

Jankowski and Sucksdorff (1996) suggested taking advantage of a disturbed day in order to maximise the effect of a set-up error. However, the global measurement noise may increase, in particular at high latitude. Indeed, the synchronisation between instruments may become critical. Additionally, a rapid change in the magnetic field may induce soil current that could affect both the DI-flux instrument and the variometer. Fortunately, as the noise is random and this is even truer during chaotic magnetic activity, it has no effect on the final results.

Equation (16) supposes a constant baseline so that a small variation will contribute to the residues. However, the use of an automatic DI-flux instrument provides a large number of measurements within a short time period. The case study has been performed during only 4 days, within which the baseline variations are reasonably considered small. Their contribution to the error can therefore be considered negligible compared to the possible scale factor and orientation parameter effects. Nevertheless, INTERMAGNET recommends performing absolute measurements with an interval ranging from daily to weekly (St Louis, 2012).

## 6 Conclusions

The baselines and absolute measurements are powerful tools for checking data quality and for highlighting possible gross errors. The present paper has demonstrated that even with a strong set-up error, it is possible to recover good magnetic data meeting the international standards. It also contributes to automatic installation and calibration of magnetic measurement systems. Future observatory deployments will be more and more complex, with automatic dropped systems in unstable environments. The challenges of tomorrow are in Antarctica, the Earth's seafloor or even Mars (Dehant et al., 2012). The application of theses methods will contribute to reaching those objectives. They will require not only automatic instruments but also regular and automatic control.

*Competing interests.* The authors declare that they have no conflict of interest.

*Special issue statement.* This article is part of the special issue "The Earth's magnetic field: measurements, data, and applications from ground observations (ANGEO/GI inter-journal SI)". It is a result of the XVIIth IAGA Workshop on Geomagnetic Observatory Instruments, Data Acquisition and Processing, Dourbes, Belgium, 4–10 September 2016.

*Acknowledgements.* We would like to acknowledge the Royal Meteorological Institute of Belgium, which allowed this research. We also acknowledge the editor and the reviewers who contributed to the improvement of this article.

Edited by: Arnaud Chulliat

## References

Alldregde, L. R.: A proposed Automatic Standard Magnetic Observatory, J. Geophys. Res., 65, https://doi.org/10.1029/JZ065i011p03777, 1960.

Dehant, V., Banerdt, B. Lognonné, P., Grott, M., Asmar, S., Biele, J., Breuer, D., Forget, F., Jaumann, R., Johnson, C., Knapmeyer, M., Langlais, B., Le Feuvre, M., Mimoun, D., Mocquet, A., Read, P., Rivoldini, A., Romberg, O., Schubert, G., Smrekar, S., Spohn, T., Tortora, P., Ulamec, S., and Vennerstrom, S.: Future Mars geophysical observatories for understanding its internal structure, rotation and evolution, Planet. Space Sci., 68, 123–145, 2012.

Eckstaller, A., Müller, C., Ceranna, L., and Hartmann, G.: The geophysics observatory at Neumayer stations (GvN and NM-II) Antarctica, Polarforschung, 76, 3–24, 2007.

Gonsette, A., Rasson, J., and Marin, J.-L.: Autodif: Automatic Absolute DI Measurements, Proceeding of the XVth IAGA Workshop on Geomagnetic Observatory Instruments, data Acquisition and Processing, 2012.

Gonsette, A., Rasson, J., Bracke, S., Poncelet, A., Hendrickx, O., and Humbled, F.: Automatic True North detection during absolute magnetic declination measurement, Geosci. Instrum. Method. Data Syst. Discuss., https://doi.org/10.5194/gi-2017-18, in review, 2017.

Jankowski, J. and Sucksdorff, C.: Guide for magnetic measurements and observatory practice, Warsaw, IAGA, 1996.

Kerridge, D. J.: Theory of the Fluxgate-Theodolite, Geomagnetic Research Group Report 88/14, Edimburgh, British Geological Survey, 1988.

Marusenkov, A., Chambodut, A., Schott, J.-J., and Korepanov, V.: Observatory magnetometer in-situ calibration, Data Science Journal, 10, https://doi.org/10.2481/dsj.IAGA-17, 2011.

Mohr, P. J., Newell, D. B., and Taylor, B. N.: CODATA recommended values of the fundamental physical constants: 2014, Review of Modern Physics, 88, https://doi.org/10.1103/RevModPhys.88.035009, 2016.

Olsen, N., Toffner-Clausen, L., Sabaka, T. J., Brauer, P., Merayo, J. M. G., Jörgensen, J. L., Leger, J.-M., Nielsen, O. V., Primdahl, F., and Risbo, T.: Calibration of the Orsted vector magnetometer, Earth Planet. Space, 55, 11–18, 2003.

Rasson, J. L.: About Absolute Geomagnetic Measurements in the Observatory and in the Field, Publication scientifique et technique no 40, Institut Royal Météorologique, 2005.

St Louis, B. (Ed.): INTERMAGNET technical reference manual: Version 4.6, 92 pp., available at: http://intermagnet.org/publications/intermag_4-6.pdf (last access: 7 February 2017), 2012.

# 13

# The MetNet vehicle: a lander to deploy environmental stations for local and global investigations of Mars

Ari-Matti Harri[1], Konstantin Pichkadze[2], Lev Zeleny[3], Luis Vazquez[5], Walter Schmidt[1], Sergey Alexashkin[2], Oleg Korablev[3], Hector Guerrero[4], Jyri Heilimo[1], Mikhail Uspensky[1], Valery Finchenko[2], Vyacheslav Linkin[3], Ignacio Arruego[4], Maria Genzer[1], Alexander Lipatov[3], Jouni Polkko[1], Mark Paton[1], Hannu Savijärvi[8], Harri Haukka[1], Tero Siili[1], Vladimir Khovanskov[2], Boris Ostesko[2], Andrey Poroshin[6], Marina Diaz-Michelena[4], Timo Siikonen[7], Matti Palin[7], Viktor Vorontsov[2], Alexander Polyakov[2], Francisco Valero[5], Osku Kemppinen[1], Jussi Leinonen[1], and Pilar Romero[5]

[1]Research Division, Finnish Meteorological Institute, Helsinki, Finland
[2]Planetary Systems Department, Lavochkin Association, Moscow, Russia
[3]Planetary Science Laboratory, Russian Space Research Center (IKI), Moscow, Russia
[4]Microelectronics Department, Instituto Nacional de Tecnica Aeroespacial (INTA), Madrid, Spain
[5]Computational Mathematics Dept, Universidad Complutense de Madrid, Madrid, Spain
[6]Dauria Ltd, Moscow, Russia
[7]Finflo Ltd, Espoo, Finland
[8]Dept of Physics, University of Helsinki, Finland

*Correspondence to:* Ari-Matti Harri (ari-matti.harri@fmi.fi)

**Abstract.** Investigations of global and related local phenomena on Mars such as atmospheric circulation patterns, boundary layer phenomena, water, dust and climatological cycles and investigations of the planetary interior would benefit from simultaneous, distributed in situ measurements. Practically, such an observation network would require low-mass landers, with a high packing density, so a large number of landers could be delivered to Mars with the minimum number of launchers.

The Mars Network Lander (MetNet Lander; MNL), a small semi-hard lander/penetrator design with a payload mass fraction of approximately 17 %, has been developed, tested and prototyped. The MNL features an innovative Entry, Descent and Landing System (EDLS) that is based on inflatable structures. The EDLS is capable of decelerating the lander from interplanetary transfer trajectories down to a surface impact speed of 50–70 m s$^{-1}$ with a deceleration of < 500 g for < 20 ms. The total mass of the prototype design is $\approx 24$ kg, with $\approx 4$ kg of mass available for the payload.

The EDLS is designed to orient the penetrator for a vertical impact. As the payload bay will be embedded in the surface materials, the bay's temperature excursions will be much less than if it were fully exposed on the Martian surface, allowing a reduction in the amount of thermal insulation and savings on mass.

The MNL is well suited for delivering meteorological and atmospheric instruments to the Martian surface. The payload concept also enables the use of other environmental instruments. The small size and low mass of a MNL makes it ideally suited for piggy-backing on larger spacecraft. MNLs are designed primarily for use as surface networks but could also be used as pathfinders for high-value landed missions.

## 1 Introduction

Significant progress in several areas of scientific investigation on Mars, such as climate circulation, water cycle, sedimentary cycle and surface–atmosphere interactions, has been made possible with spacecraft observations at Mars (Soffen, 1976; Golombek et al., 1999; Smith et al., 2008). In many investigations significant progress is contingent on good spatial coverage at several locations (MESUR study report; Chicarro et al., 1993; Harri et al., 1998; Linkin et al., 1998; Harri et al., 1999, 2007) with extended temporal and simultaneous coverage, requiring the concurrent operation of several spacecraft. Current orbital and lander observations are restricted in spatial measurements primarily due to the low number of active spacecraft available ($\sim 2$) for making simultaneous coordinated observations.

The payload mass of the launchers, and their cost, restricts the number of spacecraft and instruments that can be delivered to Mars during each launch window. Among the wide variety of science instruments and payloads relevant to Mars science and exploration some instruments and instrument types are inherently massive or sensitive, requiring relatively large and massive landing systems to enable a soft landing. Up to now large landers with multi-disciplinary and complex payloads have been favoured; Mars Science Laboratory (MSL) is perhaps the ultimate manifestation (Grotzinger et al., 2012; Gómez-Elvira et al., 2014; Taylor et al., 2008).

The use of lightweight landers would enable the delivery of an observations network to Mars possibly in a single launch. Meteorology, climate studies and seismology are areas of investigation that would benefit from a network of observations. A lightweight lander would require low-mass instruments with minimal use of resources such as power and heating, which is a requirement well suited for making atmospheric measurements. Heating requirements can be minimised by burying the bulk of the spacecraft in the regolith and so thermally isolating it from the extremes of the diurnal temperature range on Mars. Burial could be performed using the inertia of the lander as with penetrators. Keeping the lander mass low and packing density high would maximise the number of landers that could be launched towards Mars with a single launcher. This could be enabled by using inflatable aerodynamic decelerators.

This paper describes the MetNet Lander (MNL) concept, a compact and lightweight vehicle designed to deliver a set of instruments to the surface of Mars. The MNL vehicle uses a combination of lightweight inflatable aerodynamic decelerators and a penetrator-like landing system that also gives the correct final operational attitude. MNL will impact the Martian surface at a relatively lower, and hence safer, speed of around $50 \, \text{s}^{-1}$ compared to previous high-speed penetrator designs for Mars. For example, the Mars 96 and DS2 penetrators had impact speeds of 80 and $190 \, \text{s}^{-1}$, respectively (e.g. see Ball et al., 2009a). Possible uses of the MNL in Mars exploration along with programmatic and science mission aspects are also discussed.

The paper is organised as follows. In the next section previous Mars landers and their Entry, Descent and Landing System (EDLS) are reviewed in Sect. 2.1. MNL development is reviewed in Sect. 2.2. The selected MNL concept and its EDLS design are discussed and described in Sect. 2.3. Section 3 provides a more detailed description of the MNL mechanical and electrical systems. Potential mission types and

scientific applications of the MNL design are outlined and discussed in Sect. 4. Future prospects are outlined and recommendations made in Sect. 5 with Precursor missions outlined in Sect. 5.1.

## 2   Background

### 2.1   Brief overview of Mars lander technologies

The survivability of spacecraft during landing will depend largely on the spacecraft being able to absorb the impact energy without damaging its payload and critical systems. Landers can be divided into three categories with the division of these categories being defined by the landing speed, which is an indicator of the kinetic energy required to be dissipated by the spacecraft's landing system.

A soft lander typically touches down on the surface at a speed of around $1 \, \text{m s}^{-1}$ using a rocket propulsion system that is initiated at subsonic speeds to control and reduce the speed for a soft touchdown. The advantage of using a propulsion system is that manoeuvres like hazard avoidance, and pinpoint landings are possible. Examples of soft Mars landers are the Viking, Phoenix and MSL (Soffen and Snyder, 1976; Soffen, 1976; Guinn et al., 2008; Grotzinger et al., 2012) landers. Soft landing technology is required for large payloads, heavy payloads and payloads with components sensitive to high mechanical loads.

A hard lander, such as high-speed penetrators, typically impacts the surface at speeds of around $100 \, \text{m s}^{-1}$ and experiences high decelerations (up to $10^4 \, \text{m s}^{-2}$) over short time periods during the penetration of the subsurface strata. The use of high-speed penetrators for planetary science were first studied in the USA during the 1970s. The Soviet Union seemed to have initiated its studies in the 1980s (Ball et al., 2009, Chapter 19). In Europe the MarsNet mission (Chicarro et al., 1993) was the first study of a penetrator/hard lander system in the early 1990s.

Penetrators for a variety of solar system destinations have progressed to the concept stage although only two designs have actually been launched (Lorenz, 2011). These are the Russian Mars-96 penetrator (Surkov and Kremnev, 1998) and USA's Deep Space 2 Mars Microprobe (Smrekar et al., 1999). Each mission included two penetrators riding piggyback on a carrier spacecraft. None of these penetrators were successful: the Mars-96 mission failed to reach Earth's escape trajectory and the DS-2 probes' fate after deployment from the Mars Polar Lander is not known. Hard landers provide a platform to take robust science payloads to a planetary surface with a high mass efficiency. This is because the more gently a vehicle lands the more mass is needed for the EDLS to decelerate the vehicle's velocity before the touchdown on the surface.

Semi-hard landers are vehicles that impact the surface at speeds, and experience subsequent decelerations, that are be-

**Table 1.** The studied MNL EDLS concept candidates. In each concept the entry and descent phase braking devices are jettisoned to reduce decelerated mass. The concept A1 is as Mars-96 small stations (Linkin et al., 1998) and similar to the selected concept as the Mars-96 penetrators (Surkov and Kremnev, 1998), which used a rigid heat shield. The column title "Entry" refers to the hypersonic and supersonic portion of the flight. "Descent" refers to the subsonic portion of the flight. A "tension cone" refers to a type of inflatable decelerator shaped so as to contain tensile stresses; e.g. see Clark et al. (2009) for more information.

| Concept | Entry | Descent | Landing | Station type |
|---|---|---|---|---|
| A1 | rigid shell | parachute | airbags | lander |
| A2 | rigid shell | tension cone | airbags | lander |
| B1 | rigid shell | tension cone | internal shock absorber | penetrator |
| B2 | rigid shell | inflatable torus | same as descent | lander |
| B3 | inflatable | attached ballute | internal shock absorber | penetrator |
| Selected | inflatable | tension cone | internal shock absorber | penetrator |

tween those of a soft lander and a hard lander. Such landers will experience a moderate deceleration of a few hundreds of gees over the time of some tens of milliseconds. Typically low-mass Martian semi-hard landers have thus far used a combination of heat shield, parachutes and airbags (e.g. see Harri et al., 1999, and Linkin et al., 1998) for entry, descent and landing. Heavier semi-hard landers, e.g. Golombek et al. (1999), have used additional retrorockets at the end of the descent phase to decelerate down to the required impact speed. Semi-hard landers provide a practical solution for deploying planetary surface payloads that include robust geophysical instruments and are especially suited for deploying lightweight sensor systems needed to perform atmospheric science experiments.

## 2.2 MetNet Lander development history and background

The work on a semi-hard lander design for the MNL started in August 2000. Five different EDLS concepts (Table 1 and Fig. 1) were initially defined as candidates to be studied. The development of the MNL design was performed over a 7-year period from 2001 to 2008 by a team comprising of FMI, the LA and the Russian Space Research Institute IKI. The Spanish INTA joined the team in 2008. The MNL development work was funded and led by FMI. The MNL concept and key probe technologies were developed and the critical subsystems were qualified to meet the Martian environmental and functional conditions during the years 2002–2005. Development of the required system instrumentation and prototype science payloads to facilitate testing was carried out in 2004–2008.

In the initial phase of the development five different EDLS concepts were assessed from the viewpoint of finding an optimal solution for deployment of small payloads onto the Martian surface. One concept was a traditional, parachute-based approach and the remaining four utilised inflatable structures in various ways.

Comparative analysis between the five concepts, underlining and emphasising reliability, payload fraction and complexity of test programme, was carried out. The concepts were categorised into two categories. Category A contained those landers using airbags for landing and category B contained those landers using other impact shock attenuation mechanisms for landing. These categories contained a range of variants as shown in Fig. 1 whose EDLS elements are listed in Table 1. Variants A1 and B3 were selected for additional, more detailed study. This study resulted in the formation of lander concepts known as concept A and concept B. Concept A was essential variant A1, based on the Mars 96 small station, which employed a rigid heat shield, parachutes and airbags. Concept B was a new formulation of the EDLS that employed an inflatable heat shield, tension cone and penetrator to deliver the lander to the surface. Our comparative reliability analysis showed that concept B was significantly more reliable than concept A. This was due to, amongst other things, the lower amount of pyrotechnical devices required by the concept B. Penetration into the Martian regolith results in the vehicle experiencing reduced diurnal temperature variations. This could help reduce the thermal protection requirement, reducing mass, and in addition permit a wider range of qualified components for use in the vehicle. The current MNL design was chosen as it proved to best satisfy the design goals and criteria.

## 2.3 Selected Entry, Descent and Landing System concept

The selected MNL EDLS was designed to cope with relative entry speeds of slightly over $6\,\mathrm{km\,s^{-1}}$ for the current design configuration. Higher entry speeds are possible with some adjustments to the aerodynamics. The major components of the EDLS are the hypersonic inflatable braking unit (H-IBU), the transonic inflatable braking unit (T-IBU) and penetrator. The H-IBU is an inflatable heat shield designed to resist the heat during hypersonic entry into the atmosphere and decelerate the vehicle down to slightly below Mach 1. The T-IBU is an inflatable device known as a tension cone and is designed to decelerate the vehicle out of the transonic region down to fully subsonic speeds. Almost immediately after the

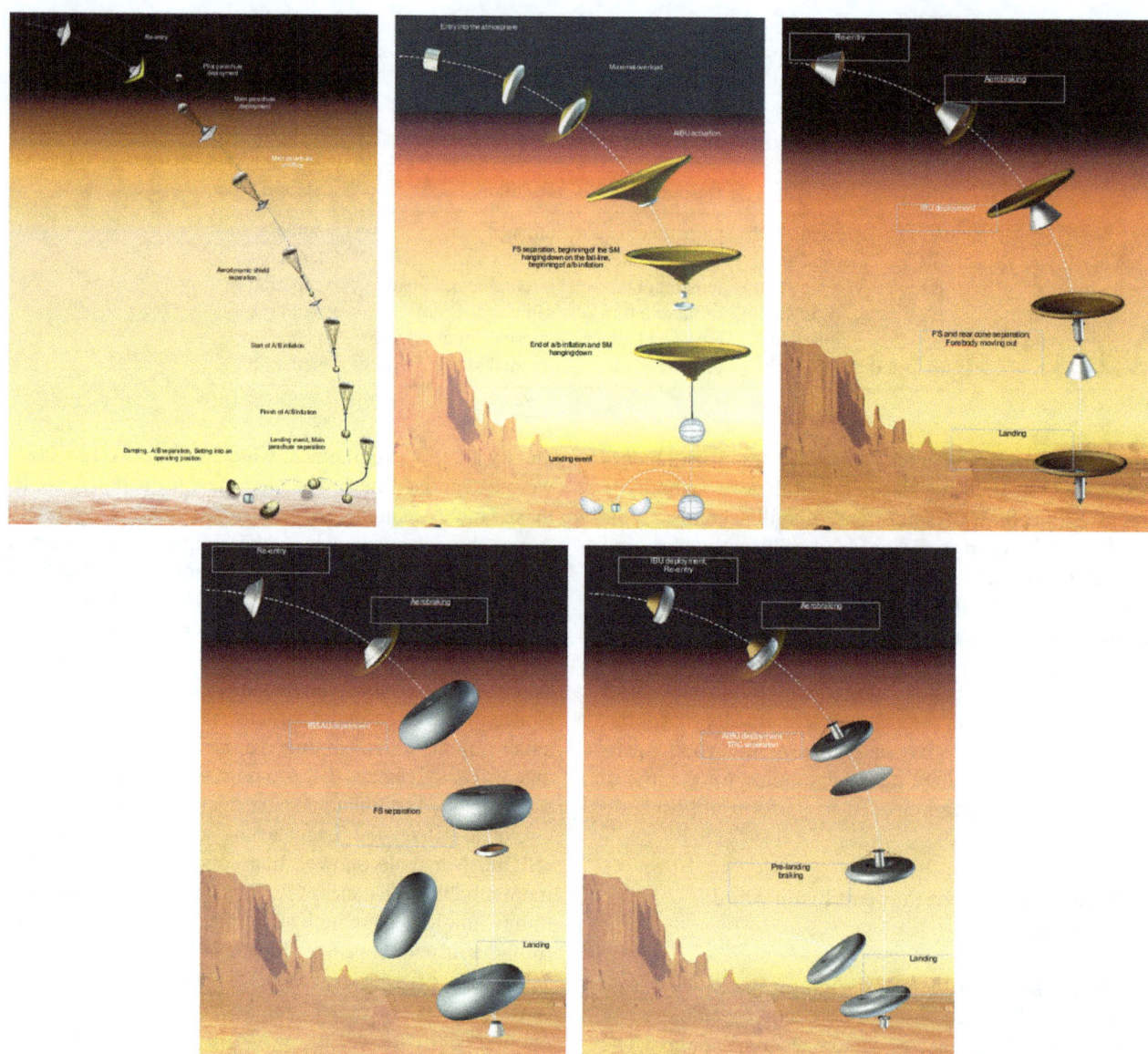

**Figure 1.** Landing schemes and designs (from left to right and top-down: A1, A2, B1, B2, B3 as in Table 1) investigated during the course of development of the MNL concept.

T-IBU is deployed the H-IBU is jettisoned. Once the H-IBU is jettisoned the forebody (FB) of the penetrator is deployed and locked into place ready for impact with the surface.

A MNL can be separated from the carrier spacecraft either directly from a Mars-approaching trajectory or from Martian orbit. Depending on the mission concept, a single carrier spacecraft may carry and deploy a single or several MNL. During the Earth–Mars cruise and possible orbital injection the carrier spacecraft provides each MNL with communications (data link) and power (for instance for health checks every few months, software upgrades, etc.) through the Carrier Spacecraft Interface and Lander Deployment System (CSI-LDS). The CSI-LDS features may vary depending on the number of MNLs carried, the mission concept and the char-

acteristics of the carrier spacecraft. A proposed MetNet mission with 16 landers (Harri et al., 2007) was made in 2007 as a study for a European Space Agency (ESA) medium class mission. Each lander was allocated a mass of 20 plus 10 kg for the spin/ejection mechanisms. The mass estimates were given with a margin of 10–20 %.

The entry, descent and landing (EDL) sequence of activities, shown in Fig. 2, begins with the separation phase from a few hours to a few days before actual separation from the carrier spacecraft. The MNL batteries (Sect. 3.2) are charged to capacity and depending on what has been performed during the preceding health check, final parameter updates to the Command and Data Management System (CDMS; e.g. software, cyclograms – see Sect. 3.3 and 3.5) may also be made.

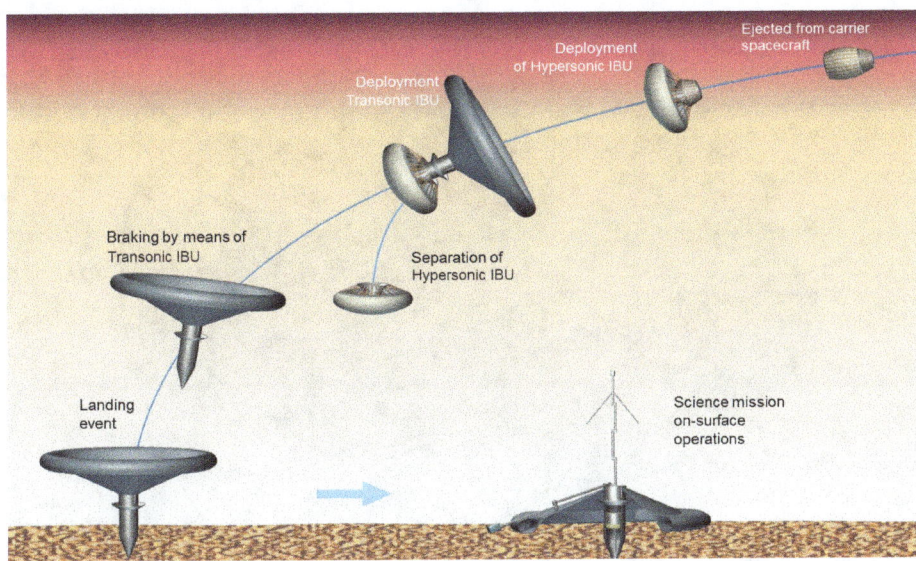

**Figure 2.** MetNet Lander (MNL) entry and landing sequence and configurations at different parts of the sequence: (1) the complete lander in stowed configuration during cruise and coast phase prior to atmospheric entry; (2) hypersonic inflatable braking unit (H-IBU) deployed for atmospheric entry; (3) transonic inflatable braking unit (T-IBU) deployed, H-IBU and rigid aerodynamic shielding (RAS) not yet jettisoned; (4) surface module (SM) in landing configuration with the forebody deployed.

Just prior to separation the MNL clock is set and the lander is spun up for stability during the entry into the atmosphere. This process takes $< 10$ min to complete.

Since the MNL itself does not have thrusters for trajectory or attitude changes, the carrier spacecraft may also need to carry out attitude change manoeuvres to release each MNL at the correct angle and at the correct time to reach its intended landing area. Stability is obtained from the aerodynamic properties of the vehicle and spinning of the MNL. The MNL is ejected and given its spin of one revolution every 6 s by a spring-loaded mechanism on the carrier spacecraft.

The behaviour of an MNL during the entire EDL is monitored by a combined three-axis accelerometer and gyroscope instrument. This diagnostic information is transmitted in packets in near-real time to the relay spacecraft (the carrier or a Mars orbiter, if one is suitably positioned during the EDL) via two dedicated beacon antennas (Sect. 3.3). The CDMS of the MNL connects the radio system first to the outer beacon antenna, after the inflation of heat shield to a second antenna and after landing and deployment of the instrument mast to the main antenna. The data packets include lander identifiers, and hence the monitoring system permits overlap or concurrence of EDL phases of multiple landers.

The entry phase begins when the MNL senses the first indications of interaction with the atmosphere and ends in the transonic (transition from super- to subsonic) speed regime. The inflatable heat shield is used during the entry phase to stabilise, decelerate the lander and protect it against excessive heat. The heat shield is inflated using a timer after release from the carrier spacecraft. The optimal range for the entry angle is $-16$ to $-18 \pm 2°$. The inflatable heat shield di-

ameter is 1 m, which decelerates the vehicle down to a Mach number of about 0.85 at an altitude of 4.5–11.0 km above the Martian datum, i.e. the point of zero elevation on Mars equivalent to the altitude where the pressure is 610 Pa, and a dynamic pressure of 95–130 $Nm^{-2}$ (both altitude and dynamic pressure depending on the angle of entry). The tension cone is fully inflated and the heat shield released 10 s later, allowing the vehicle to stabilise.

The descent phase begins when the lander speed is slightly below Mach 1, the inflatable heat shield is ejected and the tension cone is deployed. The tension cone diameter is 2 m, and is used to decelerate the MNL down to a landing speed of 47–55 m s$^{-1}$, depending on the angle of entry, at the Martian datum. The descent phase ends with the contact of the penetrator tip with the surface. Peak deceleration of the MNL payload bay during the impact will be $< 500$ g, with the outer shell experience about twice the load on the payload, and the total impact time is 20 ms. The minimum impact speed required for an operational landing is 50 m s$^{-1}$ with a maximum horizontal wind speed of 20 m s$^{-1}$.

The landing phase begins when the tip of the penetrator touches the surface and ends when the lander has come to rest on and is partially embedded in the top layers of the surface. The deceleration experienced by the payload as the lander penetrates the surface is of the order of 500 g. The structures and mechanisms involved in the final phase landing process, comprising of the shock absorbing system (SAS), which is used to reduce the g-levels on the instruments, are described in greater detail in Sect. 3.1.

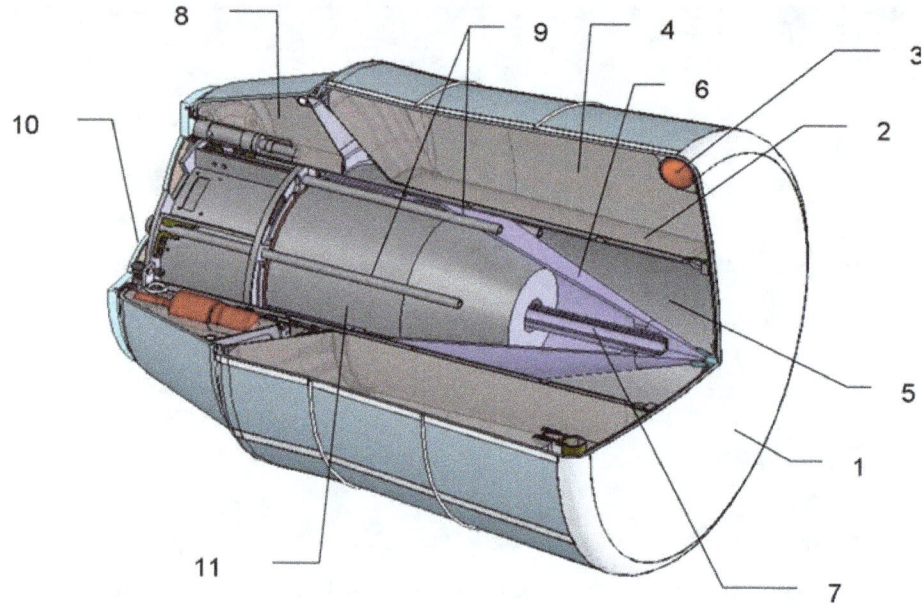

**Figure 3.** The rigid aerodynamic shielding (RAS) includes a blunt front shield plate, a toroidal pressure vessel (which stores the H-IBU inflating gas under pressure) as well as supporting structures for both the surface module and the entry and descent systems. The labels are as follows: (1) front shield (FS) with TPC; (2) body of FS; (3) H-IBD filling system; (4) H-IBD; (5) lander body; (6) telescopic cone; (7) telescopic cone drive; (8) T-IBD; (9) shock absorber; (10) cover; (11) instrument container.

## 3   Description, operation and testing of the prototype hardware

### 3.1   Structures and mechanisms

The MNL mechanisms are divided into two categories which are (a) the entry, descent and related subsystems and (b) the landing and surface operation-related subsystems. The entry and descent system consists of three subsystems:

1. rigid aerodynamic shielding (RAS) and supporting structure,

2. flexible heat protection (FHP) and

3. H-IBU (see Sect. 2.3), inflation system and load-bearing elements.

The landing and surface operation system consists of three subsystems:

1. T-IBU and gas generator,

2. surface module (SM) with a SAS and

3. equipment compartment (EC).

During cruise, entry and most of the descent phase, the EDLS-related subsystems and the SM are efficiently packed in terms of volume. This is achieved by stowing the systems telescopically inside each others where possible. The FB is stowed inside the SM cylindrical structure. When the FB is deployed the empty space provides room for the deceleration of the EC along a set of crushable rods during the impact with the surface. The FB will be deployed into a landing configuration after jettisoning the H-IBU. The stowed SM and FB are both stowed inside the mechanical support cylindrical structure of the rigid section of the front shield during entry and upper atmosphere braking phase. Figure 3 shows the complete lander with empty stowed H-IBU wrapped around it.

The SM accommodates the system electronics and payload instruments. The T-IBU is connected to and surrounds the SM. These three subsystems stay interconnected after touchdown forming the surface operating unit. The power system solar cells are attached to the upper surface of the T-IBU. Other subsystems, forming most of the EDLS, are ejected during the descent phase as shown in Fig. 2.

The SM accommodates the EC, which houses the system and payload electronics and supports external sensors (the boom with meteorological sensors, optical sensor) and telecom antenna. The SM includes a rear cover lid, which protects the module during entry and landing.

The SM includes the SAS. This system allows the equipment module (EM) to slide some tens of centimetres during the impact with the surface and thus reduce the deceleration experienced by the equipment module by a factor of around 2 compared to other rigid mechanics such as the surface module body structures. The SAS is made of six metallic (AMg3M aluminium–magnesium alloy; GOST, 1977) hollow tubes. The equipment module slides along these tubes during the impact on the surface and kinetic energy is re-

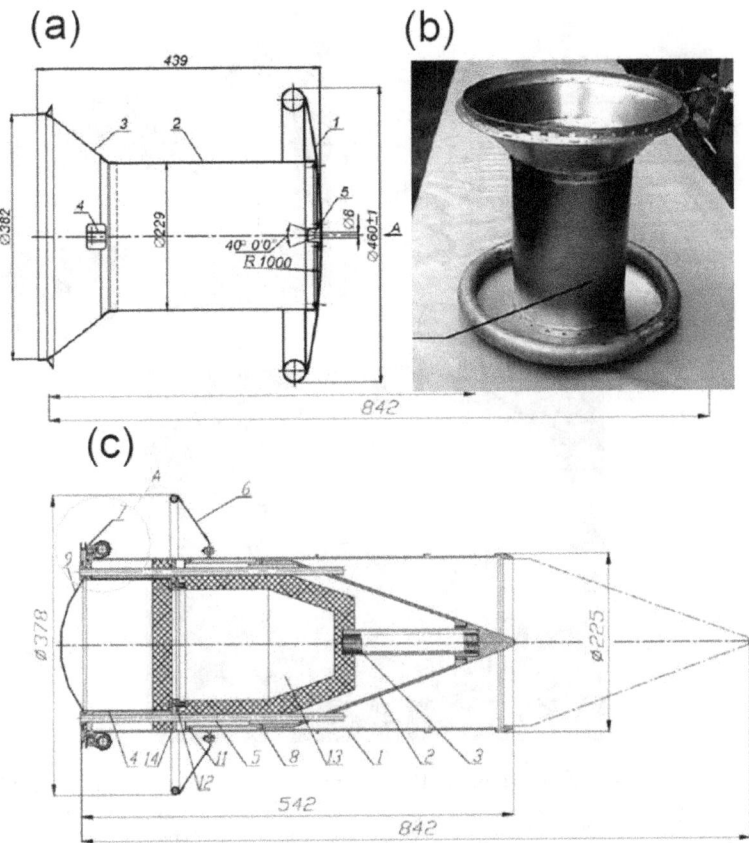

**Figure 4.** The top two figures show the rigid aerodynamic shielding (RAS), which includes blunt front shield plate, toroidal pressure vessel, which stores the H-IBU inflating gas (under pressure), and supporting structures for the surface module and entry and descent systems. Image **(c)** shows the SM in stowed configuration: (1) main body; (2) conical forebody; (3) spring to deploy forebody; (4) IB; (5) deforming tubes; (6) conical structure; (9) lid for protecting external and deployable instrumentation; (13) EM.

duced by squeezing the hollow tubes flat by squared sliding slots of the equipment module supporting adapter.

### 3.1.1 Entry- and descent-related subsystems

The RAS including structural details is shown in Fig. 4. The SM with stowed FB fits inside the cylindrical and conical structure. Figure 4c shows the SM with the FB extended. The surface module and RAS are connected and secured together by a cable and two turn-buckle devices. The RAS blunt circular front shield has radius of curvature of 1.0 m and a diameter of 0.46 m. The RAS is manufactured from AMg6, AMg6M and MA2-1 aluminium–magnesium alloys. The H-IBU interfaces with the RAS by a H-IBU inflation system (H-IBU-IS). H-IBU-IS includes a toroidal pressure vessel for storing the H-IBU inflation gas and required pyro-operated valves. The toroidal H-IBU-IS can be seen in Fig. 5 surrounding the circular front shield. The RAS has total mass of 2.31 kg.

The H-IBU consists of a toroidal inflatable wheel, shown in Fig. 5b, which supports the flexible thermal protection system (TPS), increases the frontal braking area and maintains

the stability and flight path angle during early landing phase within specifications. The H-IBU consists of 12 tubular segments, each with a diameter of 250 mm. The total diameter of the complete inflated H-IBU is 1000 mm. Inflation pressure is 63 Pa. The H-IBU consists of an internal gas-tight bladder (TPM-8 fabric), external cover fabric (aramid fibre), load bearing tapes, filling hoses as well as hardware and accommodation bag. Total mass is 1.17 kg. Figure 5c shows the inflated H-IBU.

### 3.1.2 Landing and surface operation-related subsystems

The T-IBU, shown in Fig. 6, is used during the last stages of the descent and landing. The T-IBU is deployed just before jettison of the combined H-IBU and RAS and supporting structures. The T-IBU decelerates the MNL down to subsonic speed and stabilises the lander. After landing T-IBU also supports the solar panels, which are mounted on its surface.

The T-IBU consists of toroidal shell cover and gas-tight bladder, flexible cone and inflation system which is based on a pyrotechnical gas generator. The toroidal part of the T-IBU

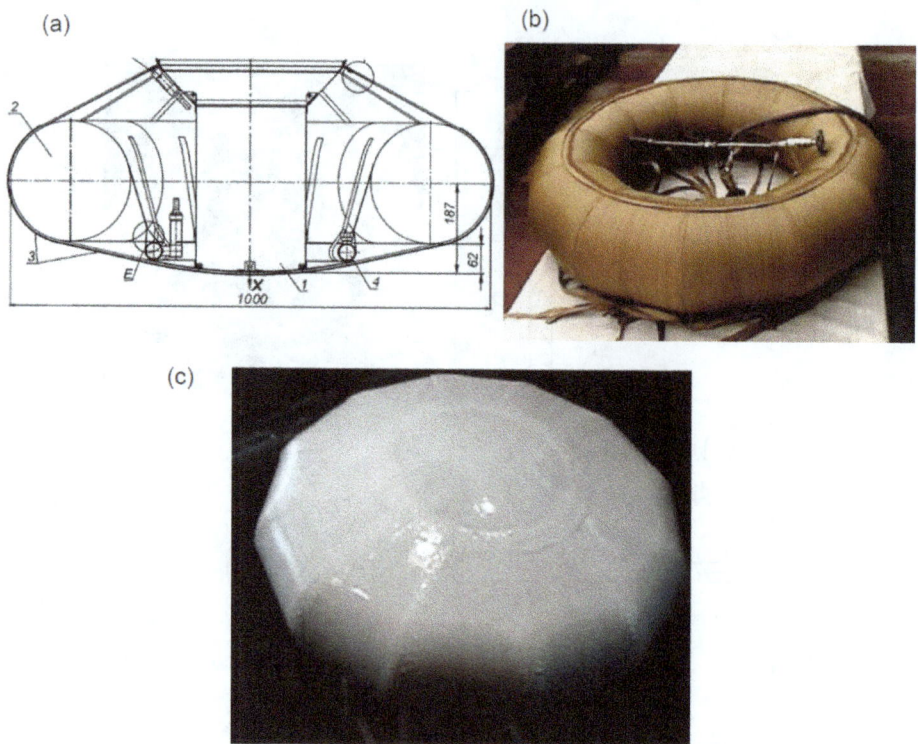

**Figure 5.** Images **(a)** and **(b)**: main inflatable braking unit (H-IBU) in the drawing on the left is a toroidal inflatable wheel (4), which supports the flexible heat protection system (3) and maintains required shape for maintaining correct attitude and flight path angle. Torus shaped pressure vessel/pressure receiver (4) provides required inflation gas. Inflated H-IBU assembly with the cover shell is shown on the right. The bottom figure shows the thermal protection shell (TPS) deployed. It is supported by the rigid section of frontal shield in the middle and toroidal inflatable H-IBU constructed from 12 segments.

**Figure 6.** Inflated additional inflatable braking unit (T-IBU) shown on the left with the surface module (SM). In the centre the cover of the T-IBU and on the right the gas-tight bladder.

is made up of 12 segments with each segment having a diameter of 200 mm. The inflated T-IBU has an overall diameter of 1800 mm. The T-IBU hardware is accommodated into a cone shaped upper part of the external body of the SM. The T-IBU has a mass of 1.06 kg.

The SM is the final stage. The surface module consists of the EM, internal body (IB), the FB and the main body (MB). They are made of AMg6 aluminium–magnesium alloy.

The FB is conical and is a telescopically extending ground-penetrating forward section of the surface module. During the last stage of deployment the forebody is locked together with the MB thus forming a unified structure. The IB is a cylindrical compartment inside the upper section of the MB. The IB is mounted together with the EM below it. The IB accommodates the external deployable instrument boom.

**Figure 7.** On the left is the surface module forebody deployed. On the right is the internal body together with equipment module mounted below it. Shock absorbing system tubes are visible. The hollow tubes absorb kinetic energy by deforming as they slide through too-small openings on the flanges of the internal body.

The EM is sealed and thermally insulated from the environment and accommodates electronics.

The SAS is installed inside the MB (Fig. 7a) and is designed to reduce g-loads on the payload compartment. The SAS is based on six deforming hollow tubes, which can be seen in Fig. 7b, mounted inside the MB. These tubes support IB and EM, which during landing impact slide along these tubes. The IB has narrower guiding slots for the tubes than their external diameter. The tubes will thus be deformed narrower accordingly and absorb the kinetic energy from the EM during the impact with the surface. The IB and EM can decelerate over a distance that is 30 cm more than the combined MB and FB. Using this method the deceleration can be limited to a maximum of 500 g. The tubes are made of aluminium–magnesium alloy AMg3M.

Figure 8 shows the configuration of the MNL's internal and external components just before and after landing. In Fig. 8b the payload compartment (EM and IB) has slid downwards along the six deforming hollow tubes.

The payload compartment consists of the EM and IB mounted together. The EM is sealed inside thermal insulation and accommodates most of the payload electronics and batteries. On the top of the EM is the IB, which accommodates the instrument boom. The boom supports temperature and humidity sensors as well as the camera and optical sensor. The boom also supports the telecom antenna. The IB features the interface with the SAS as described earlier. Figure 8c shows the payload compartment.

### 3.2 Electric power and thermal management subsystems

The primary power source for the MNL baseline design is solar energy. Flexible Si solar cells with cell dimensions 11.4 cm × 4.6 cm and total area approximately 400 cm$^2$ are placed in pockets sewn on the fabric of the upper side of the T-IBU. The cells provide daily average electrical power of

about 600 mW. Energy storage of about 40 Wh is provided by two SAFT MPS176065 Li-ion batteries connected in series inside a thermally sealed container. Originally inclusion of radioisotope thermal generators (RTG) into the basic design was investigated and consequently the design does accommodate them. The RTGs were dropped, however, due to anticipated Precursor mission options, due to difficulties related to availability of the devices and due to environmental impact assessment, political and security issues related to launching radioactive materials and devices.

This baseline non-RTG power system design limits operations to latitudes effectively between ±30° about the Equator and even within that latitude band night-time operations will be highly constrained by available power. During times of increased opacity of the atmosphere due to dust storms, during local winter time at higher latitudes or in case parts of the solar cells are covered by dust, the generated electrical energy will be reduced, limiting the operational possibilities further. Inclusion of a RTG would allow for more continuous and robust operations as well as landing sites and operations also during wintertime at higher latitudes, up to polar regions.

The rather limited amount of power (ultimately dissipated as heat necessary for thermal management) provided by the non-RTG power system is partially compensated by the passive thermal control inherent in the penetrating MNL design: after a successful landing the front part is submerged in the Martian soil and in good thermal contact with its surroundings. Since the amplitude of the temperature variations tends to decline fairly rapidly with increased depth for undisturbed material (Fig. 9). This results in smaller thermal variations for those payload components and lander subsystems housed in the front part of the lander (Paton et al., 2016). The MNL battery can operate down to temperatures of 220 K and will have its own additional thermal insulation and heaters to increase the battery temperature during charging to at least 250 K if needed. The parts and subsystems remaining above

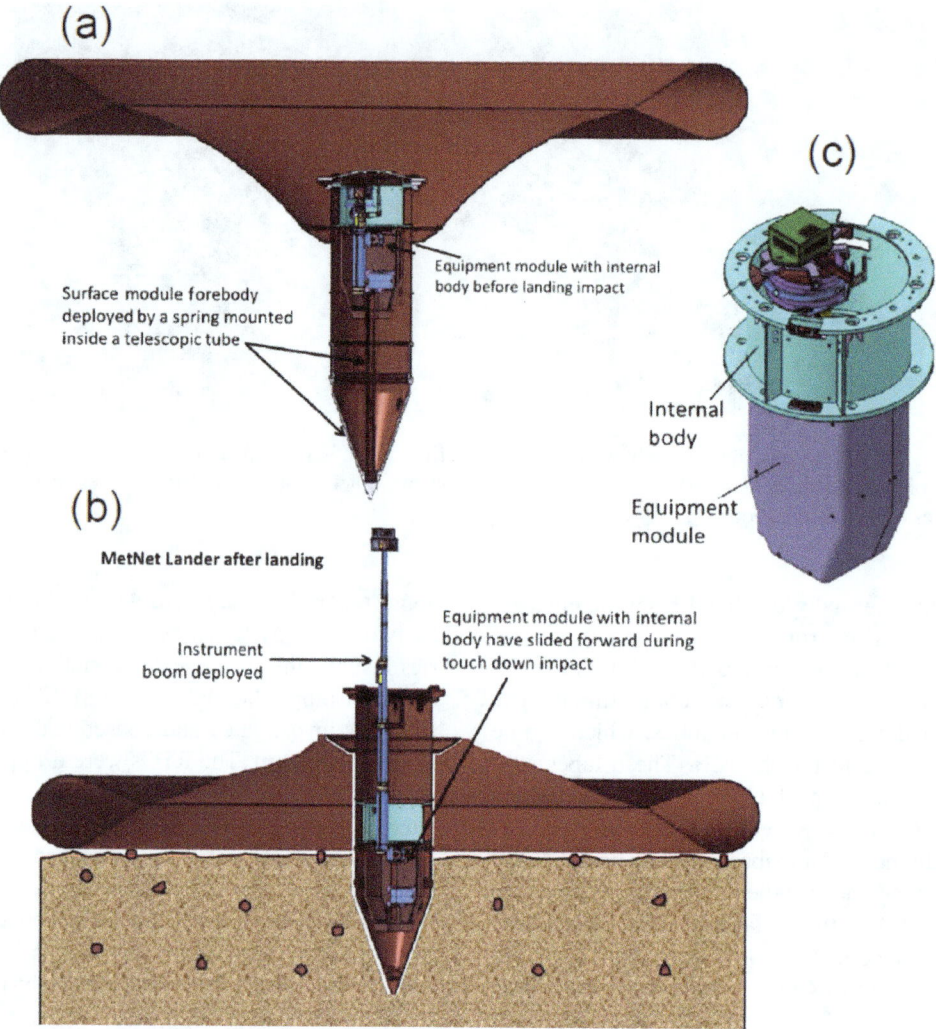

**Figure 8.** Image **(a)** shows the configuration of the external and internal components of the MNL in the last stage of the descent before impacting the surface. Image **(b)** shows the configuration of the same components after impacting the surface. Image **(c)** shows a perspective view of the internal body and equipment module.

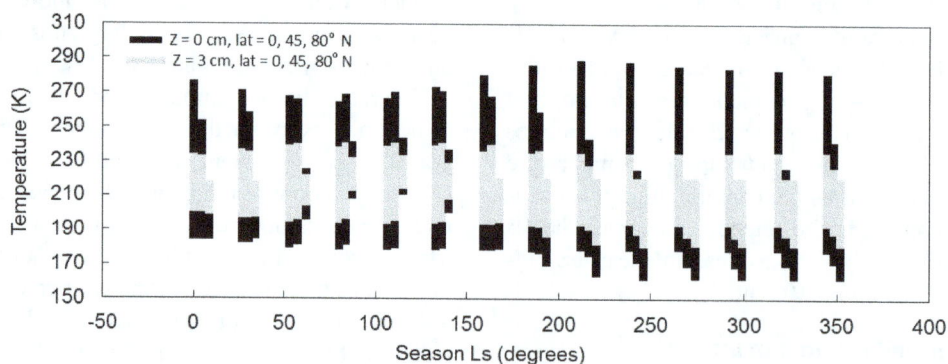

**Figure 9.** The range of temperatures experienced at different latitudes and depths on Mars over the Martian season in material undisturbed by the MNL. The light black bars represent the surface temperature and the grey bars represent the temperature at a depth of 3 cm. The bars a grouped together in bunches of three representing the three different latitudes modelled: 0, 40 and 80° N.

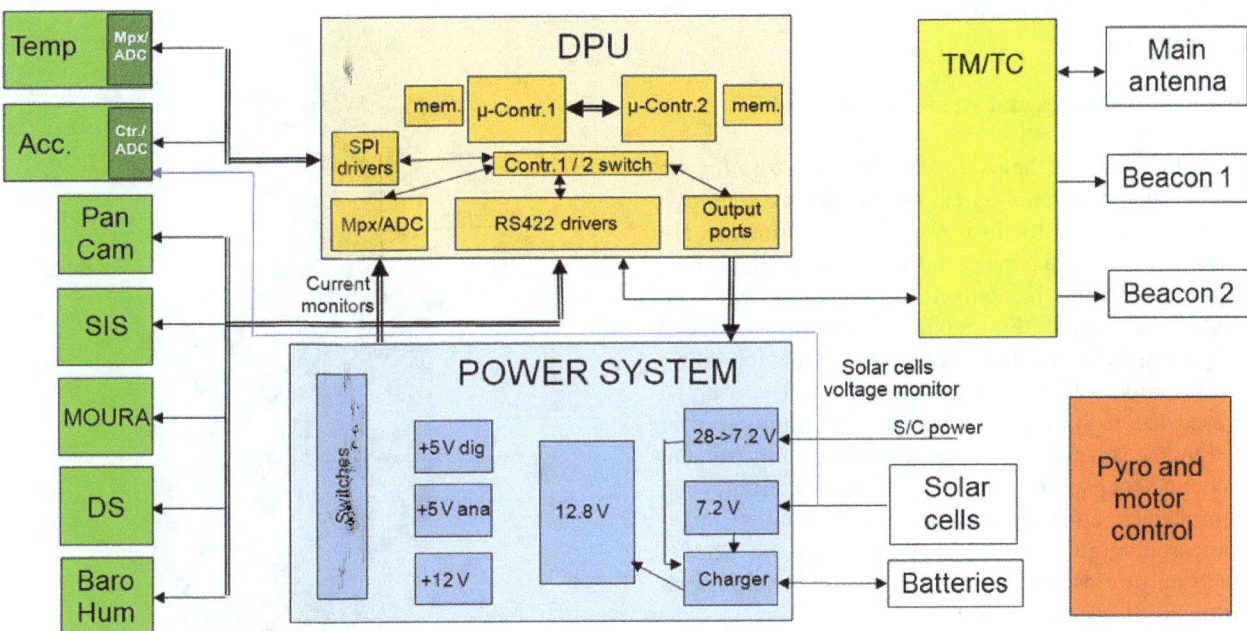

**Figure 10.** The MNL electronics consists of a hot-redundant microcontroller with duplicated interfaces, a power conditioning and a radio system. Communication with the detectors is via serial links.

the surface face comparatively much harsher thermal environment.

### 3.3 Communications, CDMS and electronics

The MNL does not include direct MNL–Earth communications capability – a relay spacecraft (either the carrier spacecraft during Earth–Mars transit or a Mars orbiter) is required. Observational and housekeeping data are preprocessed (including image compression) on board the MNL and transmitted to a relay orbiter on the UHF band. The radio system is built around the same type of micro controller used in the CDMS. Together with an field-programmable gate array it implements the Proximity 1 protocol (CCSDS 211.0-B-4) (for compatibility with the current and likely future Mars orbiting platforms). The system supports a bi-directional data link while still connected to the carrier vehicle, allowing a full system checkout as well as last-minute adjustments of operational parameters. The communication system is also capable of supporting a bi-directional link. There is also a technical capability to support software updates.

Operations are designed to make sure the transmitter does not drain the battery. The MNL goes into idle mode to save energy. The clock continues running. At preplanned times the lander waits for a hail signal from the orbiter before transmitting data.

The battery status monitor together with the system-related part of the software assures that enough energy remains available to perform the essential system tasks like telecommunication link during times of orbiter visibility, as well as time keeping. Surface to orbit link is around 16 kbps. The overall data transfer rate is expected to be low: about 0.25 to 0.75 Mb day$^{-1}$ on the average, depending on the orbital configuration.

The CDMS is built around a fully redundant micro controller system where one system is capable of autonomously detecting and correcting errors in the performance of the active controller. The micro controller type used is a free-scale micro controller MC9S12XEP100. The micro controller has 1 MB flash PROM for program, 64 kB RAM for data, 4 kB EEPROM and 32 kB D-flash. External memory used in the MetNet DPU is 2 × 128 Mbit serial flash memory. The same type of micro controller is used for DREAMS aboard Exo-Mars 2016.

All hardware interfaces and memories are duplicated so that the secondary controller can operate the system completely in case the primary one malfunctions without correction possibility. A block diagram is shown in Fig. 10. The software and even the controller hardware configuration can be updated from the operational controller via the implemented JTAG (Joint Test Action Group) input, using the own configuration as reference. The monitoring between the redundant controllers is done by a bi-directional CAN-bus interface integrated into each of the controllers. This link is also used to update cyclogram contents in the secondary controller after a commanded update.

Each of the MNL sensors are required to pre-process its observational data including image compression inside the panorama camera. The controller itself does not include any general data compression software to minimise especially energy resources.

## 3.4 Payload resources and strawman payload

With a total of 4 kg for the payload including 2.6 kg for the control system and meteorological mast including antennas, 1.4 kg are available for the instruments. As instruments are normally not operated in parallel the maximum available current from the batteries has only to be shared between the CPU and one sensor. With a maximum current of 5.8 A at $2 \times 3.5$ V about 40 W maximum power is available for a short time (of the order of 1 h; see also Sect. 3.2). The average power used has to be balanced against the average power provided by the solar cells and limits the time an instrument can be operated per sol. Instruments and their electronics can be accommodated either inside the thermally stabilised payload compartment guaranteeing temperatures above $-50\,^{\circ}$C or outside on or close to the telescopic mast. The payload compartment is illustrated in Fig. 11.

The MNL operations will be defined such that the average energy consumption does not exceed the energy provided by the solar panels. The main energy drain is the transmitter, which is used at such intervals that allow the charging of the battery between transmissions. The MNL components allow for such operational cyclograms to be defined.

The strawman payload for the Precursor mission includes sensors for temperature, pressure and humidity measurements, a four-lens panoramic camera, a multi-band spectrometer with $2\pi$ view, a three-axis magnetometer, a dust sensor and a combined three-axis accelerometer/three-axis gyrometer for descent control and monitoring.

## 3.5 MNL operations

Since commanding of a MNL is probably possible only infrequently if at all, a highly autonomous operations concept is necessary. The driving design factors constraining the vehicle design are optimal utilisation of the limited energy and availability of telemetry link time. The system also has to be able to adapt to different environmental conditions (e.g. day/night) and correct or minimise impacts of system problems. The general control scheme is illustrated in Fig. 12.

Due to the limited power supply, telemetry sessions exclude simultaneous observations. Phobos' eclipse observations (see end of this section) will also take precedence over regular observations.

The system software is implemented as a linear process, started at regular intervals of about 20 ms out of a low-power standby mode, checking first the battery status, then the availability of a telemetry link. If either the energy level is dangerously low or a telemetry link is possible, any science operation is aborted and all sensors are powered off. Telemetry remains switched on as long as relay link energy and non-transferred data are available. At other times observa-

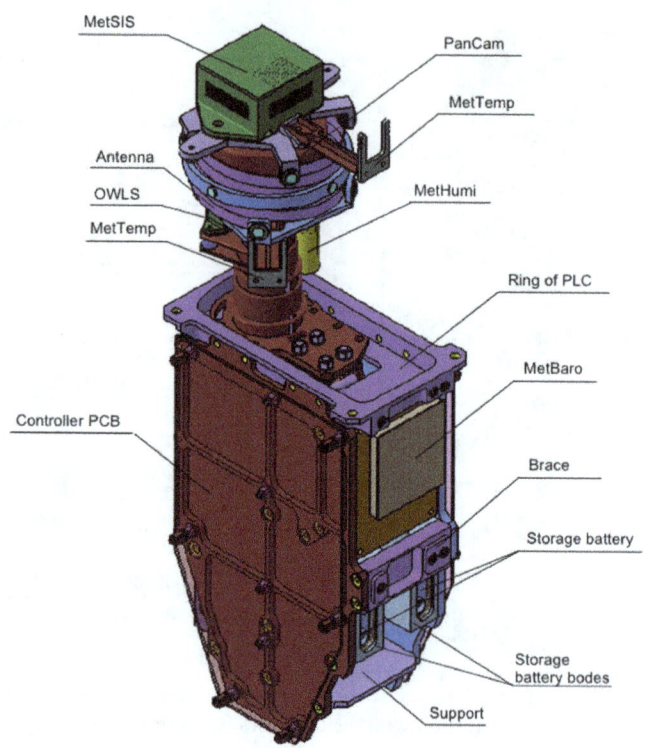

**Figure 11.** The payload compartment of the MetNet Lander slides after impact some 30 cm with deforming struts absorbing the kinetic energy remaining at the time of impact. All the instruments are packed tightly together with the CDMS and other system electronics.

tions are carried out according to a cyclograms concept. As up to 20 cyclograms – specifying detailed timing and command sequences for the sensors – can be defined, a sufficient set of scenarios can be covered. Cyclograms can be updated, if commanding is available. The concept allows consolidation and freezing of the system software at an early stage for testing while the detailed operational sequences may be optimized at least up to close to lander separation.

Concerning the electrical energy, battery charging from the carrier spacecraft or from the solar cells is completely controlled by hardware. Information flow between the controller and sensors or the telemetry system is handled by an independent processor inside the controller. This processor handles real-time tasks like sending of commands or reception of data packets, their consistency checks and acknowledgements. As conversion between internal parallel data words and external serial bit streams is also handled by autonomous hardware inside the controller, the processor is most of the time in standby mode, reducing the energy needs.

Each cyclogram is stored as a matrix with a control header, followed by many fixed-length command vectors. The header contains conditions under which the cyclogram may be used and a pointer to the next command vector to be executed. Additionally each vector defines the time delay to the fol-

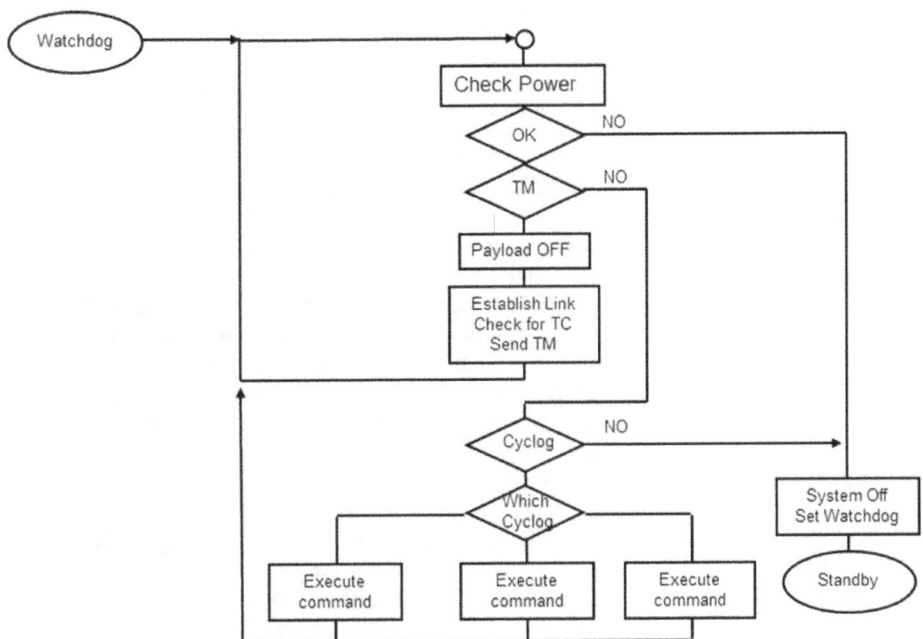

**Figure 12.** The control system consists of an infinite loop with first a sequence of resource checks, followed by the execution of the next cyclogram command if possible. In between activities the controller is in low-power wait state.

lowing vector and a set of conditions under which this vector has to be skipped. This way the same cyclogram can be used under different conditions like day or night, solar incident angles dangerous for an optical sensor or in case one of the addressed sensors is defective. The command itself contains the address of the system part or sensor, the command code itself and possibly a number of parameters associated with the command. In case the parameter list is too long to fit into the short vector space, the list is replaced by a pointer to a list at the end of the main matrix. To simplify writing and verification of the cyclogram details, the most common command combinations are hardcoded as macros and can be called directly from inside the cyclogram like a direct sensor command, shortening the cyclogram significantly.

The adaptivity of the system is based on the condition flags used to select a given cyclogram and to possibly skip a command vector inside the active cyclogram. The cyclogram control system interprets general system conditions and some of the detector measurement data to define the status of these flags. As the different cyclograms contain possible combinations of sensor operations, mainly sequential but possibly also simultaneous, an additional adaptive priority scheme ensures that each sensor gets a just share of the observation time suitable for its operation. A change in environmental or system conditions during the execution of the main control loop may result in the currently active cyclogram being aborted: the related sensors are switched off and the next possible cyclogram started. This ensures a maximum scientific return with the available resource limitations.

Criteria based on sensor measurements can be refined by programmable algorithms defining the upper and lower limit of these parameters before a condition change is initiated. The most important selection criteria are listed in Table 2 and additional criteria can easily be implemented.

A special case is observation of Phobos' eclipses, used to locate the actual landing site with high precision (Romero et al., 2011; Harri et al., 2012; Barderas and Romero, 2013). This in turn allows for correlation of the time stamps of the data with the correct local time, thereby making comparisons between observations from different landers more precise.

A time window for the passing of Phobos' shadow and for the high-time-resolution measurement with the optical sensor at the top of the mast needs to be determined and preprogrammed into the MNL control system. This can be calculated once the landing area is known based on the trajectory of the transfer vehicle and the scheduled moment of release of the lander. Only daytime passes are observed and only one diode will be used to keep the data amount as low as possible. This eclipse mode is controlled directly by the system software outside the cyclogram scheme. When the programmed absolute time is reached, the active cyclogram is aborted and the macro-like eclipse program with a finite loop is started. The data stored overwrite the not yet transferred data from other sensors if necessary. The measurement is repeated three times about 7.5 h apart to cover the theoretically expected three subsequent eclipses.

A sampling frequency of 1 Hz is used as a compromise between precision and generated amount of data inside the needed measurement window. This leads to a resolution of

**Table 2.** Environmental or system status cyclogram selection criteria.

| Criterion | Determination |
|---|---|
| Low battery status | Information comes directly from the power subsystem. Can be used to activate only sensors with low energy needs (e.g. $T$, $p$, $H$) when the battery charge is low but not yet critical. |
| Critical battery status | Information comes directly from the power subsystem. |
| Data buffer availability | The telemetry handling system keeps track of the data buffer usage per sensor. If the usage limit is reached, the sensor will be disabled and related commands skipped in active cyclograms until data have been uplinked and space made available. This limitation can be ignored to allow collection of large amounts of data under special conditions. |
| Severe detector error | To save energy and time resources, a detector is disabled if it repeatedly fails to react on commands appropriately. Recovery will be tried once per sol or after a system failure with subsequent reboot. |
| Day/night status | Determined from data from the optical sensors at the top of the mast. During the night the camera and the optical sensor will be disabled. |
| Solar incident angle | Based on accelerometer data the solar incident angle to the infrared dust sensor is calculated disabling its operation when directly illuminated. |
| Humidity | Humidity sensor data allow for selection or de-selection of the optical sensor $H_2O$ spectral band. |

approximately 3 km depending somewhat on the latitude of the actual landing side, e.g. see (Harri et al., 2012). The clock precision and stability is 5 orders of magnitude better than this, so it can be ignored. The 1 Hz sampling rate is hard-coded in the software outside the cyclogram control system to allow absolute time scheduling.

## 3.6 Aerodynamic and aerothermodynamic considerations

The aerodynamic and aerothermodynamic properties important for the design of the EDLS and trajectory have been studied in detail for MNL using laboratory tests and numerical modelling. Both the MNL with the heat shield inflated and the MNL with the tension cone deployed have been studied with wind tunnel tests and computer models to determine their aerodynamic, dynamic stability and static stability coefficients. Structural analysis of the penetrator during impact with the surface has been conducted using computer modelling and laboratory impact tests. Arc-jet tests have been made of the MNL TPS. In addition, computer modelling of the heat shield surface and payload temperatures has been performed during the descent. Computer modelling was used to assess the thermal response of the T-IBU during the descent.

The thermal protection system shown in Fig. 13 consists of three sections. Section I covers the rigid central frontal struc-

ture and the inflated torus structure around the central compartment. Section I is constructed of a double layer of KT-11 cloth, a glass-cloth-based laminate (Shalin, 1995), that can survive temperatures up to 1500 K. The exterior surface of the outer layer of the cloth is covered with a 1–1.3 mm thick layer of material that can absorb the heat experienced during atmospheric entry by sublimating. The sublimation temperature for this material is 950 K. Multi-layer insulation (MLI) is situated under the KT-11 cloth layers and is covered with a glass fabric that can survive short duration temperatures of 750–800 K. Figure 14 shows the modelled temperatures of the MNL at a speed of Mach 1.3.

Section II consists of MLI covered with a 1 mm thick layer of sublimating material that sublimates at a temperature of 950 K. Section III consists of a rigid lid covered with a 4 mm thick layer of thermal protection material that is radio-transparent.

The MNL was found to be statically and dynamically stable under conditions expected during the descent through the atmosphere from wind tunnel and computer modelling (e.g. see Figs. 14 and 15). An important system attribute of the lander is its stability during flight as this will influence the drag and lift parameters which will in turn affect the uncertainties in predicting its trajectory through the Martian atmosphere.

The drag coefficient of the heat shield, at zero angle of attack, has been determined experimentally (Heilimo et al., 2014) and computed to vary from 1.4 down to 1.0 at

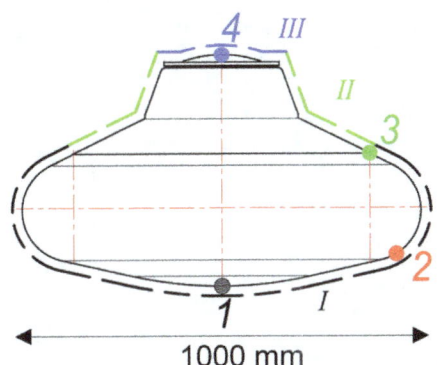

**Figure 13.** A schematic of MetNet showing the section of the heat protection sections of the inflated aeroshell. Section I covers the rigid frontal structure and inflated torus that supports the outer part of the heat shield. Section II covers the rear of the MNL with section III covering the very back.

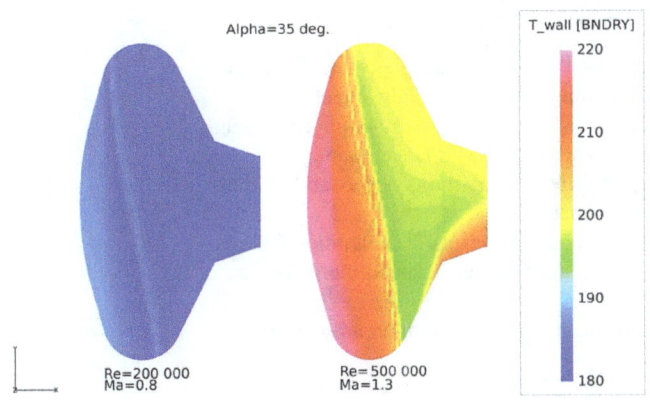

**Figure 14.** Thermal modelling of the MetNet heat shield according to Finflo Ltd simulations.

Mach 4 and 1 respectively. The drag coefficient with the tension cone deployed after the heat shield has been jettisoned has been determined to vary from 1.1 to 0.9 at Mach 0.8 and Mach 0.2 respectively. The ballistic coefficient controls the impact speed of the penetrator. With the tension cone deployed the ballistic coefficient is between 19 and $22 \, \mathrm{kg \, m^{-2}}$.

Trajectory calculations suggest the minimum entry angle for which the lander will not fly back into space is 5.8°. This is assuming a ballistic coefficient of $20 \, \mathrm{kg \, m^{-2}}$, an entry altitude of 120 km and an orbital entry speed of $4586 \, \mathrm{m \, s^{-1}}$. Minimum and maximum flight angles suitable for a landing that have been investigated are 9.5 and 13.7° respectively. It has also been calculated that the flight path angle when the tension cone is deployed will be 60° and at an altitude of 12 km, assuming the H-IBU has been flying through the atmosphere with zero angle of attack. This will allow ample altitude for the trajectory to turnover and enabling the penetrator to impact the surface vertically. The maximum surface temperature of the rigid TPS surface during the steepest trajectory has been calculated to be 523 K and a heat flux of

$190 \, \mathrm{kW \, m^{-2}}$, which is well below the short-term tolerance of the rigid TPS. See Sect. 3.1.1 for more details on the TPS.

## 4 Potential mission types

### 4.1 Overview

The MNL design features and characteristics both place constraints and open opportunities for implementation mission types and scientific investigations (Table 3). This section assesses the design from that point of view and presents some feasible mission types.

As a result of previous flight qualification activities the preparation of a flight-ready spacecraft (vehicle structure and its descent system) is estimated to take 2–4 years. With an entry mass of about 22.2 kg per unit the MNLs can be easily deployed from a wide range of transfer vehicles. The MNL structure allows the manufacturing of additional MNL units on short notice and at reasonable cost. The entry and descent systems could also be used independently from each other in other lander designs.

Planetary protection requirements are a factor in selection of possible mission types and their landing sites. Being a Mars lander any MNL-based mission is a category IV mission (COSPAR planetary protection policy). A MNL with a non-biological payload falls as a baseline into subcategory IVa with the least stringent sterilisation requirements. Since a MNL's landing involves penetration into the subsurface and depending on the payload complement and the targeted landing area, a MNL mission may also fall in the more stringent subcategories IVb and IVc.

Even if the payload does not include components aimed at studies of Martian extant life, any mission the a Martian special region (for definition and examples see COSPAR planetary protection policy) will have to comply with category IVc requirements. The hard landing also increases the probability of inadvertent exposure of the lander interior with the Martian environment and consequently for any special region mission the entire MNL would have to be sterilised to the Viking post-sterilisation biological burden. Small size of MNL facilitates comprehensive sterilisation of the entire lander.

The MNL decontamination will most likely be performed via a combination of dry heating and hydrogen peroxide treatment. Dry heating is applied for humidity sensor devices.

### 4.2 Atmospheric observation networks

A long-duration (possibly an uninterrupted time series) and global coverage in situ atmospheric science network comprising a truly significant number – from 10–20 to several tens – of observation points on the Martian surface has for a long time been the logical next step for the observational studies of dynamical features of the Martian at-

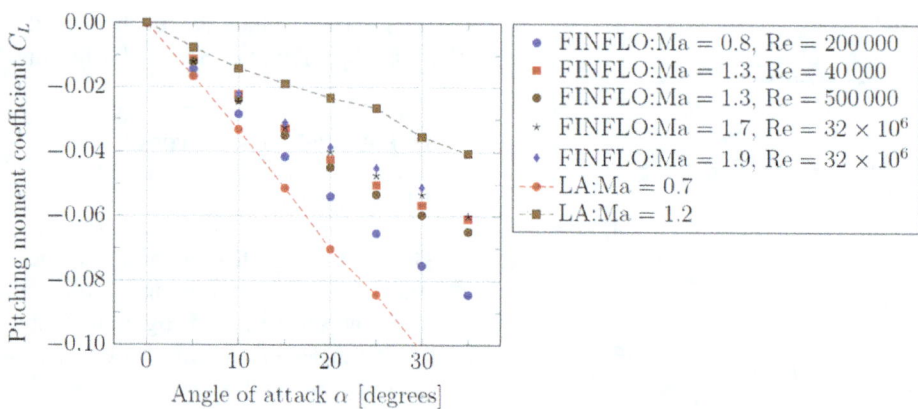

**Figure 15.** Comparison between the pitching moment coefficients for the heat shield case obtained by numerical simulations using FINFLO numerical code developed by Finflo Ltd and from LA.

**Table 3.** Key MNL design characteristics along with the associated impact (+/−) and mission design constraints.

| Characteristic | Impact | Impact description, constraints |
|---|---|---|
| Low unit mass | + | Increased launch opportunities (piggy-backing), deployment of multiple units with single launch |
| Low unit mass | + | Enables the establishment of an investigation consisting of multiple payloads on the Martian surface |
| Entry from interplanetary trajectory or parking orbit | + | Adaptability to different mission concepts and landing accuracy requirements |
| High impact acceleration | − | Only robust instrument concepts feasible, modification of the subsurface materials adjacent to the lander surface |
| Stable thermal environment in payload volume | + | Reduced instrument thermal control requirements |
| Stable near-vertical attitude after landing | + | |
| Limited control over the post-landing attitude | − | |
| Access to subsurface layers | + | Enables subsurface observations without digging or drilling |
| Limited electrical power and power storage | − | Limits instrument selection and operations |
| Both up- and down-links required | − | Constrains feasible mission concepts |
| Requirements placed on the carrier spacecraft | − | Constrains available carrier and mission concepts |
| Electromagnetic cleanliness | − | Constrains feasible instruments and their measurement accuracy |
| Planetary protection | − | Constrains types of instruments and allowed landing areas |

mosphere. Several different network concepts – differing in size and complexity of individual landers as well as number of landers forming the network – have been proposed (MESUR study report; Chicarro et al., 1993; Chicarro, 1994; Merrihew et al., 1996; Haberle and Catling, 1996; Linkin et al., 1998; Harri et al., 1999, 2007).

A spatially wide and comprehensive network alone would provide a significant leap in spatial and temporal (from diur-

nal to seasonal and up to interannual scales) characterisation of the global circulation patterns and the major climatological cycles of dust, $H_2O$ and $CO_2$ (Savijärvi et al., 2005; Harri et al., 2014b, a; Savijärvi et al., 2016). The MNL concept offers a cost-effective and hence realistic element and tool for deploying such a network. The potentially large number of observation points combined with careful selection of locations would permit analyses taking into account also corre-

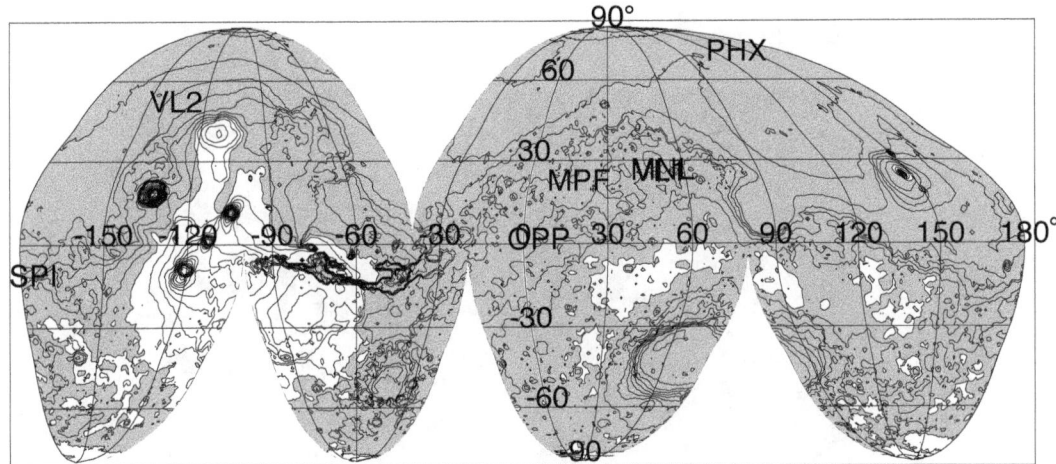

**Figure 16.** Areas with low enough surface altitudes (< 2 km above the datum) suitable as MNL landing sites are shown in grey. Shown are also landing areas of Viking Landers (VL1, VL2), Mars Pathfinder (MPF), Opportunity (OPP), Spirit (SPI), Phoenix (PHX), Mars Science Lander and a proposed MNL Precursor mission.

lations between observations. Such a set of observations has also the potential for providing sufficient constraints to become useful for assimilation into and with Mars atmospheric circulation models. Figure 16 shows the constraints on location, in terms of surface altitudes, for the MNL.

Optimal locations of observation posts depend on the total number of the network elements. If a network consists of only a few observation posts, it is worthwhile to either create a small local network or place the posts on different types of terrain and latitudes. This would encompass differences in altitude, latitude and type of surface. In all cases we should place observation posts also on the locations, where observations were previously performed by the Viking Landers (VLs), the Mars Pathfinder or the MSL. This would enable us to compare the current atmosphere at those sites to the atmospheric conditions prevailing earlier. This would be especially interesting at the VL and MSL sites, where long-duration observations are available.

Even though the initial emphasis and focus is from atmospheric science point of view naturally in achieving global and long-term coverage, in later stages one can envision deploying clusters of 3–4 landers regionally and with interstation separations of 100–1000 km to form mesoscale subnetworks (Fig. 17). Such subnetworks would be highly useful in more detailed studies of circulation patterns in regions of particular interest, e.g. the Tharsis volcano area, Valles Marineris, Hellas or perhaps the circumferences and vicinities of the permanent polar caps. The availability of observations would allow for regional models to be tuned to the characteristics of that particular region for provisions of improved understanding of the atmospheric processes and atmosphere–surface interactions, leading also to more accurate regional forecasts.

### 4.3   Joint rover–MNL atmospheric science missions

A subnetwork with interstation dimensions similar to a characteristic range of a longer-range rover (tens to a hundred km; the network "bracketing" or "interleaving" the area-of-operation of a rover) offers opportunities for having the MNLs and the rover carrying mutually complementary atmospheric science payloads with intercomparison and intercalibration opportunities. The rover could for instance carry more complex and resource-hungry instruments (e.g. Mini-TES or LIDAR type) that the MNLs could not include in their payloads. Such an arrangement would combine in the studied region stationary longer-term time series measurements with mobile, more advanced and spatially "sampling" type observations.

### 4.4   Atmospheric observations at high-risk locations

Although correlated and combined observations are qualitatively a leap forward from previous types of observations, the potentially larger numbers of deployable MNLs and acceptance of higher risk of the failure of a single vehicle (e.g. see Harri et al., 2012, and Lorenz, 2011) would permit clear advantages for studies in microscale meteorology and atmosphere–surface interactions (e.g. momentum and thermal fluxes): MNLs could be deployed to a large number of very different and also riskier locations and terrains, thus allowing for observations in and of terrains otherwise unlikely to be reachable.

### 4.5   Dedicated ground truth landers for atmospheric sounders

For best accuracy, atmospheric in-orbit sounder observations need so-called ground truth – typically independent informa-

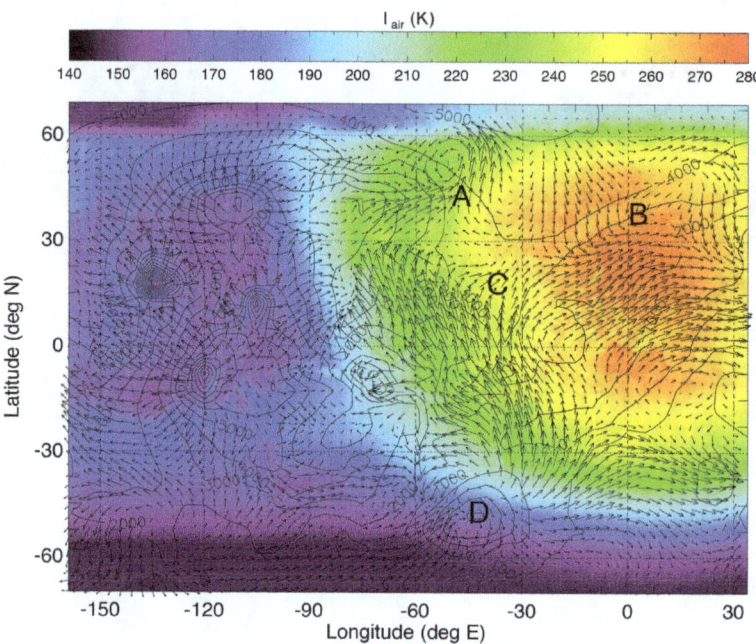

**Figure 17.** A schematic illustration of a four-lander subnetwork deployed into the equatorial region. The left figure shows the simulated near-surface temperature ($T_{air}$) and wind fields ($V$) along with the labels (A–D) of four possible landing sites. The data shown have been simulated with the Finnish Meteorological Institute/University of Helsinki Mars Limited Area Model; see e.g. Kauhanen et al., 2008).

tion on, for instance, the surface pressure. This can be provided by model estimates, but in situ observations are most reliable. Hence, having one or more atmospheric sounders in operation simultaneously with surface observation posts – especially a global surface network – would add considerable scientific value: the combination would provide and combine multipoint surface observations (providing ground truth) with spatially and temporally resolved (4-D) remotely sensed atmospheric temperature fields. However, the modest resource requirements of MNLs may enable yet another mission approach or emphasis: an advanced atmospheric remote sounder carrying with it a small number of dedicated well-placed MNLs for provision of ground truth – perhaps even forming a single mission/launch package.

### 4.6 Other science disciplines

The design offers potential uses in disciplines such as ground studies and seismology. As a MNL lander penetrates the top ground layers, one can envision for instance utilising this in investigations of the top layer of the surface – such as heat fluxes, composition or depth of permafrost. However, one needs to take into account that the energies released in the penetration process very likely modify and ground material immediately adjacent to the penetrator.

### 4.7 Pathfinders and precursors for high-value and high-cost missions

Certain Mars mission classes are inherently of extremely high value – i.e. costly in financial terms, of exceptional scientific value or may place humans at risk. Examples include components of sample return missions, in situ resource utilisation equipment or human missions. Mission assurance is hence of paramount importance and robust ability to observe the weather and atmospheric conditions are important both for planning and execution of the EDLS phase, the lift-off phase as well as surface operations.

The MNL offers an asset and tool for implementing regional weather observation infrastructure to serve high-value missions. The low cost – especially in comparison with the value of the "prime" mission – makes deployment of regional weather observation networks composed of MNLs an approach worthy of consideration. Deployment of such a network could conceivably take place during a launch window preceding a high-value mission, thus allowing for collection of a database of weather observations spanning a full Martian year. Such a data set would in turn provide observational basis for development of verified and tuned regional forecast models for the region and provision of high-fidelity forecasts to serve the prime mission components.

### 5  Discussion

We have developed a Mars lander concept – the MNL – that provides a key landing technology for the future exploration

**Table 4.** Comparison of MetNet properties and resources to a range of landers, e.g. Ball et al. (2009). In the fourth column the science payload fraction is the science payload mass divided by the entry mass. The letters W, L and H in the fifth column stand for width, length and height respectively.

| Lander | Entry mass (kg) | Science payload mass (kg) | Science payload fraction (%) | Dimensions W × L × H (cm × cm × cm) | Entry speed (m s$^{-1}$) | Impact dece- leration (g) | Generate (Wh) | Store (Ah) |
|---|---|---|---|---|---|---|---|---|
| MetNet | 22 | 1.5 | 6.8 | 22 × 22 × 84 | 6.0 | 500 | 14.6 | 40 (Wh) |
| DS2 | 2.73 | 0.15 | 5.5 | 14 × 14 × 22 | 6.9 | 60 k | 0 | 0.6 |
| Pathfinder | 584 | 8 + 10.5 | 3.2 | 90 × 100 × 100 | 7.3 | 40 (18.6) | 1200 | 40 |
| Viking | 992 | 91 | 8.2 | 100 × 200 × 50 | 4.5 | 14 | 1680 | 8 |
| MSL | 2401 | 75 | 3.1 | 270 × 300 × 220 | 5.6 | – | 2640 | 84 |

of the environment of Mars. By providing a platform for a 4 kg payload including mechanical structure the MNL is capable of serving various kinds of atmospheric science missions, as well as other kinds of environmental exploration missions

The MNL is a semi-hard penetrator utilising inflatable EDLS structures and mechanisms to improve the landed payload fraction, which is the payload fraction that could conceivably be replaced – for example if a different mission is envisaged using the MNL. The mass of the payload bay with its container and thermal insulation is 4 kg with an entry mass of 24 kg. Hence a payload fraction of 17 % based on engineering qualification hardware is an excellent number compared to earlier planned Mars landers with similar characteristics (for the Mars-96 penetrators, $F_{pl} < 7$ % (Surkov and Kremnev, 1998); for the Deep Space 2 $F_{pl}$ appears not to have been reported in open literature). The design also facilitates thermal control of the payload bay and reduces the number of pyrotechnic devices and commands needed – improving the EDLS reliability.

Table 4 compares the MNL to other soft and semi-hard Mars landers and their resources. The science payload fraction is listed here rather than the landed payload fraction. It should be noted that for older spacecraft like the Viking lander the level of integration of the instruments is low, i.e. each instrument may be self-contained rather than sharing resources, with an apparently higher payload ratio for newer spacecraft. Also small landers will tend to have a higher level of integration. The MNL compares favourable in this respect to other types of landers. The small dimensions of the MNL make it small enough to include multiple units in a carrier spacecraft.

The semi-hard nature of the entry, descent and landing system provides an excellent landed payload mass to overall mass ratio of about 0.2 facilitating an efficient use of the mass allocation of a scientific mission. The real strength of the MNL is demonstrated by atmospheric science missions requiring only modest amounts of data bandwidth, electrical energy and mass allocation for their scientific payloads. This facilitates the use of a highly versatile payload within the rel-

atively small mass allocation of the MNL vehicle. Furthermore, the MNL EDLS is inherently such that it requires less pyrotechnics (such as explosive bolts) and associated triggering commands than, for example, a traditional parachute-based landing system. This increases the overall likelihood of mission success.

A major asset of the MNL system is the eventual position, where the payload bay and its outer support structures are penetrated under the Martian surface with only the sensor boom, antenna and the outer rim above the surface. Such a position results in a favourable situation, where the payload bay will be surrounded by a natural thermal environment with temperature ranging from 230 K down to 210 K. These temperatures are still good for the electronics and other parts of the payload with the exception of batteries that need to be protected with additional thermal system. The position underneath the surface is extremely advantageous for a small probe like MNL from a thermal design point of view. At the Martian surface a small payload with low thermal inertia would require heating systems to survive the low night-time temperatures of the order of 170–190 K over a wide range of latitudes. The additional heating system would eat up a large fraction of the payload mass. Hence the MNL concept is giving both thermal shelter and a correct operational position for the payload.

The MNL EDLS allows for deployment to the Martian surface either directly from an interplanetary (hyperbolic) trajectory or from an orbit around Mars. Deployment from orbit enables more accurate landing, whereas direct deployment gives a wider selection of landing sites with the same $\Delta v$ budget. Due to fuel mass savings, direct deployment is often an appealing option – especially for atmospheric science missions for which modest landing precision is often adequate.

Presently two complete MNL EDLS systems have been manufactured and tested. They will be used on the forthcoming MetNet Precursor missions to demonstrate and validate the robustness and efficiency of the design. Prior to the launches parts of the MNL affected by shelf life, such as the fabrics of the inflatable EDLS components, will be replaced or refurbished. The Precursor landers will also carry out scientific observations and the development of two sets of atmospheric science payloads is currently under way. The payloads and their observations will be described in a separate paper.

The eventual goal of the MetNet mission concept is to create a network of MNL at the Martian surface operating simultaneously. Eventually a network is needed, for which the MNL is ideally suited. MNL could facilitate the mission with perhaps the launch of 15 units during one launch opportunity.

## 5.1 MNL concept validation: Precursor missions

The eventual validation of the MetNet Lander vehicle concept calls for an actual mission to the Martian surface and operations at Mars. The first concrete steps in this direction have already been taken. Currently an MMPM with one deployed MNL is being planned. The MMPM would perform continuous scientific observations by using a versatile set of science instruments, but the primary objective of this mission is to demonstrate the feasibility and technical robustness of the MNL concept before building the planet-wide network of observational posts.

For the Precursor missions this is extended to include also a three-axis gyroscope device. Additionally a solar incident sensor with a wide range of dedicated wavelength filters, an optical dust sensor, a three-axis magnetometer and a radiation monitor are included in the first units payload.

There exist also plans to deploy a network of some tens of MNLs furbished with atmospheric science instrumentation operating simultaneously and focused on the investigations of the Martian atmosphere. This kind of network mission has been planned for several decades but until now has never been implemented. The MNL concept provides a suitable tool to achieve this long-standing objective.

The eventual scope of the network mission is to operate the multiple scientific payloads at the Martian surface simultaneously for several Martian years. The MNL provides the means to this objective as it is designed to be operational during several Martian years.

A grand goal of creating a network of observational posts at the Martian surface can be reached either by sending a large number of MNLs to Mars onboard a single mission or by sending MNLs to Mars in successive launch windows. The latter network generation scheme requires that the lifetime of MetNet vehicles be of the order of several Martian years.

Individual or stand-alone MetNet missions can make important scientific investigations characterising the Martian environment.

## 6 Summary and concluding remarks

MNL, a small semi-hard lander/penetrator design with a payload mass fraction of approximately 17 %, has been developed, tested and prototyped. The MNL features an innovative EDLS that is based on inflatable structures capable of decelerating the lander from interplanetary transfer trajectories down to a surface impact speed of $50$–$70\,\mathrm{m\,s^{-1}}$ and a deceleration of $< 500\,\mathrm{g}$ for $< 20\,\mathrm{ms}$. The orientation of the penetrator main body into the surface strata at impact is approximately vertical and since the payload bay will be embedded in the surface materials, the bay's temperature excursions will be much less than if it was fully exposed on the Martian surface. The total mass of the prototype design is $\approx 24\,\mathrm{kg}$, with $\approx 4\,\mathrm{kg}$ of mass available for the payload. The MNL is particularly well suited for delivering meteorological and atmospheric instruments to the Martian surface. The possibility exists for sending other environmental instruments. The small size and low mass of an MNL makes it ideally suited for piggy-backing on larger spacecraft. MNLs are designed primarily for use as surface networks but could also be used as pathfinders for high-value landed missions.

*Competing interests.* The authors declare that they have no conflict of interest.

*Acknowledgements.* The authors would like to express their gratitude to the MetNet development teams at the Finnish Meteorological Institute (Finland), Lavochkin Association (Russia), the Russian Space Research Institute (IKI, Russia) and the Instituto Nacional de Técnica Aeroespacial (INTA, Spain) for their great work in the implementation of the MetNet landing vehicle. Ari-Matti Harri and Hannu Savijarvi are thankful for the Finnish Academy grants no. 132825 and no. 131723.

Edited by: G. Kargl

## References

Ball, A. J., Garry, J. R. C., Lorenz, R. D., and Kerzhanovich, V. V.: Planetary landers and entry probes, Cambridge University Press, Cambridge, UK, 2009.

Barderas, G. and Romero, P.: On the inverse problem of determining Mars lander coordinates using Phobos eclipse observations, Planet. Space Sci., 79, 39–44, 2013.

CCSDS 211.0-B-4: Proximity-1 Space Link Protocol – Data Link Layer, Consultative Committee for Space Data Systems, CCSDS Secretariat, Office of Space Communication (Code M-3), National Aeronautics and Space Administration, Washington, DC 20546, USA, 4 edn., 2006.

Chicarro, A. F., Coradini, M., Fulchignoni, M., Hiller, K., Knudsen, J. M., Liede, I., Lindberg, C., Lognonné, P., Pellinen, R., Spohn, T., Scoon, G. E. N., Taylor, F. W., and Wänke, H.: MARSNET Phase-A Study Report, Tech. Rep. SCI(93)2, European Space Agency, Paris, France, 1993.

Chicarro, A., Scoon, G., and Coradini, M.: INTERMARSNET – An international network of stations on Mars for global martian characterisation, ESA Journal, 18, 207–218, 1994.

Clark, I. G., Hutchings, A. L., Tanner, C. L., and Braun, R. D.: Supersonic Inflatable Aerodynamic Decelerators for Use on Future Robotic Missions to Mars, J. Spacecraft Rockets, 46, 340–352, 2009.

COSPAR planetary protection policy: COSPAR planetary protection policy, Policy document, Committee on Space Research (COSPAR), original dated 20 October 2002, amended 24 March 2005, 2005.

Golombek, M. P., Bridges, N. T., Moore, H. J., Murchie, S. L., Murphy, J. R., Parker, T. J., Rieder, R., Rivellini, T. P., Schofield, J. T., Seiff, A., Singer, R. B., Smith, P. H., Soderblom, L. A., Spencer, D. A., Stoker, C. R., Sullivan, R., Thomas, N., Thurman, S. W., Tomasko, M. G., Vaughan, R. M., Wänke, H., Ward, A. W., and Wilson, G. R.: Overview of the Mars Pathfinder Mission: Launch through landing, surface operations, data sets, and science results, J. Geophys. Res., 104, 8523–8554, 1999.

Gómez-Elvira, J., Armiens, C., Carrasco, I., Genzer, M., Gómez, F., Haberle, R., Hamilton, V. E., Harri, A.-M., Kahanpää, H., Kemppinen, O., Lepinette, A., Martín Soler, J., Martín-Torres, J., Martínez-Frías, J., Mischna, M., Mora, L., Navarro, S., Newman, C., Pablo, M. A., Peinado, V., Polkko, J., Rafkin, S. C. R., Ramos, M., Rennó, N. O., Richardson, M., Rodríguez-Manfredi, J. A., Romeral Planelló, J. J., Sebastián, E., Torre Juárez, M., Torres, J., Urquí, R., Vasavada, A. R., Verdasca, J., and Zorzano, M.-P.: Curiosity's rover environmental monitoring station: Overview of the first 100 sols, J. Geophys. Res.-Planet., 119, 1680–1688, doi:10.1002/2013JE004576, 2014.

GOST: Specifications, Tech. rep., 1977.

Grotzinger, J. P., Crisp, J., Vasavada, A. R., Anderson, R. C., Baker, C. J., Barry, R., Blake, D. F., Conrad, P., Edgett, K. S., Ferdowski, B., Gellert, R., Gilbert, J. B., Golombek, M., Gómez-Elvira, J., Hassler, D. M., Jandura, L., Litvak, M., Mahaffy, P., Maki, J., Meyer, M., Malin, M. C., Mitrofanov, I., Simmonds, J. J., Vaniman, D., Welch, R. V., and Wiens, R. C.: Mars Science Laboratory Mission and Science Investigation, Space Sci. Rev., 170, 5–56, doi:10.1007/s11214-012-9892-2, 2012.

Guinn, J. R., Garcia, M. D., and Talley, K.: Mission design of the Phoenix Mars Scout mission, J. Geophys. Res., 113, E00A26, doi:10.1029/2007JE003038, 2008.

Haberle, R. M. and Catling, D. C.: A Micro-Meteorological mission for global network science on Mars: rationale and measurement requirements, Planet. Space Sci., 44, 1361–1383, doi:10.1016/S0032-0633(96)00056-6, 1996.

Harri, A., Linkin, V., Polkko, J., Marov, M., Pommereau, J., Lipatov, A., Siili, T., Manuilov, K., Lebedev, V., Lehto, A., Pellinen, R., Pirjola, R., Carpentier, T., Malique, C., Makarov, V., Khloustova, L., Esposito, L., Maki, J., Lawrence, G., and Lystsev, V.: Meteorological observations on Martian surface: met-packages of Mars-96 Small Stations and Penetrators, Planet. Space Sci., 46, 779–793, doi:10.1016/S0032-0633(98)00012-9, 1998.

s Harri, A.-M., Marsal, O., Lognonné, P., Leppelmeier, G. W., Spohn, T., Glassmeier, K.-H., Angrilli, F., Banerdt, W. B., Barriot, J. P., Bertaux, J.-L., Bérthelier, J. J., Calcutt, S., Cerisier, J. C., Crisp, D., Déhant, V., Giardini, D., Jaumann, R., Langevin, Y., Menvielle, M., Mussmann, G., Pommereau, J. P., di Pippo, S., Guerrier, D., Kumpulainen, K., Larsen, S., Mocquet, A., Polkko, J., Runavot, J., Schumacher, W., Siili, T., Simola, J., Tillman, J. E., and the NetLander Team: Network science landers

for Mars, Adv. Space Res., 23, 1915–1924, doi:10.1016/S0273-1177(99)00279-3, 1999.

Harri, A.-M., Leinonen, J., Merikallio, S., Paton, M., Haukka, H., and Polkko, J.: MetNet – In situ observational Network and Orbital platform to investigate the Martian environment, Tech. rep., 2007.

Harri, A.-M., Schmidt, W., Romero, P., Vazquez, L., Barderas, G., Kemppinen, O., A. C., Vazquez-Poletti, J. L., Llorente, I. M., Haukka, H., and Paton, M.: Phobos eclipse detection on Mars: Theory and practice, Tech. rep., 2012.

Harri, A.-M., Genzer, M., Kemppinen, O., Gomez-Elvira, J., Haberle, R., Polkko, J., Savijärvi, H., Rennó, N., Rodriguez-Manfredi, J. A., Schmidt, W., Richardson, M., Siili, T., Paton, M., Torre-Juarez, M. D. L., Mäkinen, T., Newman, C., Rafkin, S., Mischna, M., Merikallio, S., Haukka, H., Martin-Torres, J., Komu, M., Zorzano, M.-P., Peinado, V., Vazquez, L., and Urqui, R.: Mars Science Laboratory relative humidity observations: Initial results, J. Geophys. Res.-Planet., 119, 2132–2147, doi:10.1002/2013JE004514, 2014a.

Harri, A.-M., Genzer, M., Kemppinen, O., Kahanpää, H., Gomez-Elvira, J., Rodriguez-Manfredi, J. A., Habserle, R., Polkko, J., Schmidt, W., Savijärvi, H., Kauhanen, J., Atlaskin, E., Richardson, M., Siili, T., Paton, M., Torre Juarez, M., Newman, C., Rafkin, S., Lemmon, M. T., Mischna, M., Merikallio, S., Haukka, H., Martin-Torres, J., Zorzano, M.-P., Peinado, V., Urqui, R., Lapinette, A., Scodary, A., Mäkinen, T., Vazquez, L., Rennó, N., and REMS/MSL Science Team: Pressure observations by the Curiosity rover: Initial results, J. Geophys. Res.-Planet., 119, 82–92, doi:10.1002/2013JE004423, 2014b.

Heilimo, J., Harri, A.-M., Aleksashkin, S., Koryanov, V., Arruego, I., Schmidt, W., Haukka, H., Finchenko, V., Martynov, M., Ostresko, B., Ponomarenko, A., Kazakovtsev, V., Martin, S., and Siili, T.: RITD – Adapting Mars Entry, Descent and Landing System for Earth, in: EGU General Assembly Conference Abstracts, vol. 16 of EGU General Assembly Conference Abstracts, 2014.

Kauhanen, J., Siili, T., Järvenoja, S., and Savijärvi, H.: The Mars Limited Area Model (MLAM) and simulations of atmospheric circulations for the Phoenix landing area and season-of-operation, J. Geophys. Res., 113, E00A14, doi:10.1029/2007JE003011, 2008.

Linkin, V., Harri, A.-M., Lipatov, A., Belostotskaja, K., Derbunovich, B., Ekonomov, A., Khloustova, L., Kremnev, R., Makarov, V., Martinov, B., Nenarokov, D., Prostov, M., Pustovalov, A., Shustko, G., Järvinen, I., Kivilinna, H., Korpela, S., Kumpulainen, K., Lehto, A., Pellinen, R., Pirjola, R., Riihelä, P., Salminen, A., Schmidt, W., Siili, T., Blamont, J., Carpentier, T., Debus, A., Hua, C. T., Karczewski, J.-F., Laplace, H., Levacher, P., Lognonné, P., Malique, C., Menvielle, M., Mouli, G., Pommereau, J.-P., Quotb, K., Runavot, J., Vienne, D., Grunthaner, F., Kuhnke, F., Mussmann, G., Rieder, R., Wänke, H., Economou, T., Herring, M., Lane, A., and McKay, C. P.: A sophisticated lander for scientific exploration of Mars: scientific objectives and implementation of the Mars-96 Small Station, Planet. Space Sci., 46, 717–737, doi:10.1016/S0032-0633(98)00008-7, 1998.

Lorenz, R. D.: Planetary penetrators: Their origins, history and future, Adv. Space Res., 48, 403–431, 2011.

Merrihew, S. C., Haberle, R. M., and Lemke, L. G.: A Micro-Meteorological mission for global network science on Mars:

a conceptual design, Planet. Space Sci., 44, 1385–1393, doi:10.1016/S0032-0633(96)00055-4, 1996.

MESUR study report: Mars Environmental Survey (MESUR) science objectives and mission description., NASA Ames Rsesearch Center study report, 1991.

Paton, M. D., Harri, A.-M., Savijärvi, H., Mäkinen, T., Hagermann, A., Kemppinen, O., and Johnston, A.: Thermal and microstructural properties of fine-grained material at the Viking Lander 1 site, Icarus, 271, 360–374, doi:10.1016/j.icarus.2016.02.012, 2016.

Romero, P., Barderas, G., Vazquez-Poletti, J. L., and Llorente, I. M.: Spatial chronogram to detect Phobos eclipses on Mars with the MetNet Precursor Lander, Planet. Space Sci., 59, 1542–1550, 2011.

Savijärvi, H., Crisp, D., and Harri, A.-M.: Effects of $CO_2$ and dust on present-day solar radiation and climate on Mars, Q. J. Roy. Meteor. Soc., 131, 2907–2922, doi:10.1256/qj.04.09, 2005.

Savijärvi, H., Harri, A.-M., and Kemppinen, O.: The diurnal water cycle at Curiosity: Role of exchange with the regolith, Icarus, 265, 63–69, doi:10.1016/j.icarus.2015.10.008, 2016.

Shalin, R. E.: Polymer Matrix Composites, Tech. rep., 1995.

Smith, P. H., Tamppari, L., Arvidson, R. E., Bass, D., Blaney, D., Boynton, W., Carswell, A., Catling, D., Clark, B., Duck, T., De-Jong, E., Fisher, D., Goetz, W., Gunnlaugsson, P., Hecht, M., Hipkin, V., Hoffman, J., Hviid, S., Keller, H., Kounaves, S., Lange, C. F., Lemmon, M., Madsen, M., Malin, M., Markiewicz, W., Marshall, J., McKay, C., Mellon, M., Michelangeli, D., Ming, D., Morris, R., Renno, N., Pike, W. T., Staufer, U., Stoker, C., Taylor, P., Whiteway, J., Young, S., and Zent, A.: Introduction to special section on the Phoenix Mission: Landing Site Characterization Experiments, Mission Overviews, and Expected Science, J. Geophys. Res.-Planet., 113, E00A18, doi:10.1029/2008JE003083, 2008.

Smrekar, S., Catling, D., Lorenz, R., Magalhães, J., Moersch, J., Morgan, P., Murray, B., Presley, M., Yen, A., Zent, A., and Blaney, D.: Deep Space 2: the Mars microprobe mission, J. Geophys. Res., 104, 27013–27030, doi:10.1029/1999JE001073, 1999.

Soffen, G. A.: Scientific Results of the Viking Missions, Science, 194, 1274–1276, doi:10.1126/science.194.4271.1274, 1976.

Soffen, G. A. and Snyder, C. W.: The First Viking Mission to Mars, Science, 193, 759–766, doi:10.1126/science.193.4255.759, 1976.

Surkov, Y. A. and Kremnev, R. S.: Mars-96 mission: Mars exploration with the use of penetrators, Planet. Space Sci., 46, 1689 – 1696, doi:10.1016/S0032-0633(98)00071-3, 1998.

Taylor, P. A., Catling, D. C., Daly, M., Dickinson, C. S., Gunnlaugsson, H. P., Harri, A.-M., and Lange, C. F.: Temperature, pressure, and wind instrumentation in the Phoenix meteorological package, J. Geophys. Res.-Planet., 113, E00A10, doi:10.1029/2007JE003015, 2008.

# 14

# Digital photography for assessing the link between vegetation phenology and $CO_2$ exchange in two contrasting northern ecosystems

**Maiju Linkosalmi**[1], **Mika Aurela**[1], **Juha-Pekka Tuovinen**[1], **Mikko Peltoniemi**[2], **Cemal M. Tanis**[1], **Ali N. Arslan**[1], **Pasi Kolari**[3], **Kristin Böttcher**[4], **Tuula Aalto**[1], **Juuso Rainne**[1], **Juha Hatakka**[1], and **Tuomas Laurila**[1]

[1]Finnish Meteorological Institute, Helsinki, Finland
[2]Natural Resources Institute Finland (LUKE), Vantaa, Finland
[3]Faculty of Biosciences, University of Helsinki, Helsinki, Finland
[4]Finnish Environment Institute (SYKE), Helsinki, Finland

*Correspondence to:* Maiju Linkosalmi (maiju.linkosalmi@fmi.fi)

**Abstract.** Digital repeat photography has become a widely used tool for assessing the annual course of vegetation phenology of different ecosystems. By using the green chromatic coordinate (GCC) as a greenness measure, we examined the feasibility of digital repeat photography for assessing the vegetation phenology in two contrasting high-latitude ecosystems. Ecosystem–atmosphere $CO_2$ fluxes and various meteorological variables were continuously measured at both sites. While the seasonal changes in GCC were more obvious for the ecosystem that is dominated by annual plants (open wetland), clear seasonal patterns were also observed for the evergreen ecosystem (coniferous forest). Daily and seasonal time periods with sufficient solar radiation were determined based on images of a grey reference plate. The variability in cloudiness had only a minor effect on GCC, and GCC did not depend on the sun angle and direction either. The daily GCC of wetland correlated well with the daily photosynthetic capacity estimated from the $CO_2$ flux measurements. At the forest site, the correlation was high in 2015 but there were discernible deviations during the course of the summer of 2014. The year-to-year differences were most likely generated by meteorological conditions, with higher temperatures coinciding with higher GCCs. In addition to depicting the seasonal course of ecosystem functioning, GCC was shown to respond to environmental changes on a timescale of days. Overall, monitoring of phenological variations with digital images provides a powerful tool for linking gross primary production and phenology.

## 1 Introduction

Phenology is an important factor in the ecology of ecosystems. The most distinctive phenomena comprising vegetation phenology are the changes in plant physiology, biomass and leaf area (Migliavacca et al., 2011; Sonnentag et al., 2011, 2012; Bauerle et al., 2012). In part, these changes drive the carbon cycle of ecosystems, and they have various feedbacks to the climate system through effects on surface albedo and aerodynamic roughness, and ecosystem–atmosphere exchanges of various gases (e.g. $H_2O$, $CO_2$ and volatile organic compounds) (Arneth et al., 2010). Besides leaf area, gas exchange is modulated by seasonal variations in photosynthesis and respiration (Richardson et al., 2013). Globally, these variations contribute to the fluctuations in the atmospheric $CO_2$ concentration (Keeling et al., 1996). In the long term, possible trends in vegetation phenology can have a systematic effect on the mean $CO_2$ level. Phenology further plays a role in the competitive interactions, trophic dynamics, reproductive biology, primary production and nutrient cycling (Morisette et al., 2009). Phenological phenomena are largely controlled by abiotic factors such as temperature, water availability and day length (Bryant and Baird, 2003; Körner and

Basler, 2010), and thus they are sensitive to climate change (Richardson et al., 2013; Rosenzweig et al., 2007; Migliavacca et al., 2012).

Several studies have reported an advanced onset of the growing season during recent decades (Linkosalo et al., 2009; Delbart et al., 2008; Nordli et al., 2008; Pudas et al., 2008). An earlier onset of growth has been observed to play a significant role in the annual carbon budget of temperate and boreal forests, while lengthening autumns have a less clear effect (Goulden et al., 1996; Berninger, 1997; Black et al., 2000; Barr et al., 2007; Richardson et al., 2009). This can be explained by the rapid C accumulation that starts as soon as conditions turn favourable for photosynthesis and growth in spring, while the opposing effect, i.e. ecosystem respiration, becomes increasingly important in summer and autumn (White and Nemani, 2003; Dunn et al., 2007).

In general, monitoring of vegetation changes by digital cameras has become feasible with the development of advanced but inexpensive cameras that produce automated and continuous real-time data. It has been shown that simple time-lapse photography can facilitate detection of vegetation phenophases and even the related variations in $CO_2$ exchange (Wingate et al., 2015; Richardson et al., 2007, 2009). This provides new possibilities for monitoring and modelling of ecosystem functioning, for verification of remote sensing products, and for analysis of ecosystem $CO_2$ exchange fluxes and related balances. Especially dynamic vegetation models and simulations of C cycle could be improved by more accurate information on the timing of budburst and leaf senescence, as simple empirical parameterizations, typically based on degree days or the onset and offset dates of C uptake, are presently used as indicators of the growing season start and end (Baldocchi et al., 2005; Delpierre et al., 2009; Richardson et al., 2013).

Digital cameras produce red-green-blue (RGB) colour channel information, from which different greenness indices can be calculated. For example, canopy greenness has been expressed in terms of the so-called green chromatic coordinate (GCC), which has been related to vegetation activity and further to carbon uptake of forests (Richardson et al., 2007, 2009; Ahrends et al., 2009; Ide et al., 2011) and peatlands (Sonnentag et al., 2011; Peichl et al., 2015). In deciduous forests, the main driver of gas exchange is leaf area that changes rapidly in spring and autumn, which is easy to detect. In evergreen conifer forests the leaf area changes are much smaller, so it is not obvious whether a similar relationship can be established for them. In a peatland environment, repeat images have been used to map the mean greenness of mire vegetation over a wide area (Peichl et al., 2015). For peatland ecosystems with a heterogeneous vegetation cover, it may be possible to simultaneously detect seasonality effects of different vegetation types. Thus digital repeat images of differentially developing vegetation types could potentially help decompose an integrated $CO_2$ flux observation into components allocated to these vegetation types.

Comparisons of phenological observations made in contrasting ecosystems are needed for highlighting the phenological features that can be extracted from camera monitoring at different sites (Wingate et al., 2015; Keenan et al., 2014; Toomey et al., 2015; Sonnentag et al., 2012). Differences in the ecosystem characteristics may also affect the ideal set-up of cameras and the interpretation of images, for example in conjunction with surface flux data.

The objectives of this study were to (1) evaluate the digital repeat photography as a method for monitoring the phenology of boreal vegetation at high latitudes, (2) investigate the differences in the phenology between two adjacent but contrasting ecosystems (pine forest and wetland) located in northern Finland, and (3) assess whether the data obtained from such cameras can support the interpretation of the micrometeorological measurements of $CO_2$ fluxes conducted at the sites.

This paper is structured as follows: Sect. 2 introduces the measurement sites, camera set-up, image analysis, and the $CO_2$ flux and meteorological data employed; Sect. 3 provides the results and related discussion, including tests of the monitoring system and an analysis of the observed phenological development in relation to $CO_2$ exchange; finally, Sect. 4 presents the conclusions emerging from this study.

## 2  Materials and methods

### 2.1  Measurement sites

The study sites were located at Sodankylä in northern Finland, 100 km north of the Arctic Circle. They represent two contrasting ecosystems, a Scots pine (*Pinus sylvestris*) forest (67°21.708′ N, 26°38.290′ E; 179 m a.s.l.) and an open pristine wetland (67°22.117′ N, 26°39.244′ E; 180 m a.s.l.). The long-term (1981–2010) mean temperature and precipitation within the area are −0.4 °C and 527 mm, respectively (Pirinen et al., 2012).

The Scots pine stand is located on fluvial sandy podzol and has a dominant tree height of 13 m and a tree density of 2100 ha$^{-1}$. The age of the trees within the camera scope is about 50 years. A single-sided leaf area index (LAI) of 1.2 m$^2$ m$^{-2}$ has been estimated for the stand based on a forest inventory in 2000. The sparse ground vegetation consists of lichens (73 %), mosses (12 %) and ericaceous shrubs (15 %).

The wetland site is located on a mesotrophic fen that represents typical northern aapa mire. The vegetation at this site mainly consists of low species (*Carex* spp., *Menyanthes trifoliata, Andromeda polifolia, Betula nana, Vaccinium oxycoccos, Sphagnum* spp.). There are no tall trees, only some *B. pubescens* and a few isolated Scots pines. Different types of vegetation are located on drier (strings) and wetter (flarks) parts of the wetland.

The physical surface structure (aerodynamic roughness length) differs between the pine forest and wetland sites.

Also, the microclimate and surface exchange of $CO_2$ and sensible and latent heat differ due to different vegetation and soil characteristics.

## 2.2  Camera set-up

The images analysed in this study were taken automatically with StarDot Netcam SC 5 digital cameras. The set-up included a weather-proof housing and connections to line current and a web server. The pictures were stored in the 8 bit JPEG format every 30 min with 2592 × 1944 resolution and transferred automatically to a remote server. The daily collecting period varied according to the time of the year roughly covering the daylight hours.

At the forest site, the cameras were mounted to a tower at two different heights: 29 m ("canopy camera") and 13 m ("crown camera"). The viewing angle of the canopy camera was 45° from the horizontal plane, while the crown camera was positioned nearly horizontally. The images of the canopy camera covered parts of the forest canopy and some general landscape. The crown camera was focused to individual trees to detect their phenological development (e.g. bud burst, shoot growth, needle shedding) more closely. At the wetland site, the camera was adjusted in an angle of 45° on top of a 2 m pole. This camera mostly observed the ground vegetation, with some *B. pubescens* and sky also visible in the images. All cameras were placed facing the north to minimize lens flare and maximize illumination of the canopy.

## 2.3  Grey reference plates

At the forest site, grey reference plates were employed to monitor the stability of the image colour channels. The plates were attached to the cameras in such a way that they are visible in every picture. The idea behind the reference plates was to detect possible day-to-day shifts in the colour balance due to changing weather conditions, such as radiation variations. The reference images should also not show any obvious seasonality (Petach et al., 2014). The grey colour of the plates was close to the "true grey" in a sense that it has an equal mix of red, green and blue colour components. To achieve this, the reference plates were painted with Tikkurila grey/1948 (RGB values: R = 95, G = 95, B = 95).

## 2.4  Automatic image analysis

The digital images were analysed with the FMIPROT software that has been designed as a toolbox for image processing for phenological and meteorological purposes (Tanis and Arslan, 2016). FMIPROT calculates the colour fractions for red, green and blue channels. In the present analysis we use the GCC defined as

$$\text{GCC} = \frac{\sum G}{\sum R + \sum G + \sum B}, \tag{1}$$

**Figure 1.** View from the wetland camera. The solid lines indicate four regions of interest defined according to vegetation types. The dashed lines indicate the region of interest that includes all vegetation types except *Betula pubescens*.

where $\sum G$, $\sum R$ and $\sum B$ are the sums of green, red and blue channel digital numbers, respectively, of all pixels comprising an image.

Within each image, it was possible to define limited subareas of regions of interest (ROIs). The ROI feature of FMIPROT makes it possible to limit the GCC calculation to an area that represents a homogeneous vegetation area. It also provides an option for analysing several subareas within the image simultaneously.

## 2.5  Selection of the region of interest

At the wetland site, GCC was calculated separately for four different, clearly identifiable vegetation types. These vegetation types were dominated by (1) bog rosemary (*Andromeda polifolia*) and other shrubs, (2) sedges (*Carex* spp.) and *Sphagnum* mosses, (3) big-leafed bogbean (*Menyanthes trifoliate*), and (4) downy birch (*Betula pubescens*) (Fig. 1). The first three ROIs also included other ground vegetation, while the fourth ROI was limited to the birch canopy. The GCC values were also analysed from a larger area that includes the three first vegetation types (Fig. 1).

The forest site had two cameras, one zoomed to the crown of a pine tree (Fig. 2) and the other providing a general view of the canopy (Fig. 3). From the general canopy image, three separate ROIs were subjectively selected with an aim to define similar homogenous areas of forest canopy (Fig. 3).

## 2.6  CO$_2$ flux measurements

The ecosystem–atmosphere $CO_2$ exchange was measured at both study sites by the micrometeorological eddy covariance (EC) method. The EC measurements provided continuous data on the $CO_2$ fluxes averaged on an ecosystem scale. The vertical $CO_2$ flux is obtained as the covariance of the

**Figure 2.** View from the pine forest crown camera. The line indicates the region of interest.

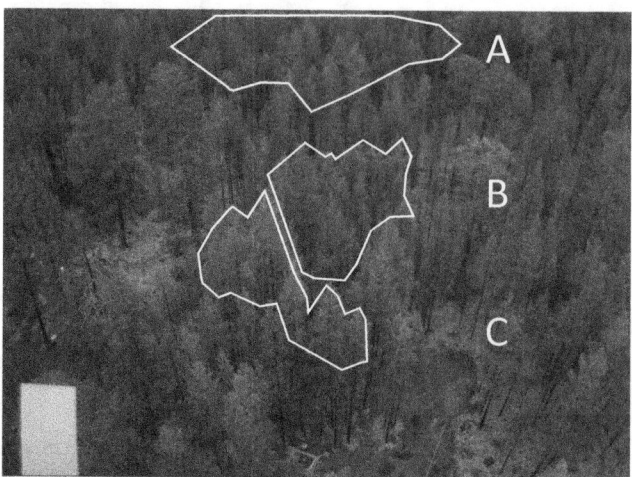

**Figure 3.** View from the pine forest canopy camera. The lines indicate three regions of interest.

high-frequency (10 Hz) fluctuations of vertical wind speed and $CO_2$ mixing ratio (Baldocchi, 2003). At both sites, the EC measurement systems consisted of a USA-1 (METEK GmbH, Elmshorn, Germany) three-axis sonic anemometer/thermometer and a closed-path LI-7000 (LI-COR, Inc., Lincoln, NE, USA) $CO_2/H_2O$ gas analyzer. The measurement systems and the data processing procedures have been presented in detail by Aurela et al. (2009).

The $CO_2$ fluxes obtained from the EC measurements represent the net ecosystem exchange (NEE) of $CO_2$, which is the sum of gross photosynthetic production (GPP) by plants and a respiration term that includes both the autotrophic respiration by plants and the heterotrophic respiration by microbes. GPP is typically derived from the NEE data by using a dedicated flux partitioning technique, for example based on nonlinear regressions with photosynthetic photon flux density (PPFD) and air temperature as predictors (Reichstein et al., 2005). Instead of performing such an explicit partitioning, we determined the daily GPP in terms of the gross photosynthesis index (GPI); for details, see Aurela et al. (2001), where a similar index was termed "PI". GPI indicates the maximal photosynthetical activity in optimal radiation conditions. It is obtained by calculating the differences of the daily averages of the daytime (PPFD $> 600 \, \mu \text{mol m}^{-2} \text{s}^{-1}$, which limit represents light saturation of photosynthesis at our sites) and night-time (PPFD $< 20 \, \mu \text{mol m}^{-2} \text{s}^{-1}$) NEE. The resulting GPI scales well with the maximal GPP obtained from a traditional NEE partitioning, despite the day–night differences in respiration. GPI provides a useful measure especially for depicting the seasonal GPP cycle, but as it is robust against missing data, it also estimates photosynthetic activity during fast changes due to short-term variations in air temperature and humidity (Aurela et al., 2001).

## 2.7 Meteorological measurements

An extensive set of supporting meteorological variables was measured at both measurement sites, including air temperature and humidity, various soil parameters (temperature, humidity, soil heat flux and water table level) and different radiation components (incoming and outgoing shortwave (SW) radiation, PPFD and net radiation). Here we used the temperature data measured at 3 m height on the wetland (Vaisala, HMP155D) and at 8 m at the forest site (Pentronic, PT100). From the SW radiation measurements (Kipp & Zonen, CM11) we calculated the surface albedo as the proportion of incident radiation that is reflected back to the atmosphere by the underlying surface. In addition, fractional cloud cover (CL) data were available from the nearby observatory.

## 3 Results and discussion

### 3.1 Testing the set-up

#### 3.1.1 Effect of environmental conditions on GCC

An accurate GCC observation requires a sufficient illumination level, which was here ensured by selecting only mid-day (10:00–14:00 local winter time) photographs for further analysis. This period was determined on the basis of the GCC of the grey reference plate in different radiation conditions (Supplement, Figs. S1–S3).

The influence of cloudiness on GCC was estimated from the data collected in July 2014. This particular month was selected for the test because July represents the peak growing season (for both radiation levels and LAI), and in July 2014 sunny and cloudy days were equally frequent. Based on the observations of fractional cloud cover (rang-

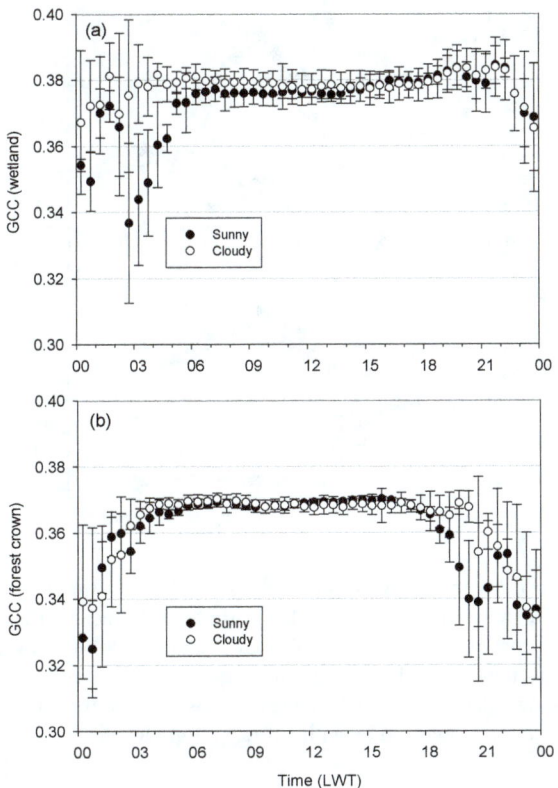

**Figure 4.** Mean (± standard deviation shown by error bars) diurnal cycle of GCC during sunny and cloudy conditions observed with (**a**) the tree crown and (**b**) the wetland cameras in July 2014.

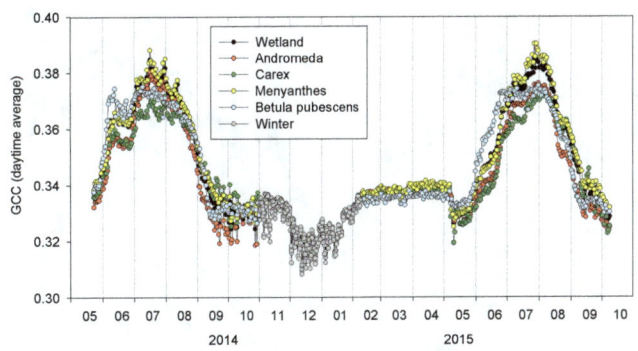

**Figure 5.** Mean daytime (10:00–14:00, local winter time) GCC of different regions of interest (vegetation types) during the measurement period of May 2014 to October 2015. Wetland refers to combined ROI shown in Fig. 5b. The grey circles indicate the wintertime data that are influenced by an insufficient light level.

ing from clear sky with $CL = 0$ to completely cloudy conditions with $CL = 8$), the images were pooled to two contrasting cloudiness groups representing sunny ($CL = 0$–1) and cloudy ($CL = 7$–8) conditions. During the daily period of 10:00–14:00, the differences in the mean GCC between sunny and cloudy conditions were statistically insignificant (Mann–Whitney U test) (Fig. 4). The mean GCC difference between the cloudy and sunny groups was 0.0014 and 0.0011 for the fen and forest, respectively. Sonnentag et al. (2012) found an equivalently small, though in part statistically significant, difference between the diurnal GCC cycles of sunny and overcast situations for their deciduous and coniferous forests.

The dependence of GCC on the solar angle with respect to ROI was also estimated from the data of July 2014 (Fig. 4). The difference between the minimum and maximum values of the hourly GCC means within the daytime window was 0.0030 and 0.0020 for sunny and cloudy cases, respectively. This is less than 5 % of the seasonal amplitude of the GCC curve (0.069 between May and July) associated with phenological greening of the fen. At the forest site, the corresponding values were 0.0022 (sunny) and 0.0012 (cloudy) and, despite the lower annual amplitude (0.024 between May

and July), the difference was less than 10 % of the seasonal variation.

### 3.1.2 Sensitivity of GCC to selection of the region of interest

The sensitivity of the GCC values to the selection of a sub-area within an image, i.e. a region of interest, was tested by comparing the GCC calculated for different vegetation patches. In particular, we wanted to examine, on the one hand, whether the forest images are homogeneous and thus insensitive to the ROI definition; on the other hand, the wetland images may provide an opportunity to simultaneously observe various microecosystems incorporated into a single image.

The GCC values of the wetland ROIs defined according to vegetation types showed significant differences in the seasonal cycle, both in the timing of the major changes in spring and autumn and in the magnitude of the maximum GCC (Fig. 5). For example, downy birch had the earliest growth onset, while the big-leafed bogbean had the largest growing-season maximum. While the seasonal patterns of the GCCs of different ROIs can be compared, the same may not be true for the absolute GCC values, which were affected by different viewing angles and distances to the target. To gain a better insight into the quantitative differences between different ROIs, these ROI-specific GCC data should be investigated in conjunction with direct vegetation analysis (LAI, biomass) and small-scale (chamber-based) $CO_2$ exchange measurements. For further analysis here we chose to use the larger ROI combining three vegetation types (Fig. 1), which matches better the areally integrating flux measurements.

The daily mean GCC values of different forest canopy ROIs remained very similar throughout the time series (Fig. 6). The GCC values determined from crown images differed from those from the camera with a general canopy view, most likely because the cameras had different viewing

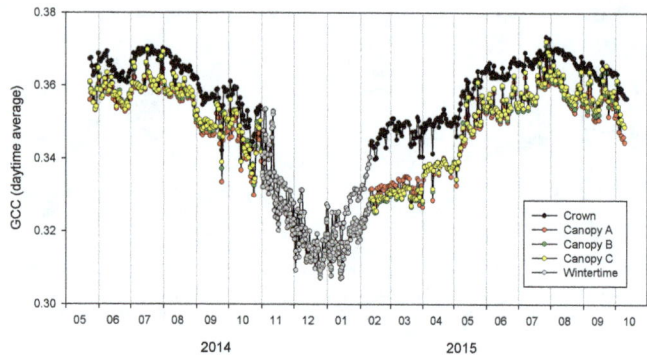

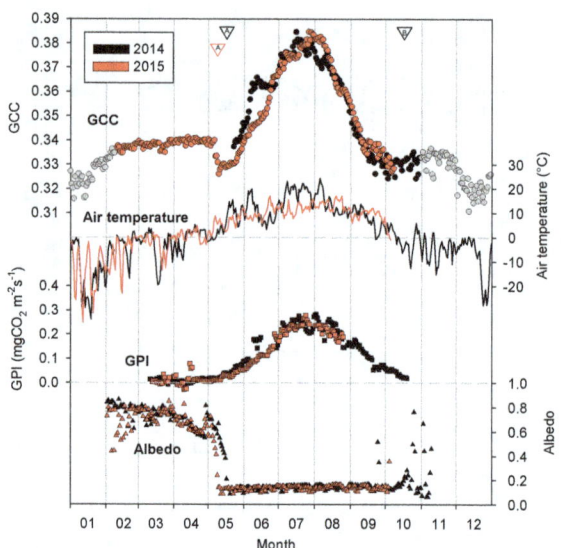

**Figure 6.** Mean daytime (10:00–14:00, local winter time) GCC values of different ROIs from two forest cameras during the measurement period of May 2014 to October 2015. The grey circles indicate the wintertime data that are influenced by an insufficient light level.

**Figure 7.** Mean daytime (10:00–14:00, local winter time) GCC (of the ROI shown in Fig. 1b) together with the daily mean air temperature, gross photosynthesis index (GPI) and albedo in 2014–2015 at the wetland site. The triangles indicate the dates of snowmelt (A) and snow appearance (B). The grey circles indicate the wintertime data that are influenced by an insufficient light level.

angles and distances to the object. The contribution of ground is mixed with the canopy signal, which partially explains why the GCC values in the distant canopy camera images were lower than in the crown camera images. In winter, there was more snow visible behind the canopy in the smaller-scale ROIs. Thus, we decided on using in further analysis only the images from the crown camera.

### 3.2 Phenological development

#### 3.2.1 Wetland site

As previously observed by Peichl et al. (2015), at the wetland the growing season is clearly discernable in the development of GCC data (Fig. 7). GCC started to increase as soon as the wetland vegetation started to assimilate $CO_2$. This growth onset took place in May after the snowmelt, for which the ground albedo provides a sensitive indicator by quantifying the proportion of incident solar radiation that is reflected back to the atmosphere. However, the onset was preceded by a short period of reduced GCC values, which were associated with the moist and dark soil.

The warm spells during late May and early June in 2014 induced a rapid emergence and growth of annual plants. Despite the later snowmelt that year, by mid-June the growing season had developed much further than in 2015. This difference is clearly visible in the GCC as well as photosynthetic activity (GPI) data (Fig. 7). The cold period in late June 2014 ceased this fast development, which is also well reflected in the GCC data that show a stabilization and even a temporary reduction during that period. GPI shows a similar pattern, highlighting the coherence between the greenness observation and the actual photosynthetic processes.

Following the earlier onset of the growing season in 2014, the peak of plant development was also observed earlier (Fig. 7). However, the magnitude of the GCC maxima during the two years was the same (0.385). From mid-August

to mid-September, the rate of GCC decline was approximately the same in 2014 and 2015. In mid-September, the slightly higher GCC in 2014 can be attributed to a warm period. By the first sub-zero values in daily mean temperatures, the GCC had decreased to its minimum value, close to the springtime minimum, and by the snowfall in mid-October it had started increasing towards the level observed for the fully snow-covered conditions in spring.

Previous observations suggest that GPP is well correlated with the GCC of wetlands, especially during spring (Peichl et al., 2015). Our results support these observations showing a strong relationship between the daily GCC and GPI data (Fig. S4), with a correlation coefficient of 0.90 and 0.92 for the snow-free period in 2014 and 2015, respectively. Especially during the springtime, the match between the GCC and GPI time series was remarkably close during both years, while in the autumn of 2014 GPI lagged slightly behind GCC.

#### 3.2.2 Forest site

Due to the closeness of the measurement sites, the meteorological conditions in forest were similar to those observed at the wetland (Figs. 7 and 8). However, the onset of photosynthetical activity differed slightly at the beginning of the growing season: the warm days of early May 2015 were not observed at the wetland as an GPI increase due to the absence of annual vegetation right after the snowmelt, while the photosynthesis of boreal trees is triggered as soon as temperature

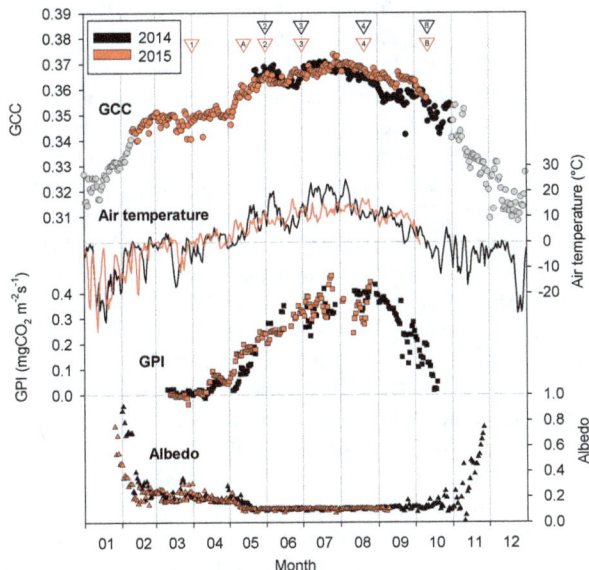

**Figure 8.** Mean daytime (10:00–14:00, local winter time) GCC (crown camera) together with the daily mean air temperature (at 18 m), gross photosynthesis index (GPI) and albedo in 2014–2015 at the forest site. The triangles indicate the start dates of visually observed phenological phases (1 – bud burst, 2 – bud growth, 3 – shoot growth, 4 – old needle browning) and snow status (A – snowmelt, B – snow appearance). The grey circles indicate the wintertime data that are influenced by an insufficient light level.

reaches a sufficient level (Tanja et al., 2003). Thus the growing season in the forest started earlier in 2015 than in 2014, while that was not the case at the wetland. Nevertheless, the warm period in late May–early June 2014 also enhanced the forest growth, and by mid-June both GCC and GPI had surpassed the corresponding level in 2015. The cold period in late June 2014 was again observed as reduced $CO_2$ uptake and even a clearer reduction in GCC than at the wetland.

Although in deciduous forest and open wetlands GCC is generally well correlated with the gross ecosystem photosynthesis during the start of the growing season (Peichl et al., 2015; Toomey et al., 2015), for evergreen needleleaf forests it has been reported that such correlation is often weaker (Toomey et al., 2015; Wingate et al., 2015). In our pine forest, however, the simultaneous development of GCC and photosynthesis was evident during the year with spring data available (Fig. S5).

Similarly to the wetland, the maximum GCC level at the forest site did not differ between 2014 and 2015, but this level was reached slightly earlier in 2014. This was probably due to the higher temperatures during the first part of the growing season. During both years, GCC started decreasing at the same time, i.e. at the end of July. This was slightly earlier than the start of the senescence detected visually (Phase 4 in Fig. 8). Similarly to the wetland, in 2014 there was a clear phase difference between GCC and GPI, the latter of which

stayed at the maximum level until the end of August. In forest, this may be due to the oldest needles, whose senescence takes place in August, while their photosynthetic capacity has diminished already earlier (Vesala et al., 2005).

In both 2014 and 2015, the photosynthetic activity continues until the end of August, but the interannual comparison is not possible here owing to the missing $CO_2$ data in 2015. Nevertheless, in both years GCC decreases to the wintertime level at the beginning of October, at the same time as the daily mean temperature decreases below 0 °C.

Our results show that the phenological development of the pine canopy could be accurately monitored with the GCC analysis, even though the GCC changes in forest were subtler than those observed for the wetland vegetation. This was confirmed by visually identifying the phenological stages of the forest from the crown camera pictures (Fig. 8). In 2014, the cameras were installed too late to detect the bud burst, but the GCC time series was consistent with the observation that the buds started their growth at the beginning of June and remained brown until the beginning of July, when they started to green.

## 4　Conclusions

We demonstrated the feasibility of digital repeat photography for assessing the link between vegetation phenology and $CO_2$ exchange for two contrasting high-latitude ecosystems. While the seasonal changes in the greenness index GCC are more obvious for those ecosystems where the vegetation is renewed every year (here an open wetland), seasonal patterns can also be observed in the evergreen ecosystems (here a coniferous forest).

We examined the illumination sensitivity of our digital camera system by analysing the images of a grey reference plate, which was included in the camera view. Limited solar radiation restricts the use of images during the wintertime as well as during the night-time. At our sites in northern Finland, the daytime radiation levels were sufficient for image analysis from February to October. During that period, a diurnal window of 10:00–14:00 (local winter time) provides stable GCC data. Our results show that the variability in cloudiness and solar zenith angle during the daytime does not play a significant role in the GCC analysis. However, it would be relevant to investigate the seasonal dependence of GCC on sun elevation, especially for the coniferous forest.

We observed a clear seasonal GCC cycle at both study sites. At the wetland, GCC correlated well with the daily photosynthetic capacity estimated from the ecosystem–atmosphere flux measurements. The interannual variation in GCC was also consistent with the observed $CO_2$ exchange and meteorological conditions. At the forest site, the seasonal GCC cycle correlated well with the flux data in 2015 but showed more deviations during the summer of 2014. For both ecosystems, the correlation between GCC and $CO_2$ exchange was highest during the spring.

In addition to depicting the seasonal course of ecosystem functioning, we showed that GCC responds to environmental changes on a shorter timescale. We observed that at both sites the increase of GCC and photosynthesis ongoing in June was ceased during a 2-week-long cold and wet period. For an unknown reason, the GCC values even slightly decreased during that period. It is possible that such a reduction is an artefact caused by wet surfaces, for example, rather than a response to an actual decrease in the chlorophyll concentration in leaves and needles.

Due to the low cost of the instrumentation involved, phenology monitoring can be established in a much larger number of locations than ecosystem–atmosphere flux measurements, thus providing a wider geographical basis for improvement of the phenological and photosynthesis components of land surface models that need more calibration and validation. The digital repeat images allow the detection of phenological events, such as shoot elongation and the start of needle growth that cannot be obtained from $CO_2$ flux measurements alone. Therefore, they should be utilized to enhance the analysis of flux data. Furthermore, as our results show, the seasonal cycle of different vegetation types within the footprint of the flux measurements can be determined. This could help decompose the integrated $CO_2$ flux observations when the distribution of the vegetation types within the area is known.

*Acknowledgements.* This work was supported by the EU: the installation of the cameras and the development of the image processing tool (FMIPROT) was done within MONIMET Project (LIFE12ENV/FI/000409), funded by EU Life+ Programme (2013–2017) (http://monimet.fmi.fi).

Edited by: M. Paton

# References

Ahrends, H. E., Etzold, S., Kutsch, W. L., Stoeckli, R., Bruegger, R., Jeanneret, F., Wanner, H., Buchmann, N., and Eugster, W.: Tree phenology and carbon dioxide fluxes: use of digital photography for process-based interpretation at the ecosystem scale, Clim. Res., 39, 261–274, doi:10.3354/cr00811, 2009.

Arneth, A., Harrison, S. P., Zaehle, S., Tsigaridis, K., Menon, S., Bartlein, P. J., Feichter, J., Korhola, A., Kulmala, M., O'Donnell, D., Schurgers, G., Sorvari, S., Vesala, T.: Terrestrial biogeochemical feedbacks in the climate system, Nat. Geosci., 3, 525–532, 2010.

Aurela, M., Tuovinen, J.-P., and Laurila, T.: Net $CO_2$ exchange of a subarctic mountain birch ecosystem, Theor. Appl. Climatol., 70, 135–148, 2001.

Aurela, M., Lohila, A., Tuovinen, J.-P., Hatakka, J., Riutta, T., and Laurila, T.: Carbon dioxide exchange on a northern boreal fen, Boreal Environ. Res., 14, 699–710, 2009.

Baldocchi, D.: Assessing the eddy covariance technique for evaluating carbon dioxide exchange rates of ecosystems: past, present and future, Glob. Change Biol., 9, 479–492, 2003.

Baldocchi, D. D., Black, T. A., Curtis, P. S., Falge, E., Fuentes, J. D., Granier, A., Gu, L., Knohl, A., Pilegaard, K., Schmid, H. P., Valentini, R., Wilson, K., Wofsy, S., Xu, L., and Yamamoto, S.: Predicting the onset of carbon uptake by deciduous forests with soil temperature and climate data: a synthesis of FLUXNET data, Int. J. Biometeorol., 49, 377–387, 2005.

Barr, A. G., Black, T. A., Hogg, E. H., Griffis, T. J., Morgenstern, K., Kljun, N., Theede, A., and Nesic, Z.: Climatic controls on the carbon and water balances of a boreal aspen forest, 1994–2003, Glob. Change Biol., 13, 561–576, 2007.

Bauerle, W. L., Oren, R., Way, D. A., Qian, S. S., Stoy, P. C., Thornton, P. E., Bowden, J. D., Hoffman, F. M., and Reynolds, R. F.: Photoperiodic regulation of the seasonal pattern of photosynthetic capacity and the implications for carbon cycling, P. Natl. Acad. Sci. USA, 109, 8612–8617, 2012.

Berninger, F.: Effects of drought and phenology on GPP in Pinus sylvestris: a simulation study along a geographical gradient, Funct. Ecol., 11, 33–43, 1997.

Black, T. A., Chen, W. J., Barr, A. G., Arain, M. A., Chen, Z., Nesic, Z., Hogg, E. H., Neumann, H. H., and Yang, P. C.: Increased carbon sequestration by a boreal deciduous forest in years with a warm spring, Geophys. Res. Lett., 27, 1271–1274, 2000.

Bryant, R. G. and Baird, A. J.: The spectral behaviour of Sphagnum canopies under varying hydrological conditions, Geophys. Res. Lett., 30, 1134–1138, 2003.

Delbart, N., Picard, G., Le Toans, T., Kergoat, L., Quegan, S., Woodward, I., Dye, D., and Fedotova, V.: Spring phenology in boreal Eurasia over a nearly century time scale, Glob. Change Biol., 14, 603–614, 2008.

Delpierre, N., Dufrene, E., Soudani, K., Ulrich, E., Cecchini, S., Boe, J., and Francois, C.: Modelling interannual and spatial variability of leaf senescence for three deciduous tree species in France, Agr. Forest Meteorol., 149, 938–948, 2009.

Dunn, A. L., Barford, C. C., Wofsy, S. C., Goulden, M. L., and Daube, B. C.: A long-term record of carbon exchange in a boreal black spruce forest: means, responses to interannual variability, and decadal trends, Glob. Change Biol., 13, 577–590, 2007.

Goulden, M. L., Munger, J. W., Fan, S. M., Daube, B. C., and Wofsy, S. C.: Measurements of carbon sequestration by long-term eddy covariance: methods and a critical evaluation of accuracy, Glob. Change Biol., 2, 169–182, 1996.

Ide, R., Nakaji, T., Motohka, T., and Oguma, H.: Advantages of visible-band spectral remote sensing at both satellite and near-surface scales for monitoring the seasonal dynamics of GPP in a Japanese larch forest, J. Agr. Meteorol., 67, 75–84, 2011.

Keeling, C. D., Chin, J. F. S., and Whorf, T. P.: Increased activity of northern vegetation inferred from atmospheric $CO_2$ measurements, Nature, 382, 146–149, 1996.

Keenan, T. F., Darby, B., Felts, E., Sonnentag, O., Friedl, M., Hufkens, K., O'Keefe, J. F., Klosterman, S., Munger, J. W., Toomey, M., and Richardson, A. D.: Tracking forest phenology and seasonal physiology using digital repeat photography: a critical assessment, Ecol. Appl., 24, 1478–1489, 2014.

Körner, C. and Basler, D.: Warming, photoperiods, and tree phenology response, Science, 329, 278–278, 2010.

Linkosalo, T., Häkkinen, R., Terhivuo, J., Tuomenvirta, H., and Hari, P.: The time series of flowering and leaf bud burst of boreal trees (1846–2005) support the direct temperature observations of climatic warming, Agr. Forest Meteorol., 149, 453–461, 2009.

Migliavacca, M., Galvagno, M., Cremonese, E., Rossini, M., Meroni, M., Sonnentag, O., Manca, G., Diotri, F., Busetto, L., Cescatti, A., Colombo, R., Fava, F., Morra di Cella, U., Pari, E., Siniscalco, C., and Richardson, A.: Using digital repeat photography and eddy covariance data to model grassland phenology and photosynthetic $CO_2$ uptake, Agr. Forest Meteorol., 151, 1325–1337, 2011.

Migliavacca, M., Sonnentag, O., Keenan, T. F., Cescatti, A., O'Keefe, J., and Richardson, A. D.: On the uncertainty of phenological responses to climate change, and implications for a terrestrial biosphere model, Biogeosciences, 9, 2063–2083, doi:10.5194/bg-9-2063-2012, 2012.

Morisette, J. T., Richardson, A. D., Knapp, A. K., Fisher, J. I., Graham, E. A., Abatzoglou, J., Wilson, B. E., Breshears, D. D., Henebry, G. M., Hanes, J. M., and Liang, L.: Tracking the rhythm of the seasons in the face of global change: Phenological research in the 21st century, Front. Ecol. Environ., 7, 253–260, 2009.

Nordli, O., Wielgolaski, F. E., Bakken, A. K., Hjeltnes, S. H., Mage, F., Sivle, A., and Skre, O.: Regional trends for bud burst and flowering of woody plants in Norway as related to climate change, Int. J. Biometeorol., 52, 625–639, 2008.

Peichl, M., Sonnentag, O., and Nilsson, M. B.: Bringing Color into the Picture: Using Digital Repeat Photography to Investigate Phenology Controls of the Carbon Dioxide Exchange in a Boreal Mire, Ecosystems, 18, 115–131, 2015.

Petach, A., Toomey, M., Aubrecht, D., Richardson, A. D: Monitoring vegetation phenology using an infrared-enabled security camera, Agr. Forest Meteorol., 195, 143–151, 2014.

Pirinen, P., Simola, H., Aalto, J., Kaukoranta, J.-P., Karlsson, P., and Ruuhela, R.: Climatological statistics of Finland 1981–2010, Reports 2012:1, Finnish Meteorological Institute, Helsinki, 2012.

Pudas, E., Leppälä, M., Tolvanen, A., Poikolainen, J., Venäläinen, A., and Kubin, E.: Trends in phenology of Betula pubescens across the boreal zone in Finland, Int. J. Biometeorol., 52, 251–259, 2008.

Reichstein, M., Falge, E., Baldocchi, D., Papale, D., Aubinet, M., Berbigier, P., Bernhofer, C., Buchmann, N., Gilmanov, T., Granier, A., Grünwald, T., Havránková, K., Ilvesniemi, H., Janous, D., Knohl, A., Laurila, T., Lohila, A., Loustau, D., Matteucci, G., Meyers, T., Miglietta, F., Ourcival, J.-M., Pumpanen, J., Rambal, S., Rotenberg, E., Sanz, M., Tenhunen, J., Seufert, G., Vaccari, F., Vesala, T., Yakir, D., and Valentini, R.: On the separation of net ecosystem exchange into assimilation and ecosystem respiration: review and improved algorithm, Glob. Change Biol., 11, 1424–1439, 2005.

Richardson, A. D., Jenkins, J. P., Braswell, B. H., Hollinger, D. Y., Ollinger, S. V., and Smith, M.-L.: Use of digital webcam images to track spring green-up in a deciduous broadleaf forest, Oecologia, 152, 323–334, 2007.

Richardson, A. D., Hollinger, D. Y., Dail, D. B., Lee, J. T., Munger, J. W., and O'Keefe, J.: Influence of spring phenology on seasonal and annual carbon balance in two contrasting New England forests, Tree Physiol., 29, 321–331, 2009.

Richardson, A. D., Keenan, T. F., Migliavacca, M., Ryua, Y., Sonnentag, O., and Toomey, M.: Climate change, phenology, and phenological control of vegetation feedbacks to the climate system, Agr. Forest Meteorol., 169, 156–173, 2013.

Rosenzweig, C., Casassa, G., Karoly, D. J., Imeson, A., Liu, C., Menzel, A., Rawlins, S., Root, T. L., Seguin, B., and Tryjanowski, P.: Supplementary material to chapter 1: Assessment of observed changes and responses in natural and managed systems. Climate Change 2007: Impacts, Adaptation and Vulnerability. Contribution of Working Group II to the Fourth Assessment Report of the Intergovernmental Panel on Climate Change, edited by: Parry, M. L., Canziani, O. F., Palutikof, J. P., van der Linden, P. J., and Hanson, C. E., Cambridge University Press, Cambridge, UK, 2007

Sonnentag, O., Detto, M., Vargas, R., Ryu, Y., Runkle, B. R. K., Kelly, M., and Baldocchi, D. D.: Tracking the structural and functional development of a perennial pepperweed (Lepidium latifolium L.) infestation using a multi-year archive of webcam imagery and eddy covariance measurements, Agr. Forest Meteorol., 151, 916–926, 2011.

Sonnentag, O., Hufkens, K., Teshera-Sterne, C., Young, A. M., Friedl, M., Braswell, B. H., Milliman, T., O'Keefe, J., and Richardson, A. D.: Digital repeat photography for phenological research in forest ecosystems, Agr. Forest Meteorol., 152, 159–177, 2012.

Tanis, C. M. and Arslan, A. N.: FMIPROT – Finnish Meteorological Institute Image Processing Tool, User manual, available at: http://monimet.fmi.fi/index.php?style=warm&page=FMIPROT, 2016.

Tanja, S., Berninger, F., Vesala, T., Markkanen, T., Hari, P., Mäkelä, A., Ilvesniemi, H., Hänninen, H., Nikinmaa, E., Huttula, T., Laurila, T., Aurela, M., Grelle, A., Lindroth, A., Arneth, A., Shibistova, O., and Lloyd, J.: Air temperature triggers the recovery of evergreen boreal forest photosynthesis in spring, Glob. Change Biol., 9, 1410–1426, 2003.

Toomey, M., Friedl, M., Frolking, S., Hufkens, K., Klosterman, S., Sonnentag, O., Baldocchi, D., Bernacchi, C., Biraud, S. C., Bohrer, G., Brzostek, E., Burns, S. P., Coursolle, C., Hollinger, D. Y., Margolis, H. A., McCaughey, H., Monson, R. K., Munger, J. W., Pallardy, S., Phillips, R. P., Torn, M. S., Wharton, S., Zeri, M., and Richardson, A. D.: Greenness indices from digital cameras predict the timing and seasonal dynamics of canopy-scale photosynthesis, Ecol. Appl., 25, 99–115, 2015.

White, M. A. and Nemani, R. R.: Canopy duration has little influence on annual carbon storage in the deciduous broadleaf forest, Glob. Change Biol., 9, 967–972, 2003.

Wingate, L., Ogée, J., Cremonese, E., Filippa, G., Mizunuma, T., Migliavacca, M., Moisy, C., Wilkinson, M., Moureaux, C., Wohlfahrt, G., Hammerle, A., Hörtnagl, L., Gimeno, C., Porcar-Castell, A., Galvagno, M., Nakaji, T., Morison, J., Kolle, O., Knohl, A., Kutsch, W., Kolari, P., Nikinmaa, E., Ibrom, A., Gielen, B., Eugster, W., Balzarolo, M., Papale, D., Klumpp, K., Köstner, B., Grünwald, T., Joffre, R., Ourcival, J.-M., Hellstrom, M., Lindroth, A., George, C., Longdoz, B., Genty, B., Levula, J., Heinesch, B., Sprintsin, M., Yakir, D., Manise, T., Guyon, D., Ahrends, H., Plaza-Aguilar, A., Guan, J. H., and Grace, J.: Interpreting canopy development and physiology using a European phenology camera network at flux sites, Biogeosciences, 12, 5995–6015, doi:10.5194/bg-12-5995-2015, 2015.

# 15

# Analysis of the technical biases of meteor video cameras used in the CILBO system

**Thomas Albin**[1,2], **Detlef Koschny**[3,4], **Sirko Molau**[5], **Ralf Srama**[1], and **Björn Poppe**[2]

[1]Institute of Space Systems, University of Stuttgart, Pfaffenwaldring 29, 70569 Stuttgart, Germany
[2]Universitätssternwarte Oldenburg, Institute of Physics and Department of Medical Physics and Acoustics, Carl von Ossietzky University, 26129 Oldenburg, Germany
[3]European Space Agency, ESA/ESTEC, Keplerlaan 1, 2201 AZ Noordwijk ZH, the Netherlands
[4]Chair of Astronautics, Technical Univ. Munich, Boltzmannstraße 15, 85748 Garching, Germany
[5]International Meteor Organisation, Abenstalstr. 13b, 84072 Seysdorf, Germany

*Correspondence to:* Thomas Albin (albin@irs.uni-stuttgart.de)

**Abstract.** In this paper, we analyse the technical biases of two intensified video cameras, ICC7 and ICC9, of the double-station meteor camera system CILBO (Canary Island Long-Baseline Observatory). This is done to thoroughly understand the effects of the camera systems on the scientific data analysis. We expect a number of errors or biases that come from the system: instrumental errors, algorithmic errors and statistical errors. We analyse different observational properties, in particular the detected meteor magnitudes, apparent velocities, estimated goodness-of-fit of the astrometric measurements with respect to a great circle and the distortion of the camera.

We find that, due to a loss of sensitivity towards the edges, the cameras detect only about 55 % of the meteors it could detect if it had a constant sensitivity. This detection efficiency is a function of the apparent meteor velocity.

We analyse the optical distortion of the system and the "goodness-of-fit" of individual meteor position measurements relative to a fitted great circle. The astrometric error is dominated by uncertainties in the measurement of the meteor attributed to blooming, distortion of the meteor image and the development of a wake for some meteors. The distortion of the video images can be neglected.

We compare the results of the two identical camera systems and find systematic differences. For example, the peak magnitude distribution for ICC9 is shifted by about 0.2–0.4 mag towards fainter magnitudes. This can be explained by the different pointing directions of the cameras. Since

both cameras monitor the same volume in the atmosphere roughly between the two islands of Tenerife and La Palma, one camera (ICC7) points towards the west, the other one (ICC9) to the east. In particular, in the morning hours the apex source is close to the field-of-view of ICC9. Thus, these meteors appear slower, increasing the dwell time on a pixel. This is favourable for the detection of a meteor of a given magnitude.

## 1 Overview and scientific objectives

Recently, several multi-station video camera systems to observe meteors have been set up, among others, in Japan (SonotaCo et al., 2010), in Canada (Weryk et al., 2013) and in the USA (Cooke and Moser, 2012; Jenniskens et al., 2011). The Canary Island Long-Baseline Observatory (CILBO) is a double-station meteor camera set-up operated by the Meteor Research Group of the European Space Agency. It is part of the video camera system of the International Meteor Organisation (Molau et al., 2015). CILBO consists of two stations, one on Tenerife and one on La Palma. A small building with an automated roll-off roof houses a set of video cameras with image intensifiers that monitor the same volume in the atmosphere for meteors. The pointing of the cameras is such that their image centres point to a height of 100 km between the two islands. Analysing the same meteor as seen from both camera stations allows the position relative to the Earth to be derived and, with that, to the cameras.

The main scientific goals of the set-up are as follows:

a. To study physical and chemical properties of meteoroids and, taking into account the modifications of the meteoroid properties during their flight in the solar system, to constrain the physical and chemical properties of their parent body.

b. To study the variability of the background dust flux in the Earth environment during a complete year.

To fulfil these goals, the following measurements are needed: (a) flux densities of the meteors, derived from the meteor numbers per time; (b) the physical properties of the meteoroids and their distribution, derived from light curves and velocity analysis; (c) meteoroid orbits, derived from the double-station observations; (d) chemical properties of the meteoroids, derived from spectra of the meteors.

A double-station set-up is very well suited to address these points. Since the distances of the meteor to the cameras can be determined, the absolute magnitude and the velocity in $\mathrm{m\,s^{-1}}$ can be computed. From this, the mass of the underlying meteoroid can be estimated (see e.g. Drolshagen et al., 2014; Ott et al., 2014; Kretschmer et al., 2015). This allows us to determine the flux density of meteoroids as a function of mass. From the triangulation of the positions, the 3-D trajectory of the meteoroid in geocentric coordinates is determined. Together with the velocity, the meteoroid path can be propagated backward and the heliocentric orbit of the meteoroid can be determined. From the magnitude profile of the meteor some physical properties of the object can be determined. To measure the spectra of the meteors, a second camera is installed on Tenerife, which has an objective grating.

To properly analyse all of these measurements, many biases have to be considered. Meteors of a given mass will generate more light the higher their velocity when entering the atmosphere. They will only be detected when they are above a certain brightness, which also depends on the distance to the observing camera. Because of the optical effects of the camera, they may be detectable in the centre of the field of view but not at the edges, where the camera sensitivity is lower. The higher the apparent velocity of a meteor, the more pixels are covered per unit time by the meteor, making it more difficult to detect it. The observing geometry will affect the observations – as we will show, a camera pointing to the east will record more meteors than one pointing west. This is because the east-pointing camera sees meteors from the apex direction with lower apparent velocity, increasing the dwell time and thus the meteors signal on a pixel.

In general, we distinguish between two effects – physical biases and biases in the detection system. Physical biases include effects that are independent of the detection system. For example, meteors that due to their orbital elements do no intersect with Earth's orbit need to be estimated for modelling purposes. This paper deals with the latter, the detection system and with geometrical aspects. This affects the detectability of meteors and biases the resulting brightness and velocity distributions depending on the camera system's set-up, settings and its pointing. The following section gives more background on the technical aspects of the system. We first describe the set-up and then summarize all the expected errors.

## 2 Set-up, data flow and methods

### 2.1 CILBO overview

A detailed overview of the set-up is given in a previous paper (Koschny et al., 2013). In this paper, we focus on the camera and the detection system, with an emphasis on their technical performance. Figure 1 shows a photograph and a block diagram of one of the cameras. It consists of the following main elements: (a) an objective lens-type Fujinon, 25 mm f/0.8; an image intensifier-type DEP1700 with a fibre-coupled 2/3″ CCD sensor read out via a Teli CS8310BCi video camera. The resulting field of view is roughly $28° \times 22°$ (H × V).

In the following, we are analysing data from two cameras, called ICC7 (on Tenerife) and ICC9 (on La Palma). "ICC" stands for Intensified CCD camera. Both cameras are identical. They point to the same volume in the atmosphere, between the two islands. Thus their pointing azimuth is roughly opposite; the pointing elevation is similar but not quite identical.

### 2.2 Data flow

The video cameras continuously record the night sky. With a field of view of approximately $600\,\mathrm{deg}^2$, CILBO covers an area of around $3000\,\mathrm{km}^2$ at an altitude of 100 km, where most meteors appear. The camera delivers a PAL (phase alternating line) video stream via a professional frame-grabber card (Matrox Meteor II) to a personal computer. The video signal is searched in real time for meteors using the software MetRec (Molau, 1999). MetRec analyses downsampled images with a resolution of $384 \times 288\,\mathrm{pixel}^2$ and 8 bit dynamical range. Later, we will show both full-resolution data and downsampled data, depending on the context.

MetRec generates a background noise image which is subtracted before the detection. The detection algorithm itself is described in Molau (1999, 2014). The software searches for brightness peaks in the background-subtracted images. It checks whether these peaks move on a great circle from one frame to the next.

For each frame of a detection, MetRec records the total digital number of the event on the detector and the position of its photometric centre. For each detected event, it stores a sum image, an animation of the event and a file containing detailed information on the event.

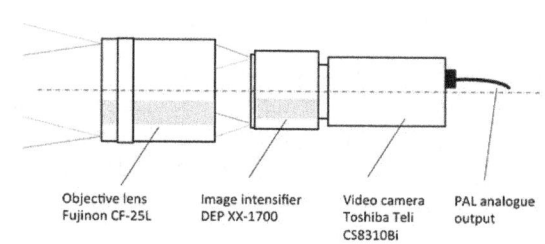

**Figure 1.** Photograph and sketch of the video cameras, called ICC (Intensified CCD camera).

For each night, MetRec saves all files in a daily directory. The data for ICC7 and ICC9 are stored in individual paths. The detailed information of each meteor is saved in an individual ASCII file with the extension *.inf, henceforth called "information file". Additionally, MetRec saves a log file that contains, e.g. the used detection parameters, the used reference file which contains the astrometric information of the stars and additional information of a recorded meteor.

The complete content of an information file is, for each frame where the meteor was detected, frame number, precise time taken from the computer clock, magnitude of the event, position of the photometric centre in coordinates relative to the detector and in celestial coordinates and fitted coordinates as described in the following paragraph. An example information file can be found in Koschny et al. (2013).

In addition to the information for each individual meteor, we use the log file entries in this paper to characterize the system behaviour. This file provides additional information for each detected meteor.

The automated event detection runs every clear night, controlled by a scheduling software as described in Koschny et al. (2013). At the end of the night, the data are uploaded to a central server for further processing. On the next day, the data for each night is visually inspected and false detections are deleted. The data are submitted on a monthly basis to the video archive of the International Meteor Observation, where a peer-review process ensures good data quality. All data are available and searchable via the Virtual Meteor Observatory (Barentsen and Koschny, 2008, http://vmo.imo.net).

MetRec allows us to manually compare a grabbed image with a star chart to produce a so-called "reference star" file. With this file MetRec can convert the relative positions together with the time of the event to right ascension and declination. The "referencing" process also generates a calibration file to convert pixel values to stellar magnitude. This process is typically done only when the camera pointing has changed.

MetRec attempts to correct any measurement errors in the position determination. It takes the originally measured right ascension and declination values and fits them to a great circle. The measured points are projected onto this great circle.

In a next step, MetRec shifts the points on this great circle to be equally spaced. For longer meteors (> 7 frames), MetRec shifts the points to match a distribution following a 2nd-order polynomial.

If a second meteor appears during the same second as a previous on, an additional log entry with the same time stamp is saved. However, the corresponding information file with the astrometric information is overwritten and lost.

### 2.3 Expected errors

### 2.3.1 Overview

In the later sections of this paper, we will present some findings on different parameters measured by the system. Then we will draw conclusions on how important the different biases are and which ones can be corrected. In summary, we expect the following errors.

### 2.3.2 Instrumental errors

a. The mechanical/thermal instability of the mounting: due to thermal effects, the precise pointing position of the camera may change. This is a systematic error affecting the position measurement of the meteor.

b. The lens and possibly also the image intensifier generate a drop-off caused by both vignetting and the tangent effect at larger distances to the centre of the field of view. This is a systematic error affecting the detectability of a meteor.

c. Due to the projection of the celestial sphere on the flat sensor surface, the system generates distortion which needs to be corrected when computing positions of the meteors. This is corrected by the 3rd-order polynomial "plate fit" performed during the measurement; however see Sect. 2.3.3 point c.

d. The sensor is read out with 25 frames per second and the readout generates noise. In addition, random noise is generated by the image intensifier. The noise statistics are estimated from a sequence of dark frames (no light entering the sensor system). It is random noise affecting all measurements.

e. The pixel resolution of the sensor does not precisely match the pixel format of the used PAL format (768 pixel × 576 pixel) and pixels may be interpolated.

f. The sensor is an interline-transfer sensor, i.e. every second physical line on the sensor is masked and used for readout. This and the previous point will reduce the quality of the position determination of the meteor.

g. (Absolute) timing errors (offset of the computer clock): this is a systematic error that only affects the position, not the velocity. A timing error of 1 s would correspond to a position error in right ascension of 1/4′.

h. Distortion of the image of a meteor close to the edge of the field of view. This effect is particularly pronounced for bright meteors and it will result in errors in the astrometric position of the meteor.

### 2.3.3 Algorithmic errors

a. Wake: during the movement of the meteor it may develop a train, which shifts the photometric centre to the opposite direction of the meteor's movement. This effect will result in an apparent change in the velocity of the meteor. Typically, trains develop towards the end of the meteor, so this effect will reduce the perceived speed of the meteor towards the end.

b. Blooming: for bright meteors, so-called blooming may occur; i.e. electrons spill over from one pixel to other adjacent pixels. The shift of the photometric centre can then go in any direction.

c. The image distortion is corrected using a 3rd-order polynomial fit. In particular, towards the edges of the field of view, a 3rd order may not be good enough to properly describe the distortion. This will introduce a systematic deviation of the measured positions with respect to the real position.

d. When determining the position of a meteor, our detection software attempts to fit the positions using a linear or quadratic equation resulting in a constant and linear equation for the velocity, respectively. Due to geometric effects this may not be sufficient to describe the position and causes a deviation between the fit and the actual measured meteor position. The effect is meteor dependent, as it is affected by the length of the meteor in number of frames. Any velocity determination error may be estimated by calculating how the velocity will really change when crossing the field of view and how good the quadratic fit is.

e. Meteor beginning and end: since the meteor will start or end at a random time during the exposure of the first or last frame, taking the photometric centre as the position of the meteor for this frame does not give the correct results. This is a systematic error that only affects the velocity.

f. Quantization error of position in the information files: the position of a meteor is stored as a relative position in the frame (from 0 to 1) with an accuracy of three decimal places only. This corresponds roughly to 0.3 pixel. If meteor positions are recomputed later in the analysis process this information is used, resulting in a quantization of the position. This is a random error which affects both position and velocity. It is meteor dependent, because meteors with more frames will be less affected.

### 2.3.4 Statistical errors

a. Statistical random error: both the position and the brightness measurements of a meteor in an individual frame are affected. This is an error due to the probabilistic nature of the event and is independent from the used instrument or its settings. It affects both position and velocity and it can be derived from the accuracy of the meteor fit that is currently investigated. It is meteor dependent, influenced by the number of frames, meteor brightness and possibly velocity.

In the following sections, we characterize the camera systems in detail. We give results on technical aspects related to camera and software (flat field effects, distortion, etc.). We then present statistics on overall distributions of different meteor characteristics (meteor length, brightness, etc.). We combine these results and provide, as a result, the means to properly de-bias the data from the cameras for scientific analysis.

## 3 Results

### 3.1 Overview

Albin et al. (2015a, b) have made a first attempt to analyse a selected number of bias effects for meteors detected simultaneously to ICC7 and ICC9. Here we expand on this work and also treat some of the data from the cameras separately. We use data from the information and the log files.

The data flow followed the description in Sect. 1. We have used a total of 51 062 and 56 951 information files and 925 and 913 log files for ICC7 and ICC9, respectively. The analysed time range was from 13 September 2011 to 31 August 2015.

In the following subsections, we describe different parameters of the measurements. These will be interpreted in the discussion section.

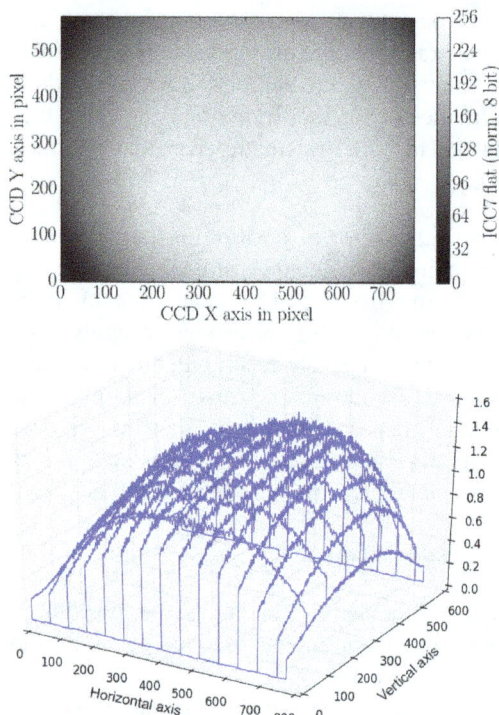

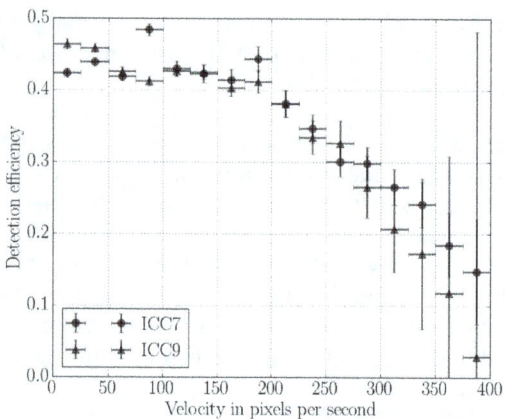

**Figure 3.** Detection efficiency vs. the downsampled velocity of a meteor in pixels per second. A detailed description of the detection efficiency can be found in Albin et al. (2015a).

**Figure 2.** 8 Bit median flat of the ICC7 camera. The $x$ and $y$ axes are not downsampled, they cover the complete PAL signal. On the left, the image is shown, with the colour bar indicating the brightness of the flat field. 256 is the maximum and can be found slightly off-centre to the right due to an offset in the optical system. The bottom panel shows a wire-mesh view of the flat field. Normalized values range from 0.3 in the corners to 1.3 in the middle.

## 3.2 Camera sensitivity

We start by analysing the detection efficiency of both cameras vs. the apparent meteor velocity in pixels per second. The detection efficiency is defined as the ratio of the theoretically expected number of meteor detections on the CCD vs. the number of actual meteor measurements on the CCD (Albin et al., 2015a). Due to vignetting and projection effects the cameras have a sensitivity drop to the edges and corners of the CCD. Thus, the number of detections decreases to the edges due to the lower signal-to-noise ratio ($S/N$) of the meteor, which results in less detections by MetRec. In other words, the detection efficiency would be 1 if a meteor of a given magnitude and velocity had the same $S/N$ over the complete field of view.

Figure 2 shows the flat field of the ICC7 system. The flat field of ICC9 looks similar. The image is an 8 bit median stack of about 10 individual images, recorded when thin fog provided a rather homogeneous sky background. The grey bar indicates the corresponding normalized brightness. It can be seen that the intensity drops to the edges and corners of the CCD. An optical system with no vignetting or projection effects would lead to a uniformly shaped distribution

and a detection efficiency of 1. To compute the theoretically expected number of measurements we take the part on the CCD with the highest detection density and extrapolate this value for the complete CCD. A detailed description can be found in Albin et al. (2015a), who also computed the detection efficiency for the CILBO system depending on the meteor brightness. They found that the detection efficiency is at around 0.55 for meteors with a brightness down to 4.5 mag and drops down to 0.45 and lower for fainter meteors. This means that the meteor cameras detect only half of the meteors which it would be able to detect for an evenly illuminated sensor.

Figure 3 shows the detection efficiency vs. the meteor velocity in pixels per second. For the analysis, we use the filtered velocity data set from the information files. The data set has been divided into bins of 25 pixel s$^{-1}$. For each bin, the theoretical and actual number of meteor detections has been computed as in Albin et al. (2015a). The plot shows the detection efficiency from 0.0 to 400 pixel s$^{-1}$. For very large velocities the number of data points decreases, increasing the shown standard deviation of the detection efficiency. It can be seen that the detection efficiency is between 0.4 and 0.5 for meteors ranging from 0.0 to 200 pixel s$^{-1}$. Then, the detection efficiency decreases approximately linearly for higher velocities.

The pixel dwell time of a meteor is inversely proportional to the apparent meteor velocity on the CCD. Consequently, a higher meteor velocity decreases the $S/N$ for a given meteor magnitude. The decreasing sensitivity to the edges and corners due to the projection effects result in a smaller effective detection area on the CCD for higher-velocity meteors. This can explain the lower detection efficiency for fast meteors.

The shown effects and the detection efficiency function as shown in Albin et al. (2015a) are necessary to de-bias the mass distribution of the meteors that is correlated to the

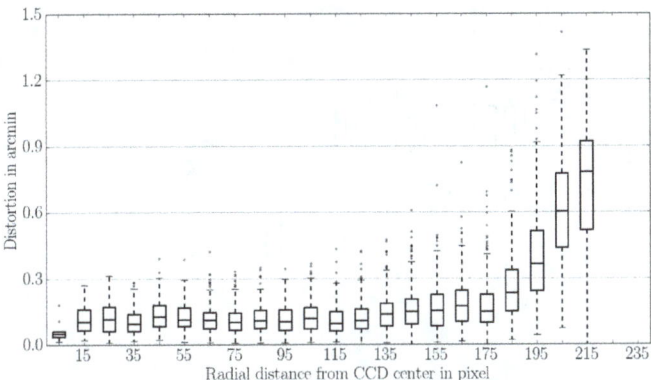

**Figure 4.** Box plot of the ICC7 distortion. The difference between actual position and CCD position is shown in arcminutes vs. the radial distance from the centre of the CCD. Each box plot contains the data of the a 10-pixel-wide bin.

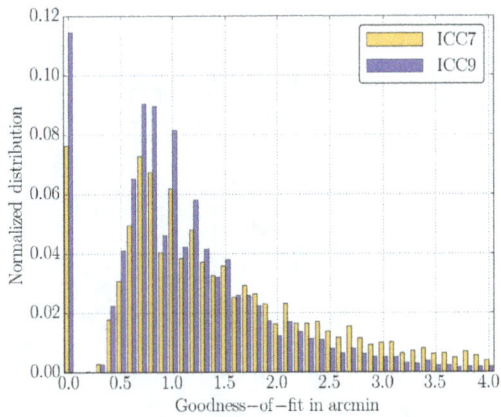

**Figure 5.** Normalized distribution of determined goodness-of-fit in arcminutes. The orange and blue bars show the distribution for ICC7 and ICC9, respectively. The bars are slightly off-centre and have an actual width of $0.1'$, e.g. the first two bins show the contribution of $[0.0', 0.1')$ for ICC7 and ICC9.

brightness measurements. Additionally, the determined flux needs to be corrected by at least a factor of 2.

### 3.3 Meteor velocity measurement bias

Albin et al. (2015b) described the velocity profiles of several simultaneously detected meteors with the CILBO camera set-up. For the analysis they used the geocentric velocity in $\mathrm{km\,s^{-1}}$ determined by the MOTS3 software package for computing trajectory data of double-station meteor cameras (Koschny and Diaz del Rio, 2002). Due to the atmospheric drag a meteoroid decelerates during the atmospheric entry. We found that 40–45 % of all meteors seem to have an increased velocity between the first and second velocity measurements. This cannot be explained by Earth's gravitational attraction. The effect is an observational bias of the camera system. Both cameras are operated with a rate of 25 frames per second and a video frame length of 40 ms respectively. The measurable beginning and ending times of a meteor do not necessarily correspond to the video frame length of 40 ms. Consequently, it may appear in the data set that the meteor covers a smaller distance at the beginning and end of a recording. The end part of the meteor overlaps additionally with the deceleration effect. Thus, to compute a proper initial geocentric velocity from a continuously operated double station meteor network, the distance between the first and second video frames should not be used for the velocity computation. The last velocity value should not be used for the same reason. As a result, no good velocity can be determined for meteors recorded on 3 frames only. To obtain two velocity measurements, the meteor has to be recorded on 5 frames.

### 3.4 Accuracy values and optical distortion

We generated optical distortion maps to determine the astrometric deviations of the real star positions relative to their expected positions according to the 3rd-order polynomial plate

fit performed by MetRec. Figure 4 shows the computed distortion distribution for the ICC7 camera. The distortion is shown by plotting the deviation of the real measured star position against its expected position determined by the plate fit. It is given in arcminutes and is plotted against the radial distance from the CCD centre in downsampled pixels. The data are summarized in bins of 10 pixels and visualized as a box plot[1]. It can be seen that the distortion remains approximately constant up to a radius of 140 pixels. The corresponding median is at around $0.1'$. With the downsampled horizontal image size of 388 pixels this corresponds to 80 % of the horizontal radius; 95 % of the horizontal radius are correct to $0.2'$. Due to the distortion of the optical system, the values worsen to the corners up to $0.75'$. In conclusion, position measurements of meteors more than about 80 % away from the field centre should be used carefully.

Since the ICC9 distribution looks similar, only the ICC7 data are shown. We will see that other astrometric errors are larger and conclude that at least for the inner 90 % of the field of view, errors due to insufficient distortion correction can be neglected.

### 3.5 Measured astrometric goodness-of-fit

For each meteor, MetRec stores a value called "accuracy" in the log file, which describes the goodness of the fit of the

---

[1] A box plot is a way to visualize non-Gaussian distributions. It uses the so-called median and the interquartile range (IQR). The median is the point where a distribution is divided into two equal-sized sets. The 25- and 75-percentile are the lower and upper limit of the IQR; the IQR contains 50 % of the data around the median. In a box plot, the median is shown as a horizontal solid line in a box; the box itself corresponds to the IQR. The dashed line has a length of $1.5 \cdot \mathrm{IQR}$. Data points outside the IQR are plotted as crosses or grey circles.

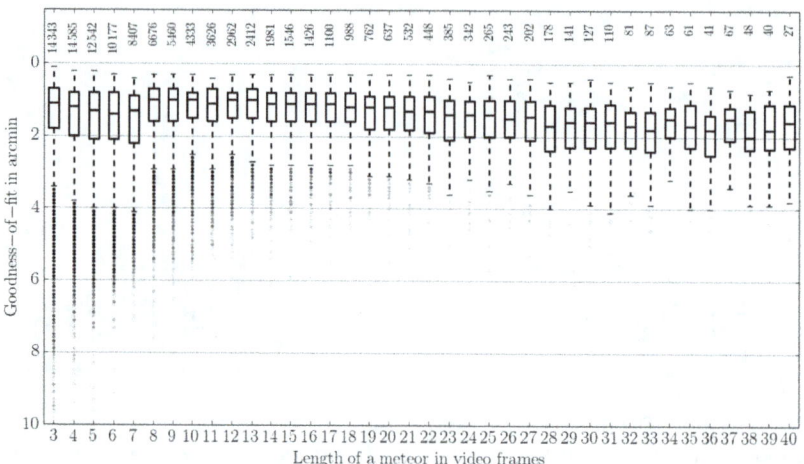

**Figure 6.** Goodness-of-fit vs. frame length. The box plots show the median, interquartile range (IQR) and 1.5 · IQR. The numbers on the top show the number of data points for each bin.

individual meteor positions relative to a great circle in the sky. We will henceforth refer to this as "goodness-of-fit". The value is given in arcminutes and is the root mean square of the deviations of individual meteor position measurements to the projections on a least-square great circle line. The smaller the value, the better the fit. This section analyses the recorded accuracies.

Figure 5 shows the normalized goodness-of-fit distribution based on all meteor observations for ICC7 (orange or bright bars) and ICC9 (blue or dark bars). "Normalized" means that the sum of all histogram bars is 1. The distribution plot is shown from 0.0 to 4.0′ with a bin width of 0.1′. This corresponds to the current accuracy resolution of MetRec. The maximum values are around 10′, but less than 3 % of the data are above 4′ (2463 values out of 73 379). We therefore decided to not display them.

It can be seen that both cameras detect a significant number of meteors with a goodness-of-fit of 0.0′. Values of 0.1 and 0.2′ are missing completely. The log files show that ICC7 has 3899 (approximately 8 %) and ICC9 has 6527 (approximately 11 %) of all measurements with values of 0. For both cameras, around 55 % of all measurements correspond to meteors with a length of 3 frames. Around 20 % correspond to a length of 4 frames, 10 and 5 % to 5 and 6 frames. The remaining 10 % correspond to longer meteors. A fraction of these can be explained with the fact that MetRec rounds the determined goodness-of-fit. However, most data points in this bin seem to have been falsely generated, otherwise the gap between the 0.0′ bin and next bin at 0.3′ cannot be explained. The following accuracy-related analysis therefore neglects these data points.

The median and IQR of the ICC7 and ICC9 accuracies are $\text{ICC7}_{\text{acc}} = 1.2^{+0.9}_{-0.5}{}'$ and $\text{ICC9}_{\text{acc}} = 1.0^{+0.5}_{-0.3}{}'$.

MetRec uses half-resolution images for the detection, i.e. 384 pixel × 288 pixel. The obtained average goodness-of-fit

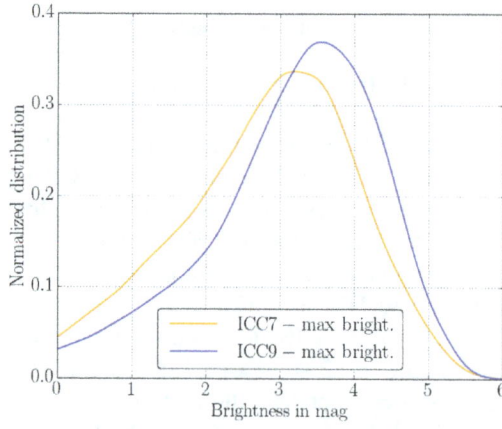

**Figure 7.** Normalized distribution of the peak brightness in magnitudes. The orange and blue curves correspond to the ICC7 and ICC9 camera, respectively.

is thus about 1/4 pixel. Taking into account that the used sensor is an interline transfer video chip and the field of view is rather large, this result is acceptable.

When using these data to compute orbits, one can use the goodness-of-fit values to estimate, via Monte-Carlo runs, the errors of the orbital elements. A Monte-Carlo-based method to compute the astro-dynamic properties of the detected meteors is described in detail in Albin et al. (2016). To simplify this procedure, it is proposed to use an average error value as derived in the following.

Figure 6 shows a box plot of the complete accuracy data of ICC7 and ICC9 in arcminutes versus the length of a meteor measured in number of frames. All goodness-of-fit values from the log files have been used with the exception of the 0.0′ data. The figure shows the distribution of meteor lengths between 3 and 40 frames and the number above each box gives the number of data points in the corresponding bin. The

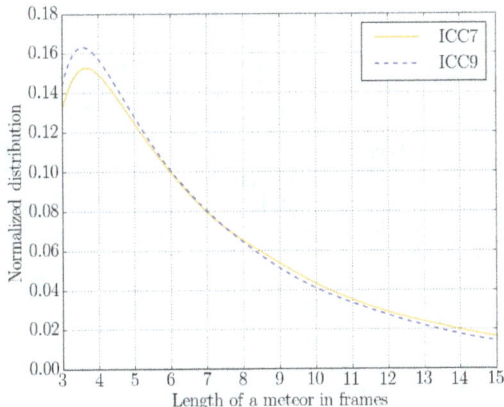

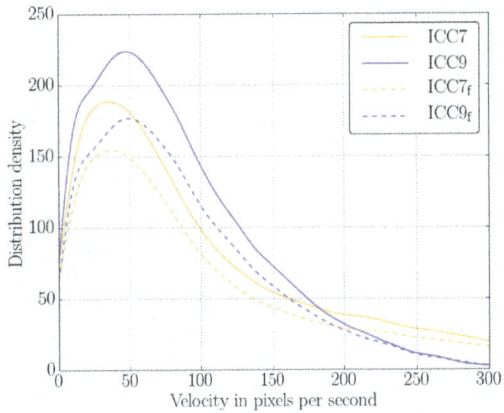

**Figure 8.** Normalized distribution of the recorded frames for ICC7 (solid curve) and ICC9 (dashed curve). Since MetRec's detection threshold is set to 3 frames, no meteors are recorded on fewer frames.

**Figure 9.** Distribution of the meteor velocities in pixel per second. The orange (bright) curves correspond to ICC7 and the blue (dark) curves show the ICC9 data. The solid distributions show the complete data set, containing all determined velocities. The dashed curves show the filtered velocity data set as explained in the text.

longest meteor recorded with CILBO is about 80 frames. For a better visualization and readability, we only show data up to 40 frames. For higher values, the total number of measurements drops further and does not allow any statistical conclusions. It can be seen that the median, the IQR and $1.5 \cdot$ IQR range increase for meteor lengths of 3 to 7 frames. The median increases from 1.1 to around 1.5'. From 7 to 8 frames, the accuracy jumps to better values: the median drops to 1.0'. This is due to a setting in the MetRec fitting algorithm. Up to 7 frames, the programme uses a constant velocity value. A meteor which is recorded on 8 or more frames is fitted with a linear velocity fit which leads to a better goodness-of-fit, as can be seen in the changing box size between frame 7 and 8. For meteors of length 8 to 40 frames, the accuracy worsens again slightly. The number of data points which lie outside the box plots decreases for higher frame numbers. The largest data scatter can be seen for meteor recorded on 3 frames. In some cases, the goodness-of-fit becomes as bad as 10', because either the linear velocity fit was insufficient for very long meteors or outlier frames caused by noise or nearby stars were not properly detected and removed.

In conclusion, we suggest assuming a typical deviation of about 1.0–1.2' to cover all uncertainties in the astrometry. This corresponds to about 1 pixel.

### 3.6   Magnitude distribution

ICC7 and ICC9 have the same technical set-up and are operated in a similar way. Items like the detection threshold and the minimum number of frames per meteor are identical. Here, we compare the measured brightness distribution of both CILBO cameras to check whether deviations in the data can be identified. For our analysis we assume that meteors appear randomly on the sky. Since some meteors either begin or end outside CILBO's field of view (FOV) or both, we consider only meteors which were completely within the FOV.

Otherwise a bias or offset in the meteors' brightness profile would affect the statistics. For the analysis we only take meteors into account that are not closer to the CCD edges than 5 % of the length and width of the CCD. Thus, the data set reduces to 49 494 meteors for ICC7 and 54 402 meteors for ICC9 which corresponds to 97 and 96 % of each individual data set, respectively.

Figure 7 shows the normalized distribution of the ICC7 and ICC9 brightness data vs. the peak brightness values in magnitudes. The orange (brighter) curve corresponds to the ICC7 data and the blue (darker) curve corresponds to the ICC9 data. The median and corresponding IQR for both cameras are $\mathrm{ICC7_{mag,peak}} = 2.92^{+0.76}_{-0.97}$ mag and $\mathrm{ICC9_{mag,peak}} = 3.32^{+0.70}_{-0.88}$ mag, respectively. This shows that ICC9 detects fainter meteors than ICC7. The brightness median difference between both cameras is 0.40 mag. We will show later that this is due to the different pointing directions of the cameras. Thus, the pointing affects the detected number of meteors for a given magnitude.

### 3.7   Distribution of the length of a meteor in frames

MetRec's detection threshold is currently set to 3 frames. With 25 frames per second this corresponds to a meteor duration of larger than 40 ms (starting at the very end of the exposure of the first frame, ending at the very beginning of the last one) to 120 ms. In some rare cases a meteor with 3 frames can also have an appearance time, e.g. of 160 ms, due to frame drops in the detection pipeline.

Figure 8 shows the normalized distribution of the length of the meteors in number of frames. The solid histogram represents the ICC7 data and the dashed histogram shows the ICC9 data. CILBO detects meteors with a length of up to 70–80 frames. For a better data readability, here we show the distributions up to a length of 15 frames, corresponding to a

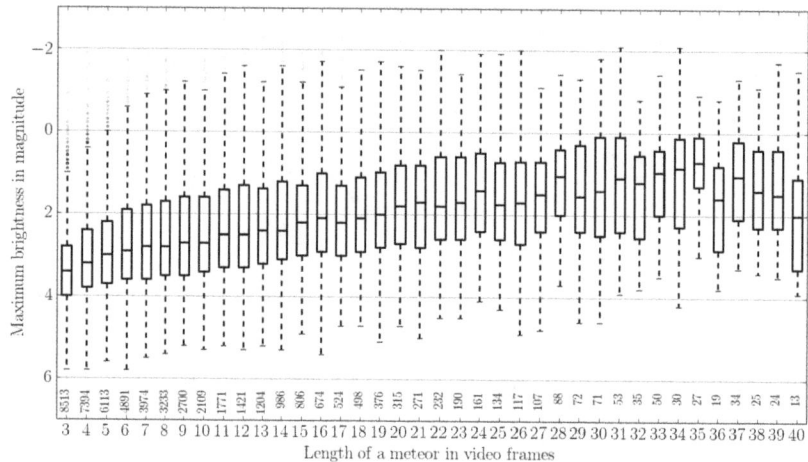

**Figure 10.** Maximum brightness in magnitude vs. the length of the meteor in frame numbers for ICC7. The box plot shows the median, IQR and $1.5 \cdot$ IQR. The number shown on the bottom indicates the number of used data points per frame bin.

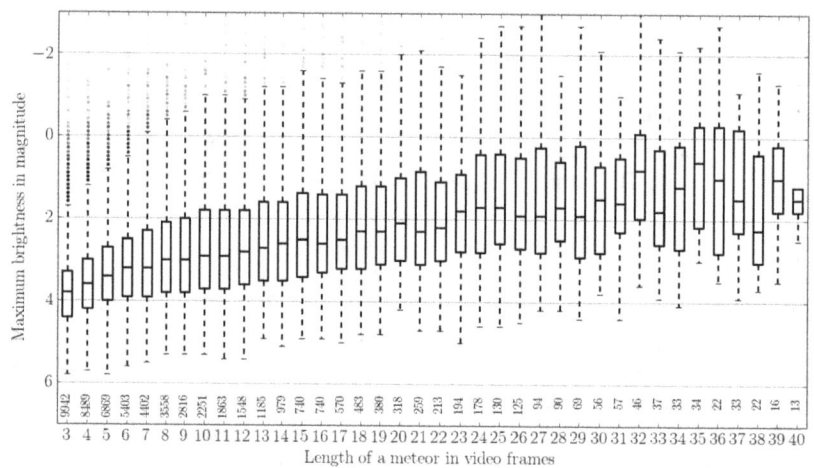

**Figure 11.** Maximum brightness in magnitude vs. length of the meteor in frame numbers for ICC9. The box plot shows the median, IQR and $1.5 \cdot$ IQR. The number shown on the bottom indicates the number of used data points per frame bin.

meteor appearance time of 0.6 s. It can be seen that the number of meteor recordings decreases for longer events. Both distributions peak at meteors with a length of 3 frames. For increasing lengths, the number of meteors decreases faster for ICC9 than for ICC7. ICC7 detects more meteors on 3 to 7 frames than ICC9. Afterwards, the ICC7 distribution is slightly above the one of ICC9.

### 3.8 Velocity distribution

The apparent velocity of a meteor is computed from its position in each frame, assuming that the frame rate is 40 ms. The position of a meteor is available in two coordinate systems: firstly, in a CCD-fixed system given as $x/y$ value pairs, corresponding to the horizontal and vertical position on the sensor, counted from the lower-left corner. $x$ and $y$ are normalized and range from 0 to 1. To convert the positions in pixels, $x$ and $y$ need to be multiplied by factors of 768 and

576, respectively, which correspond to the PAL resolution. Since MetRec downsamples both axes by a factor of two we use values of 384 pixel $\times$ 288 pixel for all detection-related aspects in this paper.

The second coordinate system in which MetRec provides the astrometry in is the equatorial coordinate system, where the meteor position is given in right ascension and declination. Due to optical distortions, the angular velocity distribution in degrees differs from the distribution given in CCD co-ordinates depending on the position in the field of view. Since this paper focuses on the technical aspects of the CILBO cameras, in the following only we consider the apparent velocity in the CCD-fixed coordinate system. For those who prefer to think in degrees per second, note that 100 pixel s$^{-1}$ will be roughly 7 deg s$^{-1}$ with the field size of our cameras.

Figure 9 shows the density distribution of ICC7 and ICC9 versus the velocity in pixels per second. The solid curves

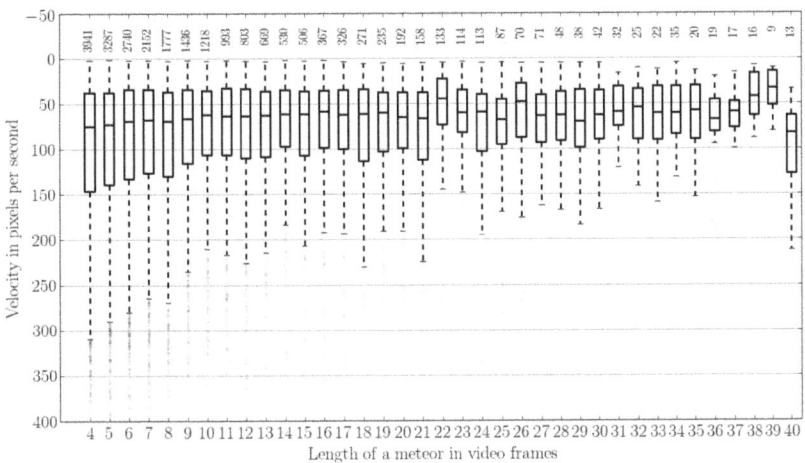

**Figure 12.** Apparent meteor velocity in pixels per second vs. the video frame length for ICC7. The box plot shows the median, IQR and $1.5 \cdot$ IQR. The number shown on the top indicates the number of used data points per bin.

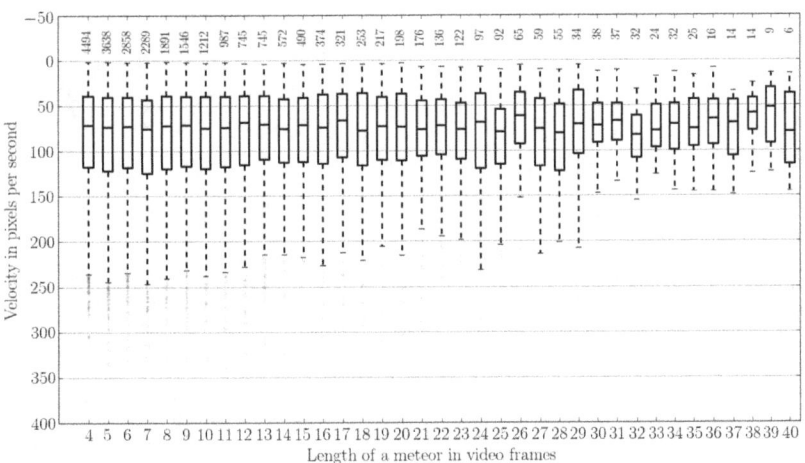

**Figure 13.** Apparent meteor velocity in pixels per second vs. the video frame length for ICC9. The box plot shows the median, IQR and $1.5 \cdot$ IQR. The number shown on the top indicates the number of used data points per bin.

are the distributions of all mean meteor velocities, where the orange (lighter) curve corresponds to ICC7 and the blue (darker) curve corresponds to ICC9 data. The velocity axis ranges from 0 to $300 \, \mathrm{pixel \, s^{-1}}$ (about $21 \, \mathrm{deg \, s^{-1}}$). It can be seen that both distributions have a similar shape; however ICC9 converges faster to 0 than the ICC7 distribution. This means that ICC7 records more fast meteors than ICC9. The curve for ICC7 is flatter and crosses that for ICC9 at $195 \, \mathrm{pixel \, s^{-1}}$. The median and IQR (given as the error values) for ICC7 and ICC9 are $\mathrm{ICC7_{vel}} = 158^{+151}_{-77}$ and $\mathrm{ICC9_{vel}} = 146^{+93}_{-66} \, \mathrm{pixel \, s^{-1}}$, respectively. This shows quantitatively that the ICC7 distribution is spread wider.

Meteors appear and disappear at some arbitrary time during the exposure time of the first and last frame of a detection (see Sect. 3.4). Thus, normally the determined photometric centres of the first and last frame are shifted towards the photometric centres determined from the second

and second-to-last video frame, respectively. To compute the velocity, the time interval between two frames is used, namely 40 ms. This means that the first and last velocity determination are typically underestimated. We leave those values and call this the filtered velocity data. The dashed curves in Fig. 9 show the filtered mean velocity data sets of ICC7 and ICC9. Both dashed curves appear similar to the solid ones. The median and IQR values for both filtered data sets are $\mathrm{ICC7_{vel,unbiased}} = 157^{+149}_{-76}$ and $\mathrm{ICC9_{vel,unbiased}} = 150^{+95}_{-67} \, \mathrm{pixel \, s^{-1}}$, corresponding to roughly $10 \, \mathrm{deg \, s^{-1}}$.

In the following sections, we only use the filtered velocity data set if not otherwise mentioned. We suggest that velocities computed from the first and last recorded frame should not be used.

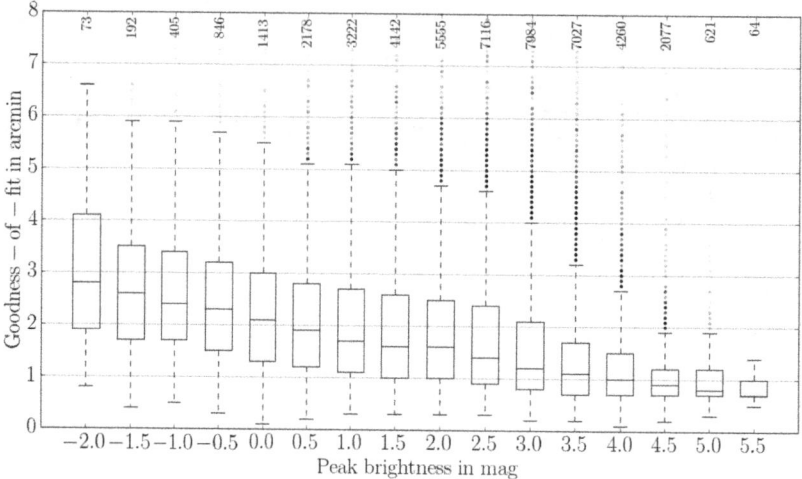

**Figure 14.** Goodness-of-fit vs. peak brightness in magnitude for ICC7. The box plot shows the median, IQR and $1.5 \cdot$ IQR. The number shown on the top indicates the number of used data points per peak brightness bin.

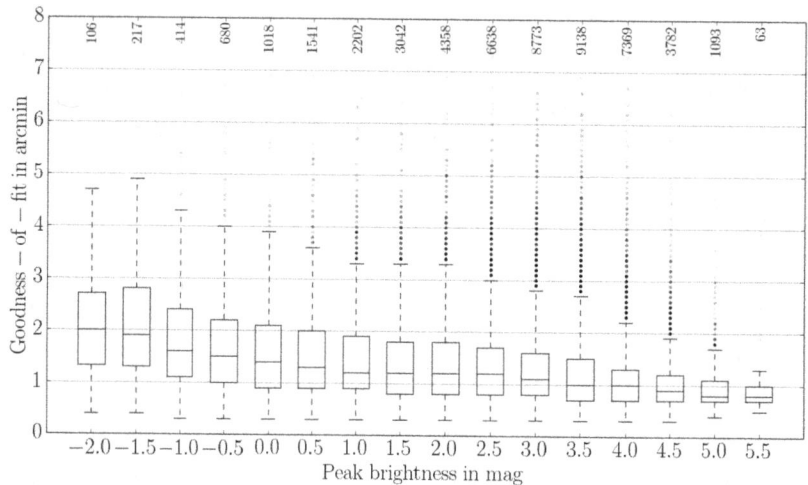

**Figure 15.** Goodness-of-fit vs. peak brightness in magnitude for ICC9. The box plot shows the median, IQR and $1.5 \cdot$ IQR. The number shown on the top indicates the number of used data points per peak brightness bin.

## 3.9 Correlation between different measurements

### 3.9.1 Overview

In Sect. 3.2 to 3.8 we showed distributions of different measured values like the accuracy or brightness of a meteor as determined by MetRec. Both ICC cameras are identical, but show deviations in the measured parameters. This section investigates possible correlations between certain measurements and parameters.

First, we describe the dependencies between the measurements and the recorded frame length. Afterwards we investigate possible detection time correlations. The last two subsections show some correlations with the measured brightness and determined velocities.

### 3.9.2 Peak magnitude as a function of meteor length and velocity

Figures 10 to 13 show box plots of the maximum brightness of a meteor in magnitudes and filtered mean apparent velocity in pixels per second for ICC7 and ICC9, respectively. The data are plotted vs. the length of a meteor in frames. Only meteors which were detected completely within the FOV of the cameras are considered.

The median and corresponding IQR of the brightness data for ICC7 and ICC9 show that the maximum brightness increases for longer meteors. Meteors with a length of 3 frames have a median and IQR of $3.4^{+0.6}_{-0.6}$ mag for ICC7 and $3.8^{+0.6}_{-0.5}$ mag for ICC9. It can also be seen that the medians and IQRs of ICC9 are shifted towards fainter meteors by a factor of around 0.2–0.4 mag, consistent with Fig. 7.

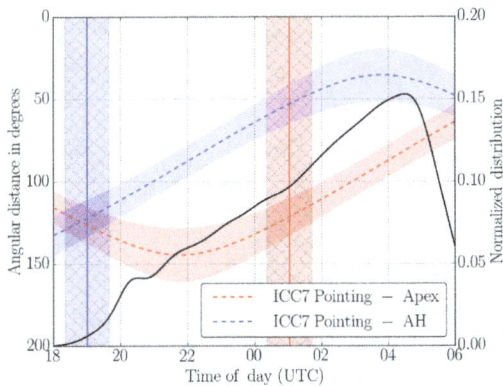

**Figure 16.** Angular distance and normalized distribution of detected meteors vs. the time of the day in UTC (ICC7). The red (upper) and blue (lower) dashed curves show the angular distances between the ICC7 boresight and the apex and antihelion directions, respectively. The coloured areas around the dashed lines show the yearly variations. The solid vertical lines indicate the rising time of the antihelion (blue, left) and the apex (red, right) radiants. The hatched area shows the yearly variations. The black curve corresponds to the right axis and gives the normalized number of all detected meteors.

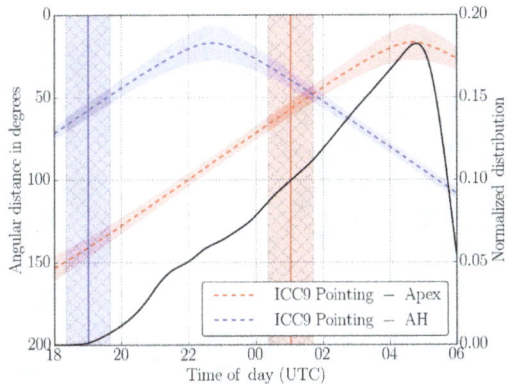

**Figure 17.** Angular distance and normalized distribution of detected meteors vs. the time of the day in UTC (ICC9). The red (upper) and blue (lower) dashed curves show the angular distance between the ICC9 boresight and the apex and antihelion directions, respectively. The coloured area around the dashed lines show the yearly variations. The solid vertical lines indicate the rising time of the antihelion (blue, left) and the apex (red, right) radiants. The hatched area shows the yearly variations. The black curve corresponds to the right axis and gives the normalized number of all detected meteors.

The box plots of the velocity distributions for ICC7 and ICC9 (Figs. 12, 13) show a slight difference. Median and IQR for ICC9 are basically constant for all shown meteor lengths. The IQR ranges between 50 and 150 pixel s$^{-1}$. ICC7, however, shows a decrease in the velocity for an increasing number of video frames. The maximum is at the beginning where the median is at around 75 pixel s$^{-1}$ and the IQR boundaries are at 40 and 140 pixel s$^{-1}$. The decreas-

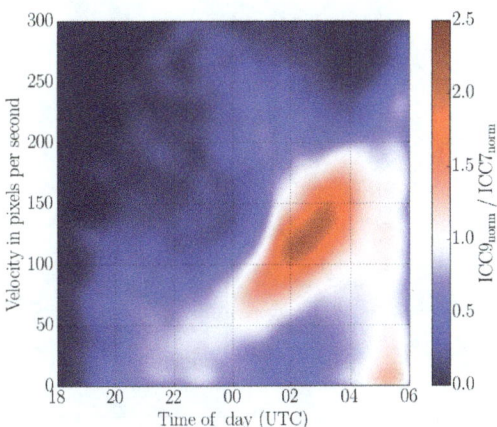

**Figure 18.** Ratio plot of the velocity in pixels per second of ICC9 divided by ICC7 vs. the detection time. The ratio is colour coded and given in the right colour bar.

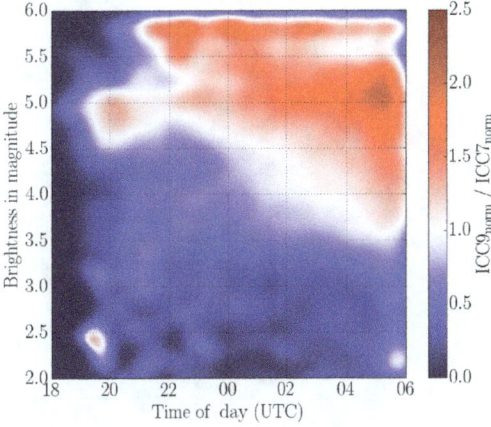

**Figure 19.** Ratio plot of the faintest brightness measurements of ICC9 divided by ICC7 vs. the detection time. The ratio is colour coded and given in the right colour bar.

ing median and IQRs converge with the ICC9 data at around frame 11.

### 3.9.3 Goodness-of-fit versus peak magnitude

Figures 14 and 15 show the measured goodness-of-fit versus the average peak brightness in mag for ICC7 and ICC9, respectively. We use all goodness-of-fit values larger than 0.0′. The shown figures show the data up to 6.0′ in a magnitude range from −2.0 to 6.0 mag. The solid line, box and the dashed lines are the median, IQR and corresponding 1.5 IQR limits. The goodness-of-fit gets smaller (i.e. better) for fainter meteors. For ICC7, the median of the goodness-of-fit at −2.0 mag is 3.0′ with an IQR of around ±1.0′. The median decreases to 1.0′ at 6.0 mag. Also, the IQR range narrows towards fainter meteors. For bright meteors, the median and IQR of ICC9 is better by around 1.0′. Median and IQR con-

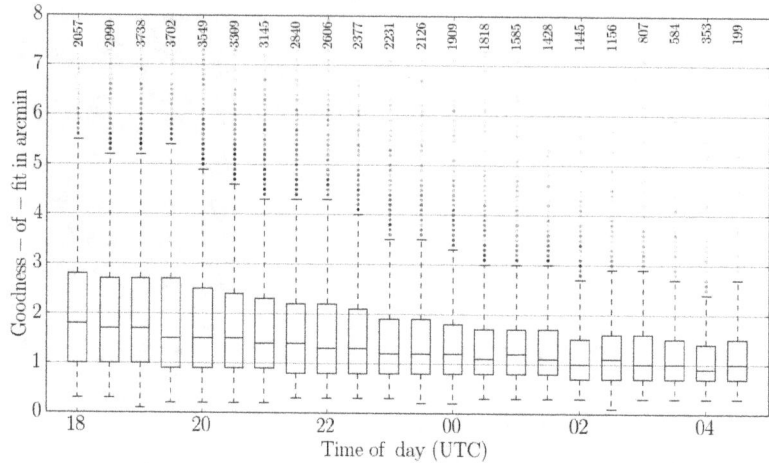

**Figure 20.** Goodness-of-fit vs. detection time for ICC7. The box plot shows the median, IQR and $1.5 \cdot$ IQR. The number shown on the top indicates the number of used data points per peak brightness bin.

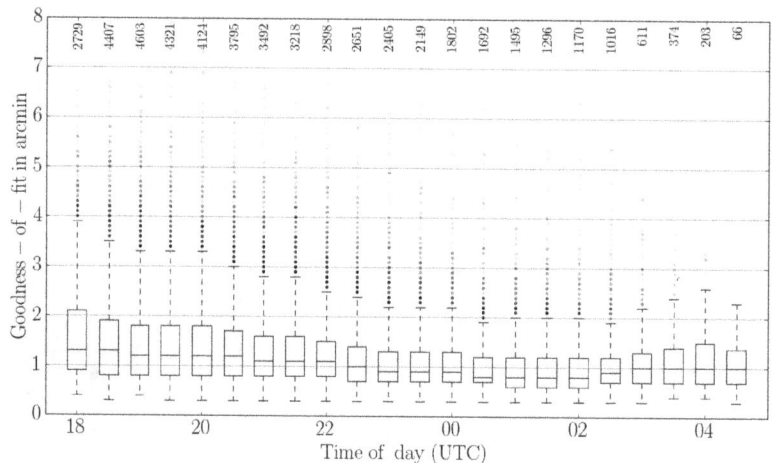

**Figure 21.** Goodness-of-fit vs. detection time for ICC9. The box plot shows the median, IQR and $1.5 \cdot$ IQR. The number shown on the top indicates the number of used data points per peak brightness bin.

verge with the ICC7 values for fainter meteors but the IQR is slightly broader.

As mentioned in Sect. 2.3.3 point b, bright meteors overexpose the CCD pixels. This leads to blooming which results in an additional broadening of the meteor on a single video frame. Another effect may be that bright meteors are more likely to display a wake (Sect. 2.3.3 point a). Due to these effects the photometric centre cannot be determined correctly, which leads to a larger position determination error for brighter meteors.

## 4 Discussion

Even though both cameras are identical from a technical point of view, ICC9 detects fainter meteors. We argue in the following that this is a geometrical effect and can be explained by the camera pointing direction.

Both camera boresights intersect between Tenerife (ICC7) and La Palma (ICC9) at an altitude of 100 km. Thus, ICC7 is pointing roughly to the west and ICC9 to the east. The elevations of the boresights with respect to the horizon are approximately 53°.

In Figs. 16 and 17 we plot the angular distance between the camera boresights and the apex and antihelion (AH) directions for the time frame 18:00 to 06:00 UTC. The red dashed line is the angular distance to the apex, the blue dashed line to the antihelion direction. The shaded areas around the lines indicate the annual variation. The black vertical lines indicate the rise times of antihelion (blue, left line) and apex (red, right line). Again, the shaded area indicates the annual variation. The thick black line is the normalized distribution of the observed meteors as a function of time during the night.

The antihelion point rises shortly after sunset, the apex direction after midnight. Since ICC7 is pointing towards the

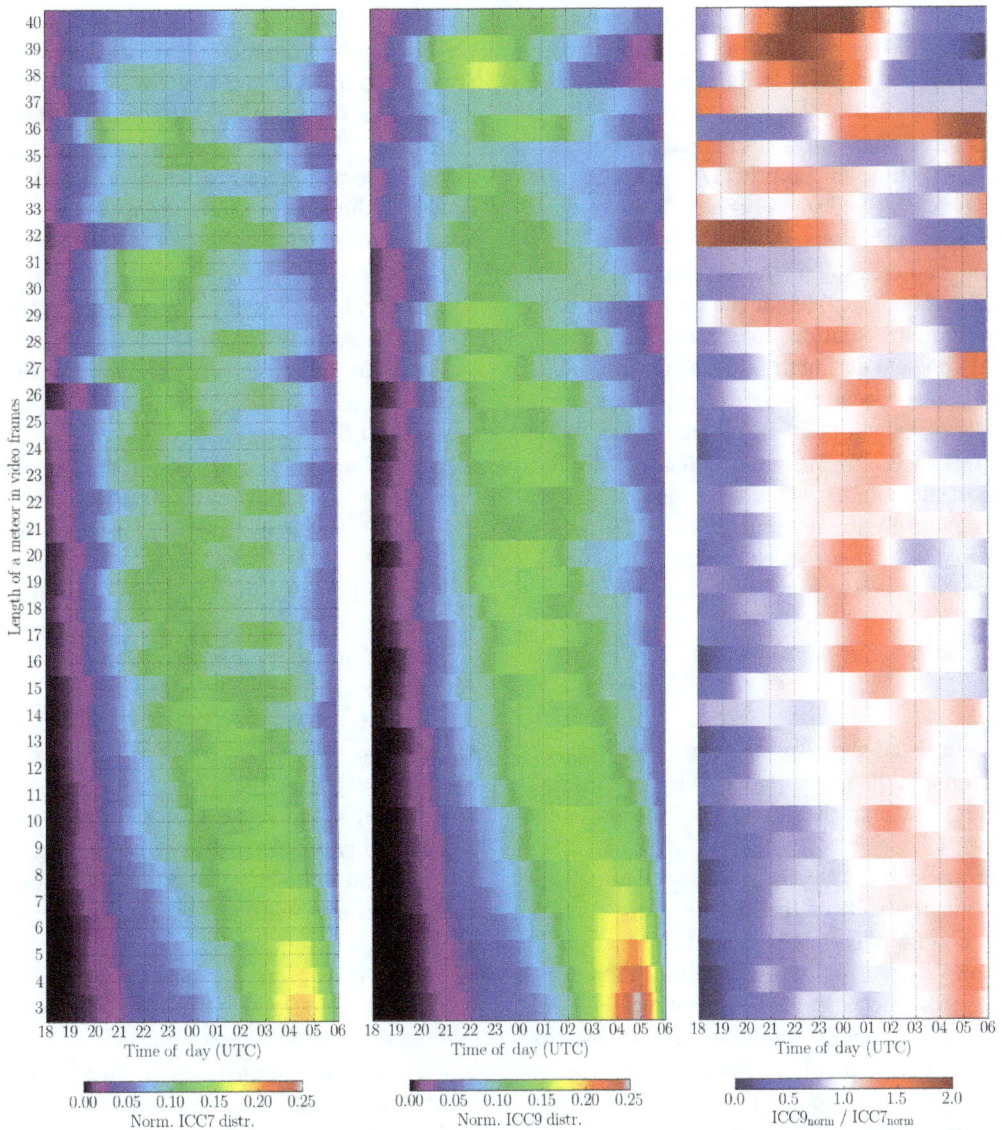

**Figure 22.** Ratio plot of the distribution of the normalized length of a meteor in frames of ICC9 divided by ICC7. Each frame distribution is shown vs. the detection time. The ratio is colour-coded, with the values given in the bottom colour bar.

west, its angular distance to the apex point is always much larger than for ICC9.

Figure 18 shows the ratio between the number of meteors for a given apparent velocity of ICC9 to ICC7 using a kernel density estimator (Pedregosa et al., 2011). This plot shows an interesting behaviour. Starting after midnight, ICC9 sees more meteors than ICC7 in the velocity range of 50 to 200 pixel s$^{-1}$. The peak moves to higher speeds during the night. After about 04:00 UTC, ICC9 also detects more meteors for low velocities. We explain this by the distance of the camera boresights to apex and antihelion sources. The apex is very close to the boresight of ICC9 in the morning hours, thus the apparent velocity of these meteors is low. Since the relative speed to the Earth is high, more meteoroids of a given mass will become visible as they generate brighter meteors.

The larger number of slow meteors in ICC9 also explains Fig. 19. Since the meteors are slower, they spend more time on a pixel and fainter meteors can be detected. This is an important finding to derive scientific conclusions, e.g. determining flux densities. The limiting magnitude determined for stars will be identical for identical systems, no matter where the camera is pointing. However, the detection threshold for meteors will be different.

In Figs. 14 and 15 we showed that the goodness-of-fit is a function of the magnitude. Since the magnitude distribution changes over the night, the goodness-of-fit also will change over night. This is illustrated in Figs. 20 and 21. The goodness-of-fit is best during the evening hours and gets worse towards the morning. The solid line indicates the median value, the dashed lines the IQRs. The values start at around 0.7' (ICC7) and 1.0' (ICC9) and decrease over the night. We claim that this is a result of the variable radiant distance and the changing magnitude distribution over the night.

Figure 22 shows three plots of the normalized length of a meteor in frames versus time for both ICC7 and ICC9, plus the ratio between two distributions. For each frame length bin, the integral of the distribution is 1. The colour map limits are the same for both cameras so that the differences between the camera systems can be visualized. It can be seen that both distributions show similar evolutions over time. Longer meteors are dominantly present during the evening and midnight hours and short meteors appear mostly during the morning hours. However, the distributions of ICC7 are spread wider than the distributions of ICC9. The ratio indicates a higher contribution of short meteors for ICC9 by a factor of up to 2. We explain this again by the apex meteors. ICC9 points closer to the apex than ICC7, in particular during the morning hours. Thus apex meteors appear shorter in ICC9.

## 5 Conclusion

In Sect. 2.3 we have listed the expected errors and biases from the instrument itself, the measurement pipeline and statistical sources. Here we map the findings of the previous section to these errors.

Mechanical/thermal stability: any mechanical/thermal instability would result in a shift of the field of view relative to an Earth-fixed direction. This would shift the measured position of a meteor. When visually inspecting the data, MetRec allows us to overlay the expected star positions with the real image. This was done regularly, and such a shift was observed in very rare cases towards the morning hours. It was typically less than 2 pixels. Since it only occurred in a few nights, it was not considered in this analysis and would deserve further study.

Brightness drop-off: the drop-off of brightness towards the edges of the optical system results in a loss of about 55 %. This will be an important effect when computing flux densities using the limiting magnitude of the system – the detected meteor numbers really are a function of the position in the field of view. The drop-off is larger than what would be expected from pure geometrical effects. It is assumed that this is an effect of the image intensifier. For non-intensified systems, we would expect this effect to be less severe.

Astrometric accuracy: the measurement accuracy of meteor positions (astrometry) is influenced by a number of the listed errors. Figure 4 shows the deviation between measured star positions and the expected position as determined by the 3rd-order polynomial plate fit performed by the detection software. It is below 0.2' up to a distance of about 90 % of the diameter of the field of view. When analysing the goodness-of-fit of individual measurement points relative to the fitted great circle of the meteor's path, errors are larger. Figures 5

and 6 show that typical errors are around 1 to 1.5', depending on the length of the meteor. We assume that these deviations come from the fact that MetRec determines the position of a meteor in a single frame by finding the photometric centre of the object. The resulting errors are listed under algorithmic errors in Sect. 2.3: a possible wake will shift the photometric centre to the back and blooming will shift the centre in an arbitrary direction, similar to the distortion of the meteor image. The possible rescaling from physical pixels to the PAL format (Sect. 2.3.3 point e) will also contribute to this result. As can be seen in Fig. 4, the deviation of the expected star positions to the real positions, based on the 3rd-order polynomial fit, stays around or below 0.2' until about 175 pixels away from the centre of the field of view. For larger values the deviation starts to increase linearly. One of the possible reasons for this could be that the 3rd order is not enough. We did not check whether a 4th-order fit would produce a better result; this will be future work.

We conclude that for our camera systems a typical error of 1 to 1.5' should be assumed.

The position measurement inaccuracies will also affect the velocity determination. In addition, the first and last frame of the meteor should not be used for velocity determination, for the obvious reason that it is not known at what time during the 40 ms exposure the meteor appears or disappears.

In a future work we will determine possible effects of daily, weekly or seasonal temperature fluctuations. Scientific projects that will derive, e.g. flux densities from the CILBO camera system, will need to consider bias effects that have been shown in this work to un-bias and derive proper scientific conclusions form the observations.

We did not do a detailed analysis of random noise affecting the measurements. We assume that since the noise is random it does not produce any bias or shift in any of the measurements, but will only increase the scatter of the data.

We find that a major contribution to the detected brightness distribution comes from the pointing direction of the cameras. The pointing direction has to be taken into account when interpreting the detected number of meteors.

*Competing interests.* The authors declare that they have no conflict of interest.

*Acknowledgements.* We acknowledge the tireless efforts of Hans Smit and Cornelis van der Luijt (ESA/Space Science Office) for keeping the cameras operational. CILBO hardware and maintenance are funded thanks to the research faculty of ESA/Space Science Office. We also acknowledge the Instituto de Astrofisica de Canarias (J. Licandro) which hosts the CILBO system and provides local support.

Edited by: L. Vazquez

# References

Albin, T., Koschny, D., Drolshagen, G., Soja, R., Srama, R., and Poppe, B.: Influence of the pointing direction and detector sensitivity variations on the detection rate of a double station meteor camera, in: Proceedings of the International Meteor Conference, edited by: Rault, J.-L. and Roggemans, P., IMO, 226–232, 27–30 August 2015, Mistelbach, Austria, 2015a.

Albin, T., Koschny, D., Drolshagen, G., Soja, R., Poppe, B., and Srama, R.: De-biasing of the velocity determination for double station meteor observations from CILBO, in: Proceedings of the International Meteor Conference, edited by: Rault, J.-L. and Roggemans, P., IMO, 214–219, Mistelbach, Austria, 2015b.

Albin, T., Koschny, D., Soja, R., Srama, R., and Poppe, B.: A Monte-Carlo based extension of the Meteor Orbit and Trajectory Software (MOTS) for computations of orbital elements, in: Proceedings of the International Meteor Conference, edited by: Roggemans, A. and Roggemans, P., IMO, 20–25, 2–5 June 2016, Egmond, the Netherlands, 2016.

Barentsen, G. and Koschny, D.: The IMO Virtual Meteor Observatory (VMO): Architectural design, Earth Moon Planets, 102, 247–252, 2008.

Cooke, W. and Moser, D. E.: The status of the NASA All-Sky Fireball network, in: Proceedings of the International Meteor Conference, edited by: Gyssens, M. and Roggemans, P., IMO, 9–12, 15–18 September 2011, Sibiu, Romania, 2012.

Drolshagen, E., Ott, T., Koschny, D., Drolshagen, G., and Poppe, B.: Meteor velocity distribution from CILBO double station video camera dat, in: Proceedings of the International Meteor Conference, edited by: Rault, J.-L. and Roggemans, P., IMO, 16–22, 18–21 September 2014, Giron, France, 2014.

Jenniskens, P. Gural, P. S., Dynneson, L., Grigsby, B. J., Newman, K. E., Borden, M., Koop, M., and Holman, D.: CAMS: Cameras for Allsky Meteor Surveillance to establish minor meteor showers, Icarus, 216, 40–61, 2011.

Koschny, D. and Diaz del Rio, J.: Meteor Orbit and Trajectory Software (MOTS) – Determining the Position of a Meteor with Respect to the Earth Using Data Collected with the Software MetRec, WGN, 30, 87–101, 2002.

Koschny, D., Bettonvil, F., Licandro, J., Luijt, C. v. d., Mc Auliffe, J., Smit, H., Svedhem, H., de Wit, F., Witasse, O., and Zender, J.: A double-station meteor camera set-up in the Canary Islands – CILBO, Geosci. Instrum. Method. Data Syst., 2, 339–348, doi:10.5194/gi-2-339-2013, 2013.

Kretschmer, J., Drolshagen, S., Koschny, D., Drolshagen, G., and Poppe, B.: De-biasing CILBO meteor observational data to mass fluxes, in: Proceedings of the International Meteor Conference, edited by: Rault, J.-L. and Roggemans, P., IMO, 209–213, 27–30 August 2015, Mistelbach, Austria, 2015.

Molau, S.: The meteor detection software MetRec, in: Proceedings of the International Meteor Conference, edited by: Arlt, R. and Knöfel, A., IMO, 9–16, 20–23 August 1998, Stará Lesná, 1999.

Molau, S.: MetRec – Meteor Recognition Software Version 5.2, available at: http://www.metrec.org/download/readme.txt (last access: 23 February 2017), 2014.

Molau, S., Kac, J., Crivello, S., Stomeo, E., Barentsen, G., Goncalves, R., Saraiva, C., Maciewski, M., and Maslov, M.: Results of the IMO Video Meteor Network – April 2015, WGN, Journal of the International Meteor Organization, 43, 115–120, 2015.

Ott, T., Drolshagen, E., Koschny, D., Drolshagen, G., and Poppe, B.: Meteoroid flux determination using image intensified video camera data from the CILBO double station, in: Proceedings of the International Meteor Conference, edited by: Rault, J.-L. and Roggemans, P., IMO, 23–29, 18–21 September 2014, Giron, France, 2014.

Pedregosa, F., Varoquaux, G., Gramfort, A., Michel, V., Thirion, B., Grisel, O., Blondel, M., Prettenhofer, P., Weiss, R., Dubourg, V., Vanderplas, J., Passos, A., Cournapeau, D., Brucher, M., Perrot, M., and Duchesnay, E.: Scikit-learn: Machine Learning in Python, J. Mach. Learn. Res., 12, 2825–2830, 2011.

SonotaCo, T., Molau, S., and Koschny, D.: Amateur contributions to meteor astronomy, European Planetary Science Congress 2010, 20–24 September, Rome, Italy, p. 798, 2010.

Weryk, R. J., Campbell-Brown, M. D., Wiegert, P. A., Brown, P. G., Krzeminski, Z., and Musci, R.: The Canadian Automated Meteor Observatory (CAMO): System overview, Icarus, 225, 614–622, 2013.

# 16

# Auroral meridian scanning photometer calibration using Jupiter

**Brian J. Jackel[1], Craig Unick[1], Fokke Creutzberg[2], Greg Baker[1], Eric Davis[1], Eric F. Donovan[1], Martin Connors[3], Cody Wilson[1], Jarrett Little[1], M. Greffen[1], and Neil McGuffin[1]**

[1]Department of Physics and Astronomy, University of Calgary, Alberta, Canada
[2]Natural Resources Canada, Geological Survey, Geomagnetism Laboratory, Natural Resources Canada Geomagnetism Laboratory, Ottawa, Ontario, Canada
[3]Department of Physics and Astronomy, Athabasca University, Alberta, Canada

*Correspondence to:* Brian J. Jackel (brian.jackel@ucalgary.ca)

**Abstract.** Observations of astronomical sources provide information that can significantly enhance the utility of auroral data for scientific studies. This report presents results obtained by using Jupiter for field cross calibration of four multispectral auroral meridian scanning photometers during the 2011–2015 Northern Hemisphere winters. Seasonal average optical field-of-view and local orientation estimates are obtained with uncertainties of 0.01 and 0.1°, respectively. Estimates of absolute sensitivity are repeatable to roughly 5 % from one month to the next, while the relative response between different wavelength channels is stable to better than 1 %. Astronomical field calibrations and darkroom calibration differences are on the order of 10 %. Atmospheric variability is the primary source of uncertainty; this may be reduced with complementary data from co-located instruments.

## 1 Introduction

Interactions between the solar wind and the terrestrial magnetic field produce a complex and dynamic geospace environment. Ionospheric phenomena, such as the aurora, are connected to magnetospheric processes by mass and energy transport along magnetic field lines. Consequently, auroral observations provide information that can be used for remote sensing of distant plasma structure and dynamics. A single ground-based instrument can only view a small part of the global system, so a combination of instruments at different locations (e.g., Fig. 1 and Table 1) is required to span larger scales. Merging multiple data sets requires accurate informa-

tion about device characteristics such as timing, orientation, and absolute spectral sensitivity.

Comprehensive calibration requires specialized equipment and skilled personnel that are typically available only at centrally located research facilities. With sufficient resources it is possible, at least in principle, to determine all device parameters that are required to convert raw instrument data numbers to physically useful quantities. Practical limitations can result in random or systematic uncertainties which may impede quantitative scientific analysis. This is particularly relevant for large networks of nominally identical instruments, where ongoing calibration of each device may be extremely challenging.

Even assuming ideal calibration at a central facility, many auroral instruments must be operated at remote field sites. Transfer between these locations requires a sequence of packing, shipping, and reassembly that is time-consuming, costly, and may unintentionally alter instrument response. Furthermore, intermittent calibration cannot distinguish between a gradual drift or sudden changes.

Extraterrestrial sources, such as planets or stars, are often used for calibration of spatially resolved optical or radio frequency data. Instrument orientation can be determined from objects whose positions are well known, while source intensity can be used to verify instrument sensitivity. Astronomical sources are often detectable in existing auroral data streams, allowing for ongoing monitoring of system response and the possibility of retrospective reanalysis of older data sets. Practical application may be restricted by instrumental limitations and complications including man-made interference, clouds, aurora, and other geophysical processes.

**Table 1.** Canadian meridian scanning photometer site information. Geographic latitude, longitude, and altitude are in degrees north, degrees east, and meters above mean sea level (WGS-84). L-shell and magnetic declination from the IGRF model.

|  | Geographic | | | L-shell | | Declination | | |
|---|---|---|---|---|---|---|---|---|
|  | Lat | Long | Alt | 1988 | 2013 | 1988 | 2013 | |
| RANK | 62.82 | 267.89 | 32 | 11.20 | 10.64 | −7.1 | −7.7 | Rankin Inlet, Nunavut |
| GILL | 56.35 | 265.29 | 99 | 6.04 | 5.83 | 2.6 | −0.5 | Gillam, Manitoba |
| PINA | 50.20 | 263.96 | 262 | 3.95 | 3.84 | 5.5 | 2.3 | Pinawa, Manitoba |
| FSMI | 60.02 | 248.05 | 205 | 6.65 | 6.58 | 24.3 | 15.8 | Fort Smith, NWT |
| ATHA | 54.70 | 246.70 | 533 | 4.50 | 4.45 | 21.1 | 15.3 | Athabasca, Alberta |

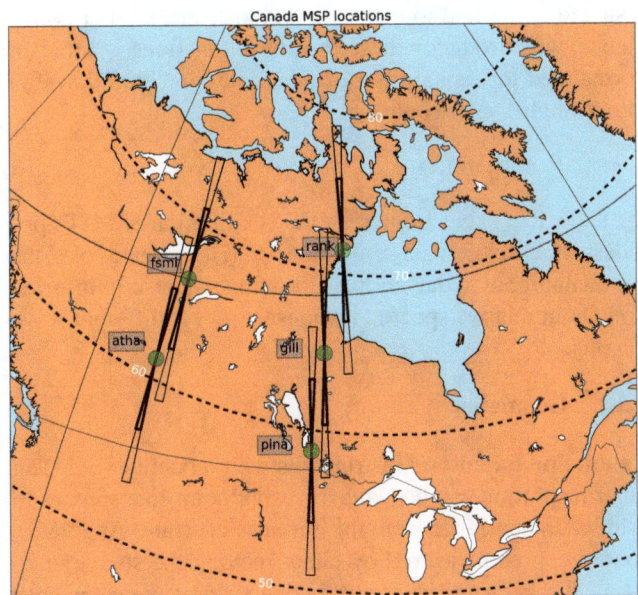

**Figure 1.** Canadian meridian scanning photometer site locations (details in Table 1). Fan shapes indicate 4° optical beam width for altitudes of 110 and 220 km at elevations of 10° above the horizon. Dashed contours indicate magnetic dipole latitude (IGRF, 2015).

There is a long history of using astronomical sources to determine the alignment of auroral instruments (Stormer, 1915; Fuller, 1931; Chapman, 1934; Kinsey, 1963). Absolute calibration using stellar spectra appears to be a more recent development (Gladstone et al., 2000; Whiter et al., 2010; Dahlgren et al., 2011; Wang, 2011; Wang et al., 2012; Grubbs et al., 2016). Detailed discussions of these topics are not always provided in the primary scientific literature, but must often be extracted from conference proceedings, technical reports, and theses.

The focus of this paper is on the field calibration of a network of four auroral photometers using Jupiter as a standard reference. Some key features of optical aurorae are provided in Sect. 1.1, Sect. 1.2 introduces key calibration concepts and results, essential astronomical topics are presented in Sect. 1.3, and atmospheric effects are briefly reviewed in Sect. 1.4. An overview of instrument details is given in

Sect. 2, data analysis and results are in Sect. 3, discussion is in Sect. 4, followed by a summary and conclusions in Sect. 5.

## 1.1 Optical aurora

In regions of geospace where magnetic field lines can be traced to the Earth, some charged particles may travel down to altitudes where neutral densities are no longer negligible. Collisions with atmospheric atoms or molecules may transfer energy which can be re-emitted as photons. Spectral, spatial, and temporal features of the optical aurora contain information about geospace plasma properties, allowing for remote sensing of magnetospheric topology and dynamics.

Auroral spectra are dominated by several relatively bright lines and bands from atomic oxygen and molecular nitrogen, with many other less intense features ranging from extreme ultraviolet through to far infrared. The intensity of auroral emission at different wavelengths depends on precipitation energy and atmospheric composition, as more energetic particles are able to penetrate to lower altitudes where constituents may be more or less abundant. Consequently, observations at multiple wavelengths can be combined to infer characteristics of the precipitating particles (Rees and Luckey, 1974; Strickland et al., 1989). These multispectral measurements can be challenging due to the wide dynamic range between very bright 558 nm green-line (1–100 kR) emissions and extremely faint 486 nm proton aurora (< 100 R).

Optical aurora typically occur within auroral ovals, roughly centered around each geomagnetic pole, extending hundreds of kilometers in latitude and thousands of kilometers in longitude (Akasofu, 1965). Luminosity can be highly dynamic over a wide range of spatial scales, but quiet-time structures generally exhibit a narrow latitudinal extent (tens to hundreds of kilometers) and relatively less longitudinal variation over hundreds or thousands of kilometers (Knudsen et al., 2001). This spatial anisotropy is one motivation for using a meridian scanning photometer (MSP; see Sect. 2) to measure auroral luminosity as a sequence of latitude profiles (keogram). As shown in Fig. 2, these data can also be used to identify other non-auroral features, such as clouds and stars.

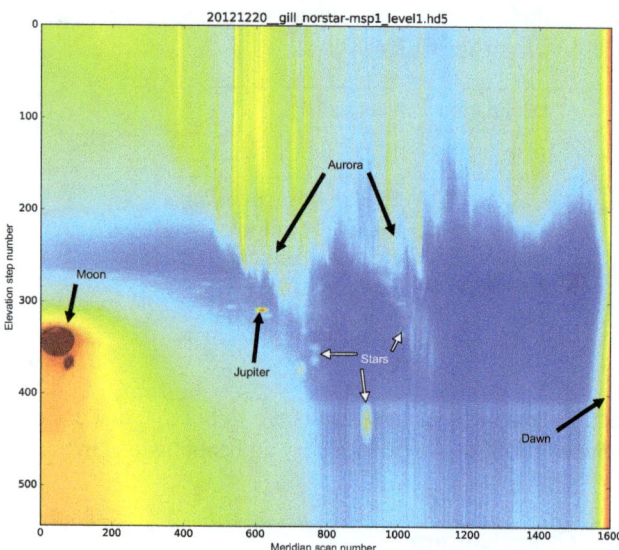

**Figure 2.** Keogram from meridian scanning photometer operating at Gillam during the night of 20 December 2012 from 00:00 to dawn at 13:20 UT. Local midnight is approximately 06:00 UT (scan number 720). Data counts have been clipped and logscaled in order to display Jupiter, stars, aurora, full moon, and dawn.

## 1.2 Instrument calibration

Optical designs can be modeled very precisely with modern software tools, but instrument calibration provides essential information about the actual performance. System response is not necessarily constant, but can change either gradually (e.g., filter bandpass drift, decreased detector sensitivity) or abruptly (e.g., damage during shipping). Such problems could be identified with calibration of instruments in the field. This process must be completely automatic, as many remote sites do not have full-time technical staff. It should be repeated frequently in order to identify abrupt changes in system response, but without interrupting or degrading normal data acquisition. A regular schedule of measurements with portable low-brightness sources (LBSs) might satisfy some of these requirements, but would involve a substantial allocation of resources for repeated site visits.

In this report, we examine some of the strengths and limitations of astronomical calibration for auroral instruments. We focus on issues related to field cross calibration of MSPs which have been used extensively for auroral research (see Sect. 2 for details). However, many of these topics can also be applied more generally to other instruments used to study the optical aurora, such as all-sky imagers (ASIs).

A single ground-based instrument may measure photons with wavelengths $\lambda$ arriving from angular locations $\theta$, $\phi$. The distribution of incident light $I$ is convolved with the instrument response function $f$ to produce a measurement $M$ with error $M_\epsilon$:

$$M(\theta, \phi, \lambda) = f(\theta, \phi, \lambda) \cdot I(\theta', \phi', \lambda') + M_\epsilon(\theta, \phi, \lambda). \quad (1)$$

For an ideal device, $f$ would be a delta function and $M$ would be equal to $I$, but any real measurement will have limited resolution. The goal of calibration or characterization is to determine the instrument response function $f$ in order to better understand the true source properties.

The general response function in Eq. (1) can be separated into a product of geometric sensitivity $f_G$ and spectral sensitivity $f_S$:

$$f_G(\theta, \phi) \times f_S(\lambda). \quad (2)$$

This approximation is not always valid (e.g., wide-angle optics coupled to a narrow-band interference filter) but can be usefully applied to many auroral instruments. For convenience, we introduce relative response functions ($\hat{f}$) that are normalized to a maximum of one, and combine all scaling into a single system constant $\mathcal{C}$:

$$\mathcal{C} \times \hat{f}_G(\theta, \phi) \times \hat{f}_S(\lambda). \quad (3)$$

We show that using Jupiter for field calibration of MSPs provides detailed knowledge about $\hat{f}_G(\theta, \phi)$, estimates of $\mathcal{C}$ that are comparable to darkroom calibration, and useful information about relative spectral response $\hat{f}_S(\lambda)$ at different wavelengths.

### 1.2.1 Geometric

Calibration for auroral instruments with moderate ($\sim 1°$) angular resolution can be achieved using point-like sources located sufficiently far from the entrance aperture. Angular response can be measured by either moving the source or rotating the instrument. The effective field-of-view (or beam shape) is often azimuthally symmetric around an optical axis with angular polar coordinates $\theta_0$, $\phi_0$, in which case relative response can be expressed in terms of off-axis angle $\gamma$:

$$\hat{f}_G(\theta, \phi) \approx \tilde{f}_G(\gamma; q_1, \ldots, q_N), \quad (4)$$

and some set of instrument parameters $q_i$ (e.g., full-width half-max).

Ideally, each instrument would arrive at a field site in exactly the same condition as it left the darkroom. It would be operated exactly as intended (i.e., perfectly level and aligned north–south) without changes for the entire design lifetime. In practice, it may be difficult to achieve desired alignment to better than a few degrees. The initial orientation may subsequently drift to some more stable state over months or years, with the possibility of more abrupt changes as the ground freezes in autumn and thaws in spring. In general, the rotation matrix **R** required to properly transform from device to local coordinates (e.g., azimuth and zenith angle) must be updated regularly in order to ensure that data are scientifically useful.

Determining Euler angles and geometric response model parameters in the field is relatively straightforward for auroral instruments that can detect and resolve at least a few of the brightest stars. Accurate GNSS site location and measurement timing can be combined with astronomical catalogs to predict the local orientation of each star. These can be converted into device coordinates and used to calculate observable quantities such as transit time and zenith angle. Discrepancies between predictions and observations can be minimized to determine optimal parameter values. A single night of good data may be sufficient to achieve subdegree accuracy, which is adequate for many auroral instruments.

Although stars are essentially point sources at infinity, other immutable properties (e.g., location, apparent motion, spectral radiance) may make them somewhat less tractable than darkroom calibration sources. Any given object will not always be visible in the night sky or pass through any specific location in an instrumental field of view. However, a substantial amount of useful information can be gathered over several days or months.

### 1.2.2 Spectral

The relative spectral response of an instrument is essential for quantitative multiwavelength analysis, such as estimating precipitation energy (Rees and Luckey, 1974; Strickland et al., 1989). Spectral response can be most effectively determined with a monochromatic source, such as

$$\int d\lambda' \hat{f}_s(\lambda')\delta(\lambda - \lambda') = \hat{f}_s(\lambda'), \tag{5}$$

which can scan through the wavelength range of interest. For narrow-band devices it may be sufficient to observe a broadband source $S(\lambda)$ with known absolute spectral flux density. If the source flux is roughly constant near some wavelength $\lambda_j$ for each device channel, as in

$$\int d\lambda \hat{f}_s(\lambda)S(\lambda) \approx \overline{S}(\lambda_k) \int d\lambda \hat{f}_s(\lambda) = \overline{S}(\lambda_k)\Delta\lambda_k, \tag{6}$$

then the throughput for each channel may be expressed in terms of the effective bandwidth $\Delta\lambda_k$.

Measurements of an absolutely calibrated LBS provide estimates of the differential sensitivity to a continuum source characterized in terms of Rayleighs per nanometer. For discrete emission lines, the effective bandwidth is also required in order to determine the sensitivity to brightness as expressed in Rayleighs. The equipment necessary for comprehensive calibration (e.g., LBS and monochromator) is not always available at remote field sites, so different methods must be established. Many stellar sources provide spectra which are apparently broad band at typical auroral instrument resolutions on the order of 1 nm. Only relatively bright stars may be above the detection threshold, and absolute flux calibrated spectra are not available for all sources. Still, in certain cases it may be possible for astronomical calibration to produce accurate and repeatable estimates of differential sensitivity.

There does not appear to be a corresponding strategy to determine effective bandwidth in the field. Most stellar spectra are essentially constant in time, so individual sources cannot be used to determine a fixed instrument response. Combining many different spectra might, in principle, allow us to distinguish between changes in effective bandwidth and total sensitivity. However, this would require nearly simultaneous observation of multiple absolutely calibrated sources with different spectral types. Low signal levels might also limit the accuracy of any estimates.

For this study, we proceed under the assumption that absolute spectral response cannot be independently determined in the field using only astronomical sources. We presume that normalized transmission integrated across each pass band,

$$\int d\lambda \hat{T}(\lambda) \equiv \Delta_\lambda \quad \hat{T}(\lambda) = [0, 1], \tag{7}$$

can be obtained in some other way, and acknowledge that simultaneous changes across multiple channels may not be detected using methods considered here. For these reasons, we shall tend to focus on the differential calibration coefficient $\dot{C}$ which can be determined using only astronomical methods. This quantity can also be directly compared to the results of darkroom calibration with an LBS. For auroral studies, data numbers $\mathcal{D}$ must be converted to Rayleighs $\mathcal{R}$, and effective bandwidth is required in order to calculate $C_{\mathcal{R}/\mathcal{D}}$.

### 1.2.3 Radiometry

At a distance $R$ from an isotropic point source with total power output $P_0$ the irradiance (intensity) $S$ will be

$$S = \frac{P_0}{4\pi R^2} \text{ W m}^2, \tag{8}$$

so that an observer at some distance $r$ will intercept an amount of power,

$$P_\delta = SA_{\text{eff}}, \tag{9}$$

proportional to the effective receiver surface area $A_{\text{eff}}$.

Power from an extended source can be expressed in terms of a volume emission rate $\rho(r, \theta, \phi)$ integrated over the entire source region weighted by the receiver angular sensitivity $G(\theta, \phi)$:

$$P_V = \oiint d\Omega \frac{LG}{4\pi} \quad 4\pi L \equiv \int_0^\infty dr \rho(r), \tag{10}$$

where the radial integral $L$ has units of radiance (W m$^2$ sr$^1$) and is often referred to as the "column emission rate". For a uniform source radiance, the total received power

$$P_V = \oiint d\Omega \frac{L(\theta, \phi)}{4\pi} A_{\text{eff}}\hat{G}(\theta, \phi) \approx LA_{\text{eff}}\Omega_0, \tag{11}$$

depends on the product of the effective area and the effective solid angle. For any signal detected from some point source there will be an equivalent volume emission which would produce the same observed power. For a uniform emission region, the relationship

$$P_\delta = P_V \rightarrow L = \frac{S}{\Omega_0} \qquad (12)$$

depends only on the effective solid angle.

Auroral intensity $\mathcal{I}$ is customarily expressed in units of Rayleighs (Hunten et al., 1956; Baker, 1974; Baker and Romick, 1976; Brändström et al., 2012) which is related to photon radiance $L_\gamma$ via

$$4\pi L_\gamma(\lambda) \equiv \mathcal{I}(\lambda) \quad 10^{10} \text{ photon s}^1 \text{ m}^2 \qquad (13)$$

where the subscript E indicates energy flux and $\gamma$ is photon number flux. For narrow-band channels

$$\mathcal{I}(\lambda) = \int \dot{\mathcal{I}}(\lambda) \approx \dot{\mathcal{I}} \Delta\lambda = 4\pi \frac{\dot{S}_E}{\Omega_0} \frac{\lambda}{hc} \Delta\lambda \qquad (14)$$

converting differential radiant spectral density $\dot{S}$ to equivalent Rayleighs per nanometer $\dot{\mathcal{I}}$ requires only the effective solid angle, which can also be estimated from observations of a point source. Working with Rayleighs requires some additional knowledge in the form of the effective bandwidth $\Delta\lambda$. As this is also true for darkroom LBS calibration, we focus here on relating $\dot{\mathcal{I}}$ in Rayleighs per nanometer to $\dot{S}$ in watts per meter squared per nanometer.

### 1.3 Astronomical sources

Extraterrestrial objects have many properties which are required for accurate calibration. Locations in the celestial sphere are known to arc-second resolution or better, which is sufficient for determining orientation and geometric response of most auroral instruments. Absolute spectral irradiance profiles are available for many sources, providing opportunities for radiometric calibration of narrow-band instruments. Total visible intensity of most sources is essentially constant, allowing for long-term monitoring of system performance. A single object can be viewed simultaneously by multiple instruments at nearby sites, facilitating quantitative intercomparisons.

Most astronomical objects are effectively point sources, and under good viewing conditions modern all-sky imagers can resolve hundreds of stars with a relatively short exposure time. Ironically, the presence of bright aurora or airglow can be a major source of error in radiometric calibration. For the MSP considered here, the total light from Vega passing through a 3 nm filter is approximately 200 Rayleighs, which is comparable to typical red-line airglow emissions. Even on a moonless night, continuum emissions can be on the order of 10 R nm$^{-1}$, equivalent to stars of magnitude 2 as observed by our MSP. Note that there are only 50 stars of magnitude 2

**Table 2.** Selected astronomical source irradiance at Earth. Energy flux is joules per s m$^{-2}$ nm$^{-1}$ and number flux is photons per s m$^{-2}$ nm$^{-1}$ Rayleighs are for a viewing solid angle of $\Omega = 0.002$ steradians (2.9° of arc).

|         | (nm) | $(J)$                  | (No.)                 | (R nm$^{-1}$)           |
|---------|------|------------------------|-----------------------|-------------------------|
| Jupiter | 486  | $4.78 \times 10^{-10}$ | $1.17 \times 10^9$    | 735                     |
| Jupiter | 556  | $5.45 \times 10^{-10}$ | $1.53 \times 10^9$    | 958                     |
| Sirius  | 556  | $1.35 \times 10^{-10}$ | $3.78 \times 10^8$    | 237                     |
| Vega    | 556  | $3.44 \times 10^{-11}$ | $9.63 \times 10^7$    | 60.5                    |
| Moon    | 556  | $4.63 \times 10^{-6}$  | $1.3 \times 10^{13}$  | $8.14\text{e} \times 10^6$ |
| Sun     | 556  | 1.81                   | $5.07 \times 10^{18}$ | $3.18 \times 10^{12}$   |

or brighter, and fewer than half of them are visible from the northern auroral zone at any given time.

Celestial source brightness spans a wide range and is usually expressed in terms of logarithmic magnitude $m$:

$$I = \sqrt[5]{100}^m \approx 2.512^m, \qquad (15)$$

so that the relative intensity of two sources can be determined from the difference of their magnitudes. Absolute flux distributions as a function of wavelength are available for most of the brightest stars, including Vega (Colina et al., 1996), Sirius (Bohlin, 2014), and Arcturus (Blackwell et al., 1975; Griffin and Lynas-Gray, 1999). Other catalogs contain many other stars (Hayes, 1985; Alekseeva et al., 1996, 1997; Bohlin, 2007, 2014), but the majority may be too dim for reliable observation by typical auroral instruments.

Conversely, the sun is so bright that direct observation will saturate detectors designed for relatively faint aurora. Thuillier et al. (2003) provide an absolutely calibrated distribution of flux vs. wavelength at 1 AU with subnanometer spectral resolution. For a nominal instrument solid angle of 2 millisteradians (3° of arc) the apparent solar brightness at 556 nm is roughly 3 teraRayleighs per nanometer (Table 2). Daytime operations are only possible for systems that respond to an extremely narrow range of wavelengths (Galand et al., 2004).

Although direct sunlight is unsuitable as a calibration source for most auroral instruments, scattering from other bodies in the solar system can provide more reasonable levels of brightness. The irradiance of an arbitrary body $x$ can be modeled by isotropic emission from the sun incident on a sphere with radius $R_x$ at distance $D_{Sx}$, followed by scattering and absorption leading to some fraction of flux traveling a distance $D_{xE}$ to arrive at the top of Earth's atmosphere. We group terms that depend on time and wavelength into $A(t)$ and $B(\lambda)$, respectively.

$$I_{xE}(\lambda, t) = A(t) \times B(\lambda) \qquad (16)$$

The wavelength-dependent term $B(\lambda)$ contains irradiance in terms of the total solar irradiance (TSI $\sim 1360$ watts m$^{-2}$) at a fixed distance of 1 AU, such as

$$B(\lambda) \equiv \text{TSI}(\lambda)\epsilon(\lambda), \qquad (17)$$

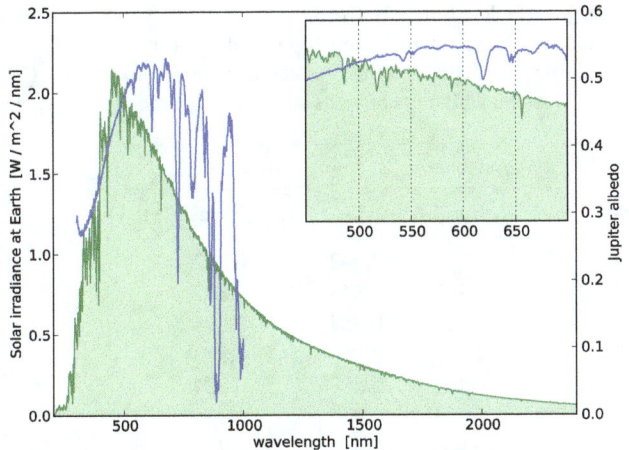

**Figure 3.** Spectra of solar irradiance (green shaded curve) from Thuillier et al. (2003) and Jupiter albedo (blue line) from Karkoschka (1998). Inset displays the same quantities for the range of wavelengths associated with most visible aurora.

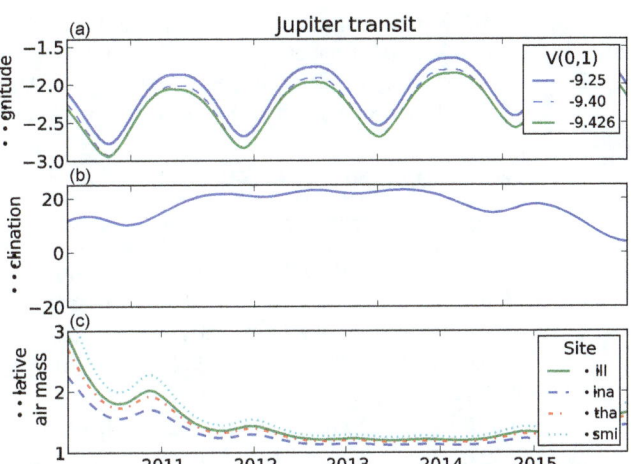

**Figure 4.** Jupiter as seen from northern auroral zone. Top panel: apparent visual magnitude (negative is brighter). Different curves correspond to results from older references ($V(1,0) = -9.25$), newer references ($-9.40$), and calculations in this study ($-9.426$). Middle panel: declination, which is effectively the same for any terrestrial observer (parallax $\approx 0$). Bottom panel: relative air mass during transit at Fort Smith, Gillam, Athabasca, and Pinawa.

where the solar power $P_s(\lambda)$ and planetary albedo $\epsilon$ are both assumed to be time independent to 1 % or less.

The time-dependent term $A(t)$ contains a phase correction factor $\Phi(\varphi)$ which accounts for any non-Lambertian scattering as a function of angle $\varphi$ between illumination and observer.

$$A(t) \equiv \frac{R_x^2 D_{SE}^2}{D_{Sx}^2 D_{xE}^2} \Phi(\varphi) \cos(\phi) \qquad (18)$$

For example, illumination from a full moon ($\phi = 0$) is reduced by a factor of $3 \times 10^{-6}$ ($m \sim 14$) relative to direct sunlight. Despite this substantial decrease, the equivalent brightness of nearly 10 megaRayleighs per nanometer (Table 2) is still a hundred times brighter than the brightest aurora. For many instruments the angular size of the moon is neither point-like nor beam-filling, requiring careful attention to details such as wavelength-dependent albedo varying across the disk (Kieffer and Stone, 2005), and making phase calculations more complicated. For these reasons, the moon is not commonly used for calibrating auroral instruments.

After the moon, Jupiter is currently the brightest celestial object that can be regularly observed well past astronomical twilight. Peak visible magnitude is nearly 4 times that of Sirius (the brightest star), making Jupiter easy to identify in the night sky. A detailed spectral distribution of Jupiter's albedo is given by Karkoschka (1998) (see Fig. 3). This can be combined with the solar spectrum of Thuillier et al. (2003) to predict the wavelength dependence of reflected light given in Table 3.

Other bodies in our solar system are less suitable as calibration sources. Mercury is only visible from Earth during the daytime when looking near the sun. Venus can often be seen near dawn or dusk, but always with excessive amounts of indirect sunlight. Mars can be visible at night for several

months in a row, but this ideal configuration only occurs on alternate years. (Fig. 5). Albedo can vary considerably during dust storms and a wide range of $\varphi$ means that the phase function $\Phi$ must be very precisely determined (Mallama, 2007). Saturn is roughly one-tenth as bright as Jupiter, with complex albedo variations due to ring geometry ($V = -0.62$ to $+1.31$) (Mallama, 2012).

The remaining outer planets are simply too dim for reliable detection by most auroral instruments.

As Jupiter and Earth each orbit around the sun, their relative motion produces significant variations in apparent magnitude and position as shown in Fig. 4. In recent years Jupiter and the Earth have been closest during winter in Northern Hemisphere, maximizing brightness during the optimal period for observations with auroral instruments. As shown in Fig. 5, Jupiter transit at Gillam Manitoba currently occurs near sunrise in early October and sunset in February. An orbital period of 11.89 years means that opposition will advance by roughly 1 month per year. Optimal configurations with transit near midnight during northern winter started in 2011, will continue until 2016, and then begin again in 2022. Previous windows of opportunity include 1988–1993 and 1999–2005. Any historical data acquired during these years could conceivably be retrospectively calibrated using Jupiter.

During this study, we identified a systematic difference between our flux calculations for Jupiter and the corresponding magnitude value provided by widely available astronomy software (Downey, 2015) using the formula

$$V = V(1,0) + 5\log_{10}(dr) + \Delta m(i), \qquad (19)$$

**Table 3.** Spectral variation of solar irradiance at Earth (Thuillier et al., 2003), albedo of Jupiter (Karkoschka, 1998), and atmospheric extinction at Cerro Paranal (Patat et al., 2011). Column 5 is the product of solar irradiance at 1 AU and Jupiter albedo (defined as $B(\lambda)$ in Eq.17) with units of watts per meter squared per nanometer. Atmospheric transmission $E_k$ at zenith is related to extinction $\kappa$ by Eq. (22). Column 8 is the product of solar irradiance, Jupiter albedo, and atmospheric transmission with units of watts per meter squared per nanometer.

| Wavelength | Solar flux ($m^{-2}\,nm^{-1}$) | | Jupiter | $B(\lambda)$ | Atmosphere | | $B(\lambda)\,Ek(\lambda)$ |
|---|---|---|---|---|---|---|---|
| (nm) | (photon $s^{-1}$) | (W) | albedo | ($W\,m^{-2}\,nm^{-1}$) | $\kappa$ | $Ek$ | ($W\,m^{-2}\,nm^{-1}$) |
| 470.9 | $1.783 \times 10^{18}$ | 2.004 | 0.446 | 0.893 | 0.187 | 0.842 | 0.752 |
| 480.0 | $1.951 \times 10^{18}$ | 2.096 | 0.454 | 0.952 | 0.179 | 0.848 | 0.807 |
| 486.1 | $1.701 \times 10^{18}$ | 1.788 | 0.455 | 0.814 | 0.171 | 0.854 | 0.695 |
| 495.0 | $2.027 \times 10^{18}$ | 2.005 | 0.470 | 0.942 | 0.160 | 0.863 | 0.813 |
| 557.7 | $2.315 \times 10^{18}$ | 1.799 | 0.515 | 0.927 | 0.127 | 0.889 | 0.824 |
| 625.0 | $2.310 \times 10^{18}$ | 1.627 | 0.495 | 0.805 | 0.101 | 0.912 | 0.734 |
| 630.0 | $2.481 \times 10^{18}$ | 1.646 | 0.520 | 0.855 | 0.097 | 0.915 | 0.782 |

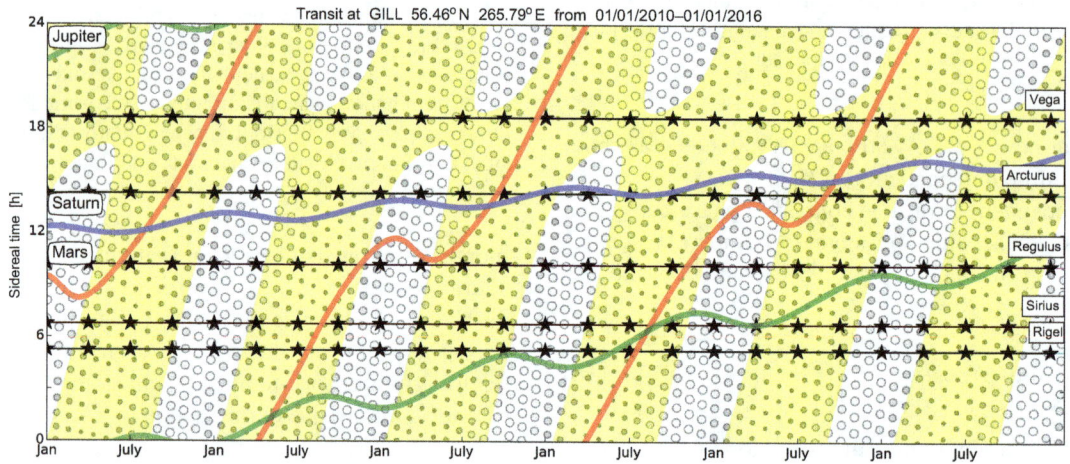

**Figure 5.** Planetary right ascension over time indicated by thick colored lines (Mars is red, Saturn is blue, and Jupiter is green). Stars indicated by thin black lines remain at constant RA. Sizes of small circles are proportional to lunar phase. Yellow shading indicates daytime extending to nautical twilight (sun 6° below horizon).

where $V(1, 0)$ is the magnitude at 1 AU with $i = 0$ and $\Delta m(\phi)$ is the magnitude phase correction. Our results were calculated by entering standard distances into Eq. (18) with irradiance and reflection from Thuillier et al. (2003) and Karkoschka (1998). We obtained equivalent values of $V(1, 0) \approx -9.426$ that were nearly 20 % larger than the standard result of $V(1,0) = -9.25$. Eventually, we discovered that the widely used lower value came from the 2nd edition of the Explanatory Supplement to the Astronomical Almanac (Seidelmann, 1992) but the most recent 3rd edition (Table 15.8, Seidelmann, 2005) now indicates $V(1, 0) = -9.40$, which differs from our results by only 2 %. This exemplifies the degree to which we attempted to cross check our results against other references. It also demonstrates that even astronomical constants may be a work in progress.

### 1.4 Atmospheric effects

Light arriving at the top of the Earth's atmosphere may undergo significant changes by the time it arrives at a ground-based observer. Gradients in the refractive index will bend ray paths, changing the apparent arrival angle. The magnitude of this effect increases with zenith angle but is only on the order of 5 arcmin at 10° elevation above the horizon. This might be important for astronomical applications, but is negligible for most optical auroral devices with precision requirements on the order of 1°.

In contrast, variations in atmospheric transmission can be important even at moderate zenith angles. Atmospheric scattering and absorption processes will reduce the radiant flux detected by a ground-based observer (Sterken and Manfroid, 1992). The decrease in apparent magnitude can be modeled as

$$\Delta m(\lambda, \zeta) = \kappa(\lambda) X(\zeta), \tag{20}$$

where $\kappa(\lambda)$ is the extinction coefficient and the relative air mass $X$ as a function of zenith angle $\zeta$,

$$X(\zeta) \approx 1 + (1 - c_1) Z - c_2 Z^2 - c_3 Z^3$$
$$Z = \frac{1 - \cos\zeta}{\cos\zeta}, \tag{21}$$

is equal to one at the zenith (i.e., $X(0) = 1$) and increases by a factor of 5 at 10° elevation above the horizon (Tomasi and Petkov, 2014). For convenience we may separate zenith angle and wavelength effects

$$E(\lambda, t) = E_k(\lambda)^{X(t)} \quad E_k \equiv 2.512^{-\kappa(\lambda)} \tag{22}$$

where $E_k$ is the transmission through one standard air mass (i.e., at zenith).

Empirical results from several nights of astronomical observations near sea level (Vargas et al., 2002) show total extinction ranging from $\kappa = 0.312$–0.604 and $\kappa = 0.180$–0.347 for standard blue and red filters, respectively. Zhang et al. (2013) found $\kappa_g = 0.69$ and $\kappa_r = 0.55$ at a low-altitude (170 m) high-humidity location. Tomasi and Petkov (2014) present an extensive review of optical air mass properties for the Arctic and Antarctic.

For this study, we use values from Patat et al. (2011) to provide a lower bound on extinction effects. The upper bound is estimated using an empirical model based on Vargas et al. (2002) and Zhang et al. (2013).

### 1.4.1 Transit zenith angle

Zenith angle at transit depends on the observer latitude $\Lambda$ and declination of the source. Consequently, two observers viewing the same source from different latitudes will be looking through different air masses. This can produce systematic differences in brightness of a few percent or more depending on the latitude offset $\Delta\Lambda$ and extinction $E_k$:

$$I_2/I_1 \propto E_k^{\Delta X} \Delta X \approx \frac{1}{\cos(\zeta_1 + \Delta\Lambda)} - \frac{1}{\cos(\zeta_1)}. \tag{23}$$

Calibration using Jupiter (or any other planet) will be further complicated by corrections for varying declination. Figure 4 shows several years' variation of air mass for Jupiter transit at the four field sites considered in this study. A significant transition occurs between large latitude-dependent extinction before 2011 to relatively uniform low levels afterward. The consequences for this study are only on the order of a few percent, but are clearly evident in results presented in Sect. 3.3. This provides some assurance that our analysis procedures are accurate near the 1 % level. Of course, calculating the effects of varying declination requires atmospheric extinction coefficients that may not be very well known. This is a challenge, but also an opportunity to test which extinction models produce the best agreement with observations.

**Table 4.** MSP filter wheel sequence.

| Wavelength (nm) | Description | Filter wheel position | |
|---|---|---|---|
| | | Calgary | CANOPUS |
| 470.9 | $N_2^+$ energy flux | 5 | 6 |
| 480.0 | blue background (1) | 1 | 2 |
| 486.1 | $H_\beta$ (1) | 2 | 3 |
| 486.1 | $H_\beta$ (2) | 3 | 4 |
| 495.0 | blue background (2) | 4 | 5 |
| 557.7 | OI green line | 6 | 7 |
| 625.0 | red-line background | 7 | 0 |
| 630.0 | OI red line | 0 | 1 |

Declination differences can even alter the intensity ratio between two different wavelengths (heterochromatic extinction in Sterken and Manfroid, 1992),

$$I_2/I_1 \propto \Delta E_k^{X(\zeta)} \quad \Delta E_k \equiv 2.512^{\kappa(\lambda_1) - \kappa(\lambda_0)}, \tag{24}$$

because extinction is a nonlinear function of air mass. This effect is considered in Sect. 3.4 and found to be significant.

## 2 Meridian scanning photometer

Auroral luminosity is often spatially anisotropic, with latitude structuring on scales of 1–100 km and longitudinal features extending from hundreds up to thousands of kilometers. Consequently, some instruments are designed with reduced azimuthal coverage in exchange for improved sensitivity along a latitude profile. These systems may be referred to as meridian imaging spectrographs (MISs) or meridian scanning photometers (MSPs) depending on the technology used for spectral discrimination and photon detection. In this paper, we explore issues related to field cross calibration of a specific MSP design that has been used extensively for auroral research in Canada. Many of these topics can also be applied more generally to other auroral optical devices.

Data used for this study were obtained from a network of four multispectral auroral meridian scanning photometers. These systems were based on the meridian scanning photometer array (MPA) component of the CANOPUS project (Rostoker et al., 1995) which operated MSPs at three sites in a latitude chain: Rankin Inlet, Gillam, Pinawa (the Churchill line), and a fourth auroral zone site 2 h to the west in Fort Smith. The primary goal was to detect proton aurora at 486.1 nm and electron aurora at several wavelengths (see Table 4) in order to determine precipitation species, characteristic energy, and energy flux. The array was operated continuously for nearly 20 years, producing a large high-quality data set which was the foundation for important research on topics including substorms (Samson et al., 1992), the polar cap boundary (Blanchard et al., 1995, 1997), poleward boundary intensifications (Lyons et al., 1999; Zesta et al., 2000), and the B2i isotropy boundary (Donovan et al., 2003).

**Table 5.** Characteristics of three sets of nominally identical narrow-band filters. Passband is integral of transmission profile, 90 % bandwidth is the range between 5 and 95 % points of the cumulative transmission.

| (nm) | Passband (nm) | | | Peak transmission (%) | | | 90 % bandwidth (nm) | | |
|---|---|---|---|---|---|---|---|---|---|
| 470.9 | 2.483 | 2.362 | 2.355 | 82.94 | 79.36 | 78.28 | 3.60 | 3.40 | 3.50 |
| 480 | 2.592 | 2.418 | 2.661 | 85.59 | 78.60 | 87.74 | 3.10 | 3.20 | 3.20 |
| 486.1 | 2.605 | 2.587 | 2.615 | 88.26 | 85.95 | 87.57 | 2.90 | 3.00 | 3.00 |
| 486.1 | 2.572 | 2.222 | 2.509 | 84.71 | 74.21 | 83.44 | 3.10 | 3.00 | 3.10 |
| 495 | 2.607 | 2.525 | 2.584 | 88.40 | 85.74 | 87.27 | 3.30 | 3.40 | 3.50 |
| 557.7 | 1.788 | 1.728 | 1.920 | 82.93 | 78.93 | 88.53 | 4.60 | 4.90 | 4.30 |
| 625 | 1.624 | 1.632 | 1.588 | 84.46 | 87.37 | 86.09 | 4.20 | 3.20 | 2.80 |
| 630 | 1.597 | 1.590 | 1.558 | 86.67 | 84.83 | 83.81 | 2.40 | 2.60 | 2.40 |

Due to bandwidth limitations, most raw instrument output was downsampled by averaging in space and time in order to produce a uniform data stream for real-time transmission. Full high-resolution data were available over a serial campaign port. In later years, data loggers were used at some sites to record the full resolution data; several years of high-res MSP data are available for retrospective recalibration. The more extensive low-res data set is averaged into 17 latitude bins per scan, which is adequate for auroral science, but diminishes the ability to resolve elevation from individual star transits.

The original CANOPUS MSPs were built by an industrial contractor (Johnston, 1989) based on a series of instruments developed at the National Research Council of Canada (NRCC). Calibration of the prototype was carried out in 1985 by NRCC and the University of Saskatchewan, the results of which led to several design modifications. The first field system was commissioned at Gillam in February 1986, with all four units operational by early 1988. By the late 1990s it was increasingly obvious that the instruments were nearing the end of their lifespan. The primary concern was the mirror motors which had driven several billion steps, but many other issues (e.g., data acquisition, high voltage supplies, photomultiplier tubes) were also causing problems. Eventually, a lack of spare parts resulted in significant failures and data loss.

An MSP revitalization project was carried out at the University of Calgary starting in 2007. The goal was to provide replacement systems with equivalent functionality. System design was based closely on the original instruments in order to minimize risk, with legacy mechanical and optical components reused where possible. Initial development was carried out on the legacy system at Rankin Inlet, which was broken beyond repair. The detector was replaced with a new PMT, high-voltage supply, and pulse-counting circuit. Anti-reflection coatings were added to several optical elements, with system throughput optimized with predictions from optical modeling software and confirmed with quantitative testing. All of the old filters were replaced, as was the filter wheel motor. The scanning mirror assembly was upgraded to provide 0.09° elevation steps (4000 steps per 360°). Thermal and power control systems were completely replaced. Low-level timing and synchronization is now coordinated by an FPGA, with a Linux PC-104 responsible for data acquisition and overall system control.

After darkroom calibration and local field trials, the new prototype system was deployed at Gillam and operated adjacent to the legacy system which was still functioning intermittently. The original Gillam system was then upgraded and sent to Fort Smith (2009), the old Fort Smith system upgraded and installed at Pinawa (2010), and the old Pinawa system upgraded and moved to a new site near Athabasca (2011). Additional improvements were implemented in later systems, motivating a round of upgrades in 2012 to the Gillam and Fort Smith units. The entire rebuild process took more than 4 years and involved multiple personnel at the University of Calgary. Despite careful attention to tracking changes, there are still some functional differences between the first and last refurbished systems. Many of these issues have been identified with internal calibration procedures, but astronomical sources provide useful insight about comparative instrument performance.

The new Calgary MSPs use the same filter wheel design as CANOPUS to acquire data from eight spectral channels, with 486.1 nm duplicated in order to increase SNR for faint proton aurora. Accurate radiometry of rapidly varying aurora requires effective simultaneous measurements of background and signal. This is accomplished by rotating the filter wheel at 1200 RPM (20 Hz) and gating the detector to provide successive 12.5 ms sample spacing. Some details about filter sequencing is given in Table 4; for simplicity, all subsequent multichannel data will be ordered by increasing wavelength (blue to red).

Interference filter transmission and blocking as a function of wavelength were provided by the manufacturer and summarized in Table 5. Results were very close to specifications: FWHM of 3 nm for the blue filters and 2 nm for green and red. Transmission peaks were broad and flat with maxima around 80 %, which is important for optimizing detection of narrow emission lines. An effective passband,

**Table 6.** Fort Smith MSP channel sensitivity $\mathcal{C}_{\dot{\mathcal{R}}/\mathcal{D}}$ (Rayleighs/nanometer/count) determined by darkroom LBS calibration.

| Site date device | 471.0 | 480.0 | 486.0 | 486.0 | 495.0 | 557.7 | 625.0 | 630.0 |
|---|---|---|---|---|---|---|---|---|
| FSMI 20100112 msp-02 | 0.2712 | 0.2570 | 0.2446 | 0.2474 | 0.2666 | 4.5029 | 0.8188 | 0.8598 |
| FSMI 20141121 msp-02 | 0.2935 | 0.2756 | 0.2647 | 0.2685 | 0.2900 | 5.8267 | 0.9180 | 0.9742 |
| FSMI 20141122 msp-02 | 0.2942 | 0.2735 | 0.2645 | 0.2687 | 0.2904 | 5.9789 | 0.9258 | 0.9833 |

$$\Delta\lambda_j = \int d\lambda \hat{T}_j(\lambda), \qquad (25)$$

is the relevant quantity for broad-band calibration sources, i.e., converting from Rayleighs per nanometer to Rayleighs. These data suggest typical passband and transmission variations on the order of 5 % between different sets of filters.

Light which passes through the filters is detected by a photomultiplier tube (PMT) with photocathode quantum efficiency ranging from 20 % at 400 nm to 2 % at 750 nm; this response was selected to maximize response for the faint $H_\beta$ emissions. A dynode chain amplifies each electron to produce a cascade which triggers a pulse-counting circuit. The high-voltage power supply required for this process is quite stable over short intervals under ideal conditions, but may change during extended field operations. Photocathode aging and high-voltage drift are likely to be the primary causes of any long-term reduction in system sensitivity.

The dead time of PMTs produces a nonlinear response at high count rates. This pulse pile-up effect can be largely removed if the time resolution $\tau$ of the system is known and is not significantly longer than the signal count interval. For the PMTs used in this study, nonlinearity only becomes important for count rates greater than $10^5$ photons per second. These rates can be produced by very bright aurora but are not a problem for any astronomical sources except the sun and moon.

Meridian scans are achieved with a 45° tilted mirror and a stepping motor. Many MSPs rotate the mirror at a fixed rate in order to produce data from evenly spaced elevations. Both the original and refurbished systems considered here instead utilize a sequence of variable steps chosen to produce nearly constant exposure times as a function of linear distance at auroral altitudes. This detail is relevant to this study because Jupiter transit profiles will be measured with different resolution depending on transit elevation. The effects are expected to be small, but must be kept in mind when considering multiyear variability.

## 2.1 System sensitivity

The relationship between incident photon flux $\mathcal{P}(\lambda)$ and measured channel count rate $\mathcal{D}_k$,

$$\mathcal{D} = A_{\mathrm{eff}} M_x \Delta t \int d\lambda \mathcal{P}(\lambda) T_k(\lambda) Q(\lambda), \qquad (26)$$

depends on the effective aperture allowing photons into the system ($A_{\mathrm{eff}}$), channel multiplexing efficiency ($M_k$), filter transmission ($T_k$), measurement interval ($\Delta t$), and the detector efficiency $Q(\lambda)$.

For wide-band input through narrow-band filters, the process can be written in terms of filter peak transmission $T_k$ and bandwidth $\Delta\lambda_k$:

$$\mathcal{D}(\lambda_i) \approx \mathcal{P}(\lambda_k) A_{\mathrm{eff}} M_k \Delta\lambda_k T(\lambda_k) \Delta t Q(\lambda_i), \qquad (27)$$

from which we can isolate a coefficient of response $_k\mathcal{C}_{\#1/\#2}DP$ for each channel,

$$
\begin{aligned}
_k\mathcal{C}_{\#1/\#2}DP &= \frac{\mathcal{D}(\lambda_k)}{\mathcal{P}(\lambda_k)} \\
&= A_{\mathrm{eff}} M_x \Delta\lambda_k T(\lambda_k) \Delta t Q(\lambda_k),
\end{aligned} \qquad (28)
$$

in terms of measured $\mathcal{D}$ and predicted $\mathcal{P}$ for each filter wavelength. In principle, this equation could be used to calculate coefficients in terms of fundamental properties of each instrument. In practice, calibration coefficients are often estimated empirically by measuring sources with known brightness. For auroral applications the goal is to determine the differential sensitivity $\mathcal{C}_{\#1/\#2}D\dot{R}$ relating data numbers to Rayleighs per nanometer.

## 2.2 Darkroom calibration

All systems have been calibrated at the University of Calgary using a low brightness source (LBS) with spectral radiance measured by the Canadian Institute for National Measurement Standards. Several sets of calibration results for one instrument at different times are shown in Table 6. Results from two successive days (21 and 22 November 2014) agree to 1 % or better, suggesting that the calibration process is highly repeatable. Earlier results from 2010 indicate that the system was about 5 % more sensitive in all channels, but with only two measurements over more than 4 years, it is impossible to determine whether this corresponds to a gradual decline or an abrupt change at some time during shipping or field operations.

## 3 Data analysis

In this section, we present methods for extracting useful calibration information from Jupiter transits in MSP data.

There are five topics organized by which parameter is under consideration and what supporting measurements are required. Results range from precise and absolute to uncertain and relative. Optical field of view is considered in Sect. 3.1, device orientation in Sect. 3.2, magnitude variation in Sect. 3.3, spectral ratios in Sect. 3.4, and absolute sensitivity in Sect. 3.5.

Each of the MSPs considered in this study repeats a sequence of scans from the northern to southern horizon. Every scan consists of multiple steps through a 160° elevation range, with measurements acquired through multiple filters at each step. The resulting data stream has units of counts or simply data numbers ($\mathcal{D}$) and can be represented by a $[K, M, N]$ array of 16-bit numbers where $K = 8$ is the number of filters, $M = 544$ is the usual number of elevation steps for the rebuilt MSPs, and $N = 120$ scans are acquired during each hour (30 s cadence).

Ephemeris software (Downey, 2015) was used to calculate the time and elevation corresponding to the transit of Jupiter through the local meridian containing the zenith and terminated by the celestial poles. To start, we assumed that instruments were perfectly level and had azimuths pointing directly north in order to obtain a starting point for identifying actual transits. A keogram subregion centered on the predicted transit was used to fit a two-dimensional generalized Gaussian model:

$$\mathcal{D}(x, y) = D_0 Exp\left[-\left|\frac{\overline{x}}{\sqrt{2}\alpha_x}\right|^{\beta_x} - \left|\frac{\overline{y}}{\sqrt{2}\alpha_y}\right|^{\beta_y}\right] \tag{29}$$
$$+ B_0\left\{1 + B_x\overline{x} + B_y\overline{y} + B_{xy}\overline{x}\overline{y}\right\},$$

where $D_0$ and $B_0$ are signal and background; $\overline{y} = y - y_0$ and $\overline{x} = x - x_0$ are the elevation and time relative to the transit peak; $x_0$, $y_0$, and $\alpha_{x,y}$ are profile widths; and $\beta_{x,y}$ are scaling parameters. Jupiter transit profiles were initially modeled with a simpler bivariate Gaussian ($\beta_x = \beta_y = 2$) which could usually achieve model–data differences on the order of 10 %. The more general representation in Eq. (29) was introduced in an attempt to ensure that model error would not be a limiting factor for analysis at the 1 % level. We subsequently found that the coefficients also provided a useful measure for classifying transit quality, and more clearly identified minor azimuthal asymmetry in the optical response.

The polynomial background model is effective for mitigating effects from dawn/dusk gradients and scattered moonlight. This significantly increases the number of transits which could be used for estimating orientation and field of view, although relatively few of these additional events are suitable for radiometric calibration. Figure 7 shows Gillam transit times obtained over several winter field seasons. Sequences of good transits correspond to cloudless nights and gaps correspond to periods of poor visibility near full moon.

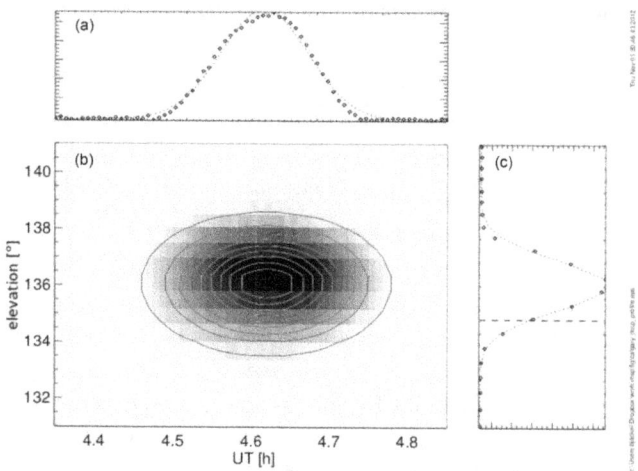

**Figure 6.** Gillam MSP observations of Jupiter on 22 November 2011. Shading in central panel corresponds to counts for each scan and step (higher data numbers (DN) are darker, ranging from 0 to 1500 DN), contours indicate best 2-D Gaussian fit. Right panel is elevation profile obtained by averaging over time (symbols) and best fit Gaussian (dotted line). Top panel is time profile obtained by averaging over elevation. Dashed lines indicate the predicted transit time (off scale) and elevation for ideal north–south scan.

### 3.1 Field of view

Stars and distant planets are effectively point sources when viewed with a single pixel detector (PMT) through optics with angular resolution on the order of 1°. Each MSP elevation sweep over an astronomical source will produce a profile that corresponds to the vertical optical angular response. Similarly, a time sequence of observations from a fixed elevation should provide a complementary measure of horizontal optical beam shape. This is illustrated in Fig. 6 with a full two-dimensional (elevation and time) distribution of observed counts along with corresponding elevation and time profiles. Each profile is approximately Gaussian, and the combined two-dimensional pattern is fairly well modeled by the bivariate generalized Gaussian in Eq. (29). A complete transit profile extends over 10 min, during which time viewing conditions may change considerably. In contrast, each elevation sweep over Jupiter lasts for only a few seconds.

Fitted horizontal (time) and vertical (elevation) beam widths from the Gillam MSP are plotted in Fig. 8. There is a cluster of points near $\sigma \sim 1.1°$ that presumably corresponds to the actual beam shape. Other points are generally associated with suboptimal viewing conditions (e.g., clouds or aurora). Seasonal average estimates of horizontal and vertical beam width for Gillam and Fort Smith sites are presented in Table 7. Results are consistent with all instruments having similar horizontal and vertical widths: $\sigma \approx 1.07°$ (FWHM $\sim 3.0°$). Average beam widths have standard deviations less than 0.05° and standard errors less than 0.01°; typ-

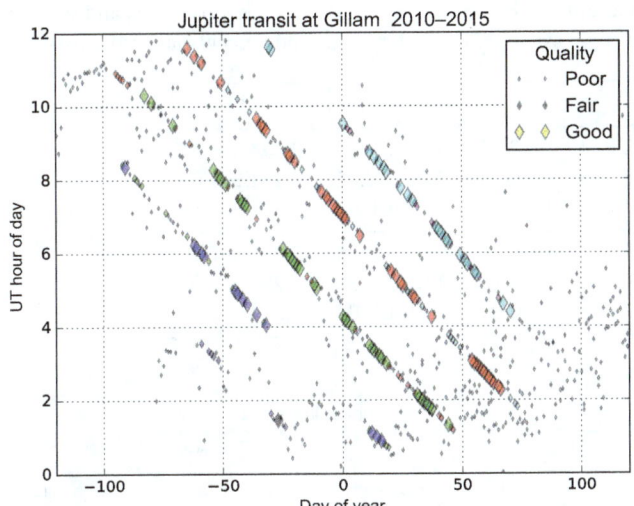

**Figure 7.** Jupiter transit time (UT) observed at Gillam during 2011–2014 Northern Hemisphere winters. Each symbol corresponds to a single night; larger symbols indicate higher-quality transits.

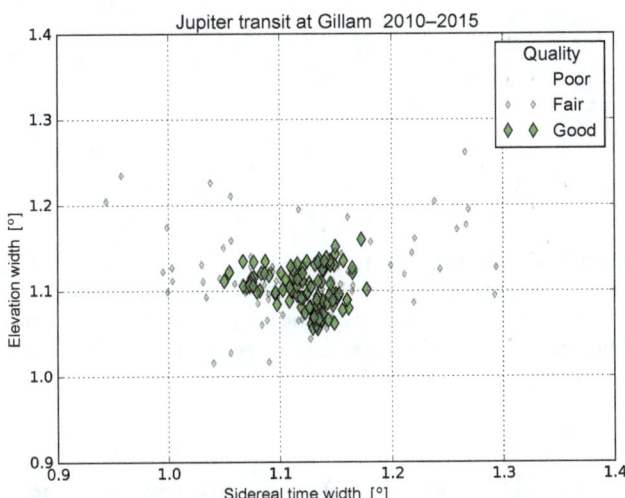

**Figure 8.** Optical beam width determined by fitting a generalized Gaussian to observations of Jupiter by an MSP at Gillam over three winters.

ical beam solid angles are approximately $2.30 \times 10^{-3}$ steradians with uncertainties of a few percent.

The effective solid angle $\Omega_0$ is essential for comparing flux from distant point sources to distributed auroral emissions. For several years of Fort Smith data, the average value was 2.07 msr with standard deviation of 0.12, and standard error of the mean was less than 1 %.

## 3.2 Orientation

An ideal MSP would be aligned to produce scans with predetermined azimuth and elevation. For outdoor installations at remote field sites, it can be difficult to reduce leveling er-

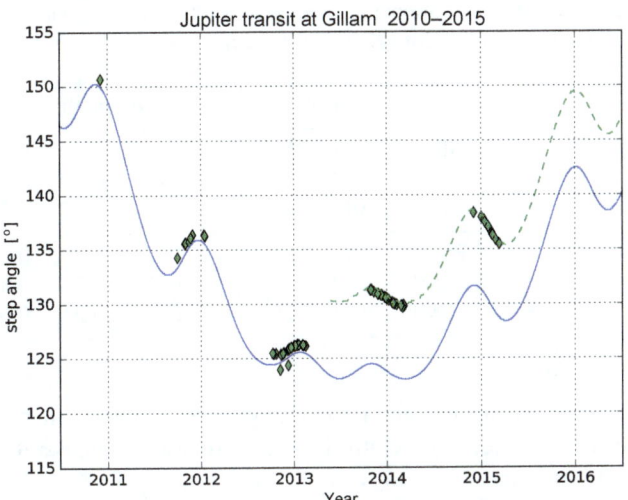

**Figure 9.** Nominal stepping mirror elevation of Jupiter as observed by Gillam MSP. Each symbol corresponds to one transit during a single night. Solid line indicates the actual elevation, while the dashed line is shifted by 7.5 °. Instrument alignment was quite good before summer 2013, after which an unplanned tilt is evident.

rors below a few degrees. Geographic azimuth may be difficult to precisely determine unless a detailed site survey is available. Alignment with magnetic north can also be challenging unless the site is magnetically clean and there are no geomagnetic disturbances. Over longer periods the magnetic declination may change significantly (see Table 1) due to secular variation in the geomagnetic field.

Fortunately, it is possible to accurately determine instrument orientation from transit observations. Starting with site locations obtained using GPS, observed transit times were used to calculate the actual elevation and azimuth of Jupiter for each night. These were interpreted in terms of two device angles. First, azimuth offset was attributed to horizontal orientation of a level instrument. Second, the difference between nominal mirror elevation and actual target elevation was attributed to instrument tilt from level.

Results for several seasons of azimuth estimates at Gillam (not shown) are extremely stable over time, with jitter $< 1°$ and no apparent drift. Tilt estimates are shown in Fig. 9. The first two seasons are generally stable, although there appears to be a small jump in early November. Examination of results from the other three sites (not shown) finds a similar feature at Fort Smith (FSMI), a smaller shift at Pinawa (PINA), and no obvious change at Athabasca (ATHA). These results are consistent with frost heave occurring in early winter as moisture in the soil freezes. The lack of this effect at ATHA may be due to better foundations for the instrument platform. The large change in tilt at Gillam during summer 2013 occurred around the same time as a maintenance trip. This shift could not have been detected in real time due to the lack of Jupiter transit data during limited observing hours during

**Table 7.** Instrument orientation and beam width from all good transits at Gillam and Fort Smith during each winter. Averages and standard deviations in degrees for azimuth, tilt, beam width, and beam height. Solid angle average in millisteradians and percent standard error.

| Site | Year | $N$ | Azimuth | Tilt | $\sigma_h$ | $\sigma_v$ | $\Omega$ |
|------|------|-----|---------|------|-----------|-----------|----------|
| GILL | 2011–2012 | 73 | $6.65 \pm 0.16$ | $0.52 \pm 0.32$ | $1.04 \pm 0.07$ | $1.12 \pm 0.06$ | $2.224 \pm 1.5\%$ |
| GILL | 2012–2013 | 67 | $6.62 \pm 0.14$ | $0.54 \pm 0.33$ | $1.10 \pm 0.05$ | $1.08 \pm 0.04$ | $2.281 \pm 1.0\%$ |
| GILL | 2013–2014 | 46 | $4.81 \pm 9.25$ | $6.52 \pm 0.77$ | $1.10 \pm 0.07$ | $1.09 \pm 0.06$ | $2.301 \pm 1.8\%$ |
| FSMI | 2011–2012 | 64 | $10.35 \pm 0.16$ | $0.59 \pm 0.29$ | $1.06 \pm 0.07$ | $1.11 \pm 0.05$ | $2.257 \pm 1.4\%$ |
| FSMI | 2012–2013 | 57 | $10.00 \pm 0.26$ | $0.87 \pm 0.19$ | $1.12 \pm 0.10$ | $1.07 \pm 0.06$ | $2.282 \pm 2.0\%$ |
| FSMI | 2013–2014 | 54 | $10.50 \pm 0.24$ | $0.66 \pm 0.22$ | $1.12 \pm 0.09$ | $1.06 \pm 0.04$ | $2.274 \pm 1.6\%$ |

summertime operations. Fortunately, once the problem has been identified, it is relatively straightforward to make the necessary corrections to scientific data products.

A yearly summary of orientation parameters for two sites is presented in Table 7. For cases with 30 or more good transits, the standard deviations are less than 0.5° and uncertainties in the average (standard errors) are less than 0.1°. This level of accuracy allows data to be accurately mapped into other coordinates (i.e., geographic); even minor changes to instrument alignment can be easily identified.

### 3.3 Magnitude variation

The signal intensity during each transit will depend on source brightness, instrument sensitivity, and atmospheric effects. This is complicated for Jupiter, as the apparent visual magnitude varies due to changes in distance from Earth. Figure 10 illustrates the importance of this effect, with predicted variation in apparent brightness following the upper bound of observations. The lower set of events typically corresponds to apparent transit profile widths that are significantly different than the best-case values, and are likely due to non-ideal atmospheric transmission (e.g., clouds or ice crystals). There are usually several dozen good transits per season; subsequent analysis will focus on these events.

Effects due to variation in source brightness can be removed by normalizing all measured $\overline{\mathcal{D}}$ cases to magnitude $m = 0$:

$$\mathcal{D}_0 = \overline{\mathcal{D}} \times \sqrt[5]{100}^{m_J}, \qquad (30)$$

where $m_J$ is the apparent visual magnitude of Jupiter predicted by the ephemeris. The resulting distribution of normalized magnitude at Gillam (not shown) has a fairly narrow peak with a sharp higher cutoff and a long tail of lower values corresponding to non-ideal viewing conditions. The 90th percentile was found to be a simple and robust estimator of peak normalized brightness, while average and standard deviation are used to estimate uncertainty in seasonal averages. Results for Gillam and Fort Smith are presented in Table 8.

Normalized brightness for all Gillam transits over 3 years is shown in Fig. 11. Linear fits to the data give a slight positive slope of roughly 2 % per year, but with statistical un-

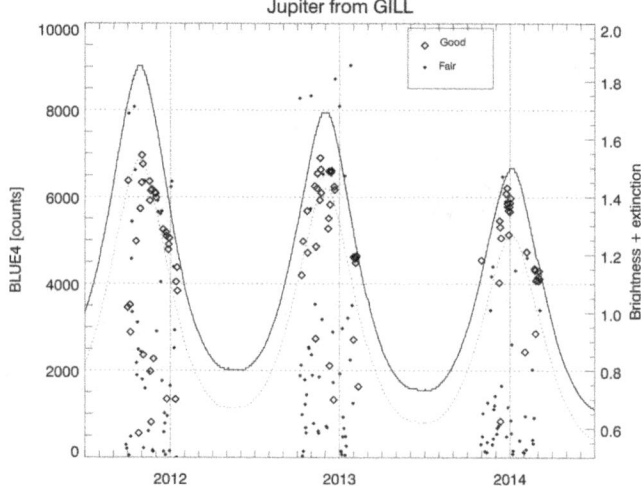

**Figure 10.** Peak counts from Jupiter at Gillam over three winters. Large symbols are transits with narrow widths, small symbols are noisier profiles. Solid line is variation in apparent visual magnitude of Jupiter, dashed line indicates the change in extinction due to doubling air mass ($\Delta \kappa = 0.15$).

certainty that includes zero. This is consistent with a stable system response at blue wavelengths, although variations on the order of 5 % cannot be excluded.

If the linear trend were significant, this would mean the instrument was becoming slightly more sensitive over time, which seems unlikely. Closer examination of the data found that most of the variation is due to a 5 % jump between 2012 and 2013 after which the signal levels remain essentially constant. The jump did not correspond to any system maintenance or modifications. A nearly identical pattern was observed at Fort Smith, further suggesting that the underlying cause was not instrumental.

In fact, this appears to be an example of atmospheric effects as discussed in Sect. 1.4.1. The apparent declination of Jupiter increased from $+5°$ in 2011 to roughly $+15°$ in 2013 and 2014. This reduced the transit zenith angle at Gillam from $\zeta = 57$ to 44°, and effective air mass from $X = 1.84$ to 1.39. For a nominal extinction coefficient $\kappa = 0.17$ at blue wavelengths with zenith transmission $K = 85.5\%$ the change

**Table 8.** Magnitude normalized intensity and self-normalized spectral sensitivity for Gillam and Fort Smith. Column 3 is the number of good transits available from each winter season. Column 4 is the 90th percentile of intensity. Column 4 is the source-normalized brightness (Eq. 30). Remaining columns are channel brightness normalized to average of two 486 nm observations.

| Site | Year | $N$ | 90 % | $\mathcal{D}_0$ | 471 | 480 | 486 | 486 | 495 | 558 | 625 | 630 |
|------|------|-----|------|-----------------|-----|-----|-----|-----|-----|-----|-----|-----|
| GILL | 2011–2012 | 73 | 530.3 | $425 \pm 143$ | 0.914 | 1.054 | 0.997 | 1.003 | 1.052 | 0.087 | 0.415 | 0.382 |
| GILL | 2012–2013 | 67 | 572.7 | $474 \pm 133$ | 0.927 | 1.033 | 1.007 | 0.993 | 1.073 | 0.087 | 0.399 | 0.359 |
| GILL | 2013–2014 | 46 | 582.6 | $487 \pm 141$ | 0.914 | 1.042 | 0.997 | 1.003 | 1.061 | 0.018 | 0.376 | 0.366 |
| FSMI | 2011–2012 | 64 | 844.9 | $651 \pm 224$ | 0.915 | 1.063 | 1.000 | 1.000 | 1.055 | 0.071 | 0.395 | 0.379 |
| FSMI | 2012–2013 | 57 | 873.2 | $732 \pm 199$ | 0.933 | 1.043 | 1.009 | 0.991 | 1.066 | 0.102 | 0.400 | 0.341 |
| FSMI | 2013–2014 | 54 | 877.0 | $715 \pm 228$ | 0.907 | 1.056 | 1.003 | 0.997 | 1.049 | 0.053 | 0.387 | 0.347 |

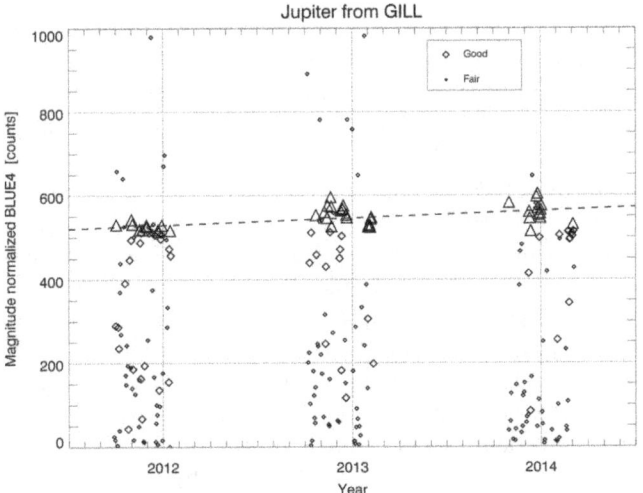

**Figure 11.** Gillam transit events from Fig. 10 normalized to magnitude 0 using Eq. (30).

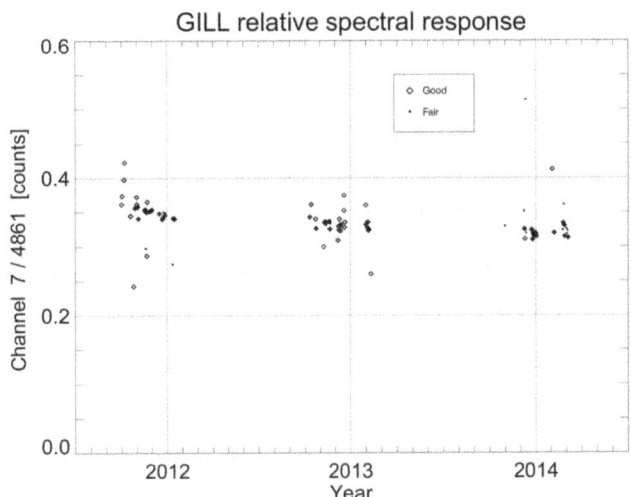

**Figure 12.** Ratio of 630.0 nm to average of two 486.1 nm channels vs. time. Large symbols correspond to good transits and small symbols to noisier events.

in declination corresponds to transmission differences of 74.9 vs. 80.4 %. Adding this correction to normalized brightness reduces the linear trend to zero, although with considerable uncertainty.

### 3.4 Spectral ratio

Absolute radiometric calibration with Jupiter is complicated by variability in observed brightness, and absolute spectral sensitivity is similarly challenging. Working with relative spectral response removes changes in source brightness, allowing us to focus on instrumental and atmospheric effects. In order to reduce statistical uncertainty, we have normalized all channels to the average of the twin $H_\beta$ channels. Results are summarized in Table 8.

Factoring out external brightness variation provides useful information about internal stability of different wavelength channels. Averages for normalized blue channels are essentially constant to within 1 % year to year. This result provides some reassurance about relative filter stability, but cannot exclude the possibility of any change which might produce

identical changes in all channels (e.g., high-voltage supply drift, optical defocusing).

Red channels exhibit more variability on both shorter and longer timescales as shown in Fig. 12. One notable feature is a clear drop after the first season, followed by 2 years of relative stability. This might be attributed to some wavelength-dependent change in sensitivity such as photocathode aging or filter delamination. However, exactly the same pattern is observed at all four sites, suggesting a cause that is external rather than instrumental.

As noted in Sect. 1.4.1, apparent changes in wavelength ratios can also be produced by variations in source declination. Extinction at zenith will have a larger effect on shorter wavelengths, thus increasing the red-to-blue ratio. This effect becomes larger as zenith angle increases with largest red-to-blue ratios observed near the horizon. From 2012 to 2013, Jupiter's declination increased by roughly 10° and transit zenith angle decreased from 50 to 40°. Assuming that observed changes in wavelength ratio are caused by this effect, a simple log-linearized regression,

**Table 9.** Calibration coefficient $\mathcal{C}_{\#1/\#2}PD$ estimated at Gillam using a single transit on 11 November 2011. Atmospheric effects are neglected.

| | | |
|---|---|---|
| 486.1 | (nm) | Channel wavelength |
| 1501 | (DN) | Peak data number |
| $5.191 \times 10^{17}$ | (photon)/(m$^2$ s$^{-1}$nm$^{-1}$) | Solar photon flux at 1 AU |
| $5.328 \times 10^{-10}$ | | Geometric factor $A(t)$ |
| 0.455 | | Jupiter albedo |
| $1.061 \times 10^{9}$ | (photon)/(m$^2$ s$^{-1}$ nm$^{-1}$) | Jupiter photon flux at Earth |
| $7.067 \times 10^{5}$ | (photon)/(m$^2$ s$^{-1}$ nm$^{-1}$ $\mathcal{D}$) | Calibration coefficient $\mathcal{C}_{\#1/\#2}PD$ |
| 0.799 | | Extinction at $\zeta = 45.6°$ |

$$\log I_1/I_2 + \log(2.512) - x\,(\kappa_1 - \kappa_2) = \log D_1/D_2, \qquad (31)$$

gives a slope of $\kappa_{\text{red}} - \kappa_{\text{blue}} \approx 0.38$, which is generally consistent with other results considered in Sect. 1.4. Since this estimate is produced by combining a large number of transits obtained during a wide range of atmospheric conditions, we do not place too much weight on the precise value. The important result is that spurious trends in wavelength ratios can be modeled well enough to allow detection of real changes on the order of 5 %.

### 3.5 Absolute sensitivity

System sensitivity defined in Eq. (28) provides a measure of the data count rate $\mathcal{D}$ produced by one Rayleigh per nanometer $\dot{\mathcal{R}}$ of extended luminosity. This can be related to the differential irradiance of an ideal point source using Eq. (14). Losses due to atmospheric effects can be modeled with Eq. (20). The combination of these three equations,

$$\mathcal{C}_{\#1/\#2}D\dot{R} = 10^{10}\frac{\mathcal{D}}{\dot{S}_\gamma}\frac{\Omega_0}{4\pi}2.512^{+\kappa X}, \qquad (32)$$

gives an expression for calibration coefficients in terms of five physical quantities (see also p. 42 of Wang, 2011). Three of these terms are easily estimated, while the other two present some challenges.

The differential number flux $\dot{S}_\gamma$ of solar photons scattered from Jupiter and arriving at the top of the Earth's atmosphere is only subject to uncertainties in the solar spectrum and Jupiter's albedo, both of which are known to 1 % or better. The effective air mass $X(\zeta(t))$ depends on the apparent zenith angle which can be calculated for any arbitrary time. The effective solid angle $\Omega$ is either known a priori or can be estimated from transit profiles; from Sect. 3.1 the uncertainty of an unbiased estimate will be less than 1 %, but systematic bias on the order of 5 % is also a possibility.

The extinction coefficient spectrum $\kappa(\lambda)$ can be highly variable, can have a major effect on received signal levels, and cannot be accurately estimated from the MSP data. In the absence of other information, the best we can do is identify an upper envelope containing the brightest events and assume that they correspond to the minimum possible extinction values. This approach seems to produce intrinsic variability less than 5 %, but does not address the issue of systematic bias.

Each transit could potentially provide a measured value for $\mathcal{D}$. A simple calculation of Poisson uncertainty for the entire profile would be on the order of 1 % assuming good transits with peaks in excess of 2000 counts. This result may be overly optimistic given the complicated nature of many transits. An alternative approach is to examine sequences of transit profiles, focus on clusters of bright events in the top quartile or decile, and assume that they provide an overestimate of the intrinsic variability. This approach produces estimated uncertainties ranging from 1 to 5 %.

Data from a single transit can be scaled by model flux density from Eq. (18) to obtain an empirical estimate of the system calibration coefficient $\mathcal{C}$. An example is provided in Table 9 for the 22 November 2011 transit at Gillam using the pair of nominally identical 486 nm channels as an example. Fitting a two-dimensional Gaussian model to each channel separately produced very similar peak values: 1501.14 and 1501.54 DN. Appropriate model values from Table 3 can be used to predict input photon flux (neglecting atmospheric effects) and estimate a system calibration coefficient relating flux from a point source to measured data numbers.

Calculation up to this point has consisted of multiplying several quantities, each with relative uncertainty of a few percent or less. These errors are negligible in comparison to atmospheric variability. The 486.1 nm extinction factor at zenith could vary between 0.73 and 0.84 for fair to good visibility, and 0.64–0.78 at $\zeta = 45°$. Lower elevations and worse viewing conditions will further attenuate incoming flux. Neglecting extinction will provide a lower bound for empirical sensitivity, as reduced flux requires higher sensitivity in order to produce the same observations.

Including more events should provide some combination of additional information and increased variability. We attempt to focus on a subset of high-quality transits that presumably correspond to good atmospheric viewing conditions. Events are first classified according to beam widths

**Table 10.** Sensitivity for each channel in data numbers (counts) per Rayleigh per nanometer $C_{\#1/\#2} DR$. Asterisks for Gillam 2012 green line correspond to a calibration without the standard neutral density filter.

|  | Year | N | 471 | 480 | 486 | 486 | 495 | 558 | 623 | 630 |
|---|---|---|---|---|---|---|---|---|---|---|
| GILL | 2011 | 59 | 0.2478 | 0.1816 | 0.2507 | 0.2427 | 0.2002 | 1.6296 | 1.0857 | 0.9721 |
| GILL | 2012 | 60 | 0.2114 | 0.1603 | 0.1698 | 0.1702 | 0.1718 | – | 0.8244 | 0.8764 |
| GILL | 2013 | 39 | 0.1434 | 0.1280 | 0.1169 | 0.1174 | 0.1365 | 1.3462 | 0.5802 | 0.5037 |
| FSMI | 2011 | 51 | 0.0734 | 0.0707 | 0.0615 | 0.0611 | 0.0704 | 1.5203 | 0.3267 | 0.3525 |
| FSMI | 2012 | 47 | 0.1239 | 0.1182 | 0.1096 | 0.1113 | 0.1292 | 3.6000 | 0.6915 | 0.6828 |
| FSMI | 2013 | 52 | 0.1316 | 0.1222 | 0.1164 | 0.1164 | 0.1307 | 3.3278 | 0.5578 | 0.3469 |

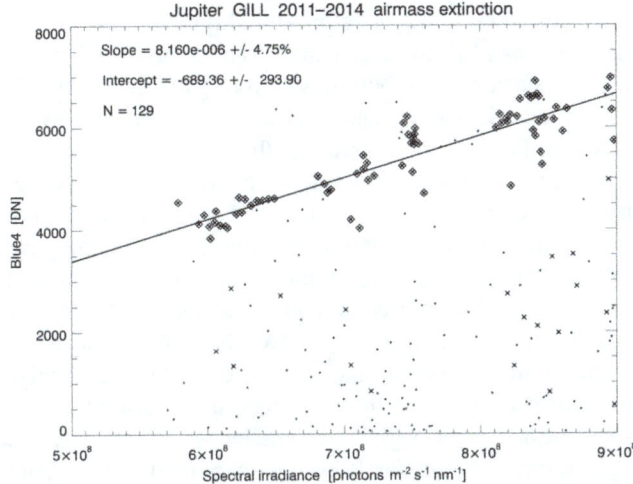

**Figure 13.** Total counts in four blue channels (excluding 470.9 nm) as a function of predicted photon flux density. The small "+" signs indicate all cases, medium "x" signs indicate good beam widths, and large squares indicate nearness to robust fit line. Flux model includes solar spectrum, illumination geometry, Jupiter albedo, and terrestrial atmospheric extinction as in Table 3.

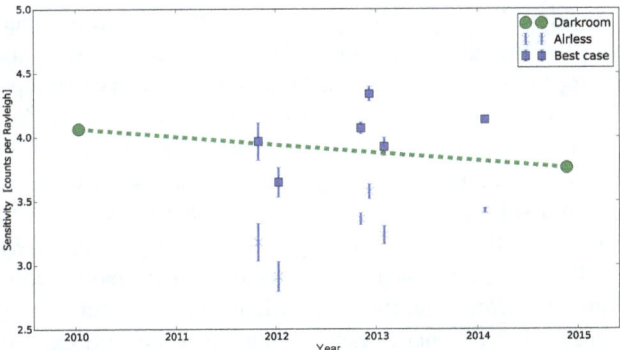

**Figure 14.** Sensitivity for the FSMI MSP 486.1 nm channels. Green circles are values obtained during darkroom calibration in 2010 and 2014; nominal linear trend of $-2\,\%\,\mathrm{yr}^{-1}$ indicated by dashed line. Blue symbols are values obtained by averaging three best values over 10-day intervals and standard deviation indicated with error bars; $x$ values are without any atmospheric correction; squares are with clear sky model.

(Sect. 3.1). Most points cluster near a common linear trend, but there are also quite a few low-brightness outliers. A robust (least absolute deviation) linear model provides a plausible fit that is insensitive to outliers. Points within a generous range around the robust fit are classified as high quality and used for subsequent analysis, including standard least squares estimates of intercept and slope $C_{D/P}$. Figure 13 shows classification and fitting results for the combined blue channel data. This automated process produces reasonable results for all the data considered in this study. Channel calibration coefficients for Gilliam and Fort Smith are presented in Table 10, and Fort Smith results plotted in Fig. 14. More sophisticated algorithms for further studies could explicitly include the asymmetric nature of extinction, i.e., hard upper bound on theoretical maximum.

## 4 Discussion

When auroral instruments operate unattended for long periods of time at remote locations, frequent comprehensive on-site calibration may not be feasible. If celestial objects can be identified in standard data streams then these may serve as the basis for alternative independent calibration procedures.

Stars are essentially point sources when viewed using auroral instruments with angular resolution on the order of 1°. They are stationary in celestial coordinates, and follow predictable paths as the Earth moves during each day and over the course of a year. Absolute flux spectra are increasingly available, although more generally for faint stars that cannot be reliably detected by most auroral devices. Even the brightest stars are only comparable to low-intensity aurora with correspondingly high statistical uncertainty. Light from extraterrestrial sources must also travel through the Earth's atmosphere before arriving at a detector. The resulting wavelength-dependent reduction in photon flux depends critically on atmospheric properties that may not be well known. Of course, auroral light is also subject to the same atmospheric effects.

Jupiter's peak intensity is greater than the brightest star, but less than the moon, so there is no risk of saturating most auroral detectors. It is effectively point-like, has a predictable trajectory, and absolute spectral flux can be calculated from existing albedo and solar irradiance measurements. Unlike stars, planets are not fixed in celestial coordinates, meaning that transit altitude is not constant. This minor complication actually provides an opportunity to study the effects of changing zenith angle on atmospheric extinction.

### 4.1 Atmospheric effects

Atmospheric transmission is likely to be the largest source of uncertainty for high SNR applications. Reducing this uncertainty will require estimation of extinction coefficients that are appropriate for each transit. Our preliminary attempts to determine these parameters using multispectral MSP data were not successful, but this problem may yield to more sophisticated analysis. In principle, extinction coefficients can be found simply by measuring the apparent magnitude of a single star at a given wavelength over a range of different zenith angles. Improved precision can be achieved by combining data from multiple stars. Many auroral observatories include all-sky camera systems which can image dozens or hundreds of stars. However, the optical response (flat field) of these systems is also a strong function of axial angle, which for an ASI is usually directed towards the zenith. Accurate flat fields will be essential for accurate extinction estimates. Recent work by Duriscoe et al. (2007), Olmo et al. (2008), and Román et al. (2012) might be adapted for auroral applications.

It is tempting to avoid the complexity of atmospheric variation by using only a small number of good days to determine calibration parameters. One obvious limitation of this approach is that it cannot reliably detect short term changes in instrument response. More importantly, all auroral observations are subject to exactly the same atmospheric issues. A constant emission feature moving from the horizon to zenith will appear brighter even after accounting for viewing geometry (i.e., van Rhijn correction) simply due to the reduction in total air mass between auroral altitudes and a ground-based observer. Atmospheric effects may be negligible when looking directly upward through clear skies, but critically important at low elevations and non-ideal viewing conditions. These effects would be even more pronounced at shorter wavelengths (e.g., 427.8 and 391.4 nm) often used in auroral studies.

### 4.2 Retrospective calibration

Some auroral instruments only acquire data during short-term campaigns, but many are operated in support of longer-term science objectives. Not all devices are fully calibrated before being deployed and few are calibrated on a regular basis. Even when the resulting data overlap in space and time,

quantitative comparison may not be possible. Astronomical observations of bright sources such as Jupiter can provide a basis for retrospective cross calibration of historical data sets.

The original CANOPUS meridian scanning photometer array (MPA) is a good example. Digital low-resolution binned data are available starting in early 1988 and continuing until spring 2005. Some higher-resolution data are available for the transition period from 2005 to 2010, after which all refurbished instruments were operating in the same high-resolution mode. The 16 years of low-res data alone extend well beyond one solar cycle and could span more than two if merged with newer data.

However, certain kinds of quantitative analysis are limited by the lack of radiometric calibration. Some key parameters (e.g., filter bandwidth and channel sensitivity) were determined for each system, but the supporting documentation is very limited. Mechanical and electrical subsystems were regularly maintained and repaired, but there was no corresponding recalibration schedule. Some terminal calibration procedures were carried out during the 2005–2010 transition, but by this point the instruments were often not functioning reliably. In order to confidently identify long-term geophysical trends in these data, it is essential to have some sense of how instrument performance changed over the same timescales.

A preliminary survey of the CANOPUS MPA data archive has confirmed the feasibility of astronomical calibration and also identified some significant challenges. First, only the brightest few stars are visible even with optimal viewing conditions. Jupiter can be clearly identified, but at count rates much lower than obtained by the newer systems, and consequently with much greater uncertainty. Elevation steps are combined into 17 latitude bins which effectively removes the ability to determine instrument tilt. More generally, it eliminates virtually all information about the optical beam shape in that direction, including that required to confidently estimate the effective solid angle $\Omega_0$. Finally, the decreased scan cadence of one per minute will slightly reduce the accuracy of azimuth estimates. Despite these limitations it should still be possible to estimate absolute sensitivity using Jupiter transits during extended intervals at both ends of the project: 1989–1993 and 1999–2005. Other bright stars or planets might be used to fill in the intervening period.

## 5 Conclusions

In this study, we have demonstrated the feasibility of using Jupiter to calibrate a network of auroral meridian scanning photometers. During times when Jupiter is visible in the night sky, it can be easily distinguished from other astronomical sources. Statistical uncertainty may be a limiting factor even for bright stars, so the increased signal from Jupiter is highly advantageous. Addition precision can be achieved by combining results from multiple days with good viewing conditions.

For geometric calibration, this approach provides an estimate of instrument orientation for each transit with even marginal viewing conditions. Changes of less than 1° between successive transits can be easily identified. Absolute orientation can be determined to at least 0.1°, which exceeds most application requirements. Angular optical response (beam shape) can be estimated to roughly 1 % precision by combining several dozen transits.

Relative spectral calibrations (ratios of different channels) can also be obtained with precisions on the order of 1 % during a single field season. Absolute radiometric calibration for individual channels is significantly less precise. This is due primarily to the difficulty of obtaining and identifying perfectly clean transits. Even results from apparently ideal transits can differ by 5–20 %, likely due to uncertainties in the true atmospheric extinction parameters.

The merits of Jupiter as a calibration source also apply to other types of auroral instruments. Utility of stellar calibration for all-sky imagers has been demonstrated (Wang, 2011; Wang et al., 2012; Grubbs et al., 2016) and these methods would be even more effective with a brighter source. Given the complexities of absolute calibration, it might be helpful if observations were presented in some standard format, e.g., data numbers normalized to source magnitude $\mathcal{D}_0$ as defined in Eq. (30). This, along with estimates of solid angle $\Omega_0$ and bandwidth $\Delta\lambda$, would greatly facilitate the intercomparison of different data products, which would be beneficial for both instrument operators and end-users of scientific data products.

In principle, astronomical calibration could be extracted from almost any auroral data set. In practice, this process is typically applied on a case-by-case basis and requires a considerable amount of human intervention and instrument-specific knowledge. The essential next step is to develop automated software tools which can be applied more broadly. This will significantly increase the utility of optical auroral observations for quantitative scientific analysis.

*Acknowledgements.* Funding for MSP refurbishment and ongoing operation was provided by the Canadian Space Agency under Go Canada initiative contract 13SUGOHSTO for the H STORM project. Field operations support is provided by SED systems and Athabasca University.

Edited by: M. Genzer

# References

Akasofu, S. I.: Dynamic morphology of auroras, Space Sci. Rev., 4, 498–540, 1965.

Alekseeva, G. A., Arkharov, A. A., Galkin, V. D., Hagen-Thorn, E. I., Nikanorova, I. N., Novikov, V. V., Novopashenny, V. B., Pakhomov, V. P., Ruban, E. V., and Shchegolev, D. E.: The Pulkovo spectrophotometric catalog of bright stars in the range from 320 to 1080 nm, Baltic Astron., 5, 603–838, 1996.

Alekseeva, G. A., Arkharov, A. A., Galkin, V. D., Hagen-Thorn, E. I., Nikanorova, I. N., Novikov, V. V., Novopashenny, V. B., Pakhomov, V. P., Ruban, E. V., and Shchegolev, D. E.: The Pulkovo spectrophotometric catalog of bright stars in the range from 320 to 1080 nm – A supplement, Baltic Astron., 6, 481–496, 1997.

Baker, D. J.: Rayleigh, the unit for light radiance, Appl. Optics, 13, 2160–2163, 1974.

Baker, D. J. and Romick, G. J.: The rayleigh: interpretation of the unit in terms of column emission rate or apparent radiance expressed in SI units, Appl. Optics, 15, 1966–1968, 1976.

Blackwell, D., Ellis, R., Ibbetson, P. A., Petford, A. D., and Willis, R. B.: The continuum flux distribution for Arcturus, Mon. Notic. Roy. Astron. Soc., 171, 425–439, 1975.

Blanchard, G. T., Lyons, L. R., Samson, J. C., and Rich, F. J.: Locating the polar cap boundary from observations of 6300 Å auroral emission, J. Geophys. Res., 100, 7855–7862, 1995.

Blanchard, G. T., Lyons, L. R., and Samson, J. C.: Accuracy of using 6300 Å auroral emission to identify the magnetic separatrix on the nightside of Earth, J. Geophys. Res., 102, 9697–9703, doi:10.1029/96JA04000, 1997.

Bohlin, R. C.: HST Stellar Standards with 1 % Accuracy in Absolute Flux, in: The future of photometric, spectrophotometric, and polarimetric standardization, vol. CS-364, edited by: Sterken, C., ASP Conference Series, Blankenberge, Belgium, 315–333, 2007.

Bohlin, R. C.: HST Calspec flux standards: Sirius (and Vega), arXiv, 2014.

Brändström, B. U. E., Enell, C.-F., Widell, O., Hansson, T., Whiter, D., Mäkinen, S., Mikhaylova, D., Axelsson, K., Sigernes, F., Gulbrandsen, N., Schlatter, N. M., Gjendem, A. G., Cai, L., Reistad, J. P., Daae, M., Demissie, T. D., Andalsvik, Y. L., Roberts, O., Poluyanov, S., and Chernouss, S.: Results from the intercalibration of optical low light calibration sources 2011, Geosci. Instrum. Method. Data Syst., 1, 43–51, doi:10.5194/gi-1-43-2012, 2012.

Chapman, S.: A mechanical-optical method of reduction of pairs of auroral plates, Terr. Magnet. Atmos. Elect., 39, 299–303, 1934.

Colina, L., Bohlin, R., and Castelli, F.: Absolute flux calibrated spectrum of Vega, Tech. rep., Space Telescope Science Institute, 1996.

Dahlgren, H., Gustavsson, B., Lanchester, B. S., Ivchenko, N., Brändström, U., Whiter, D. K., Sergienko, T., Sandahl, I., and Marklund, G.: Energy and flux variations across thin auroral arcs, Ann. Geophys., 29, 1699–1712, doi:10.5194/angeo-29-1699-2011, 2011.

Donovan, E. F., Jackel, B. J., Strangeway, R. J., and Klumpar, D. M.: Energy Dependence of the Latitude of the 1–25 KeV Ion Isotropy Boundary, Sodankyla Geophys. Obs. Publ., 92, 11–14, 2003.

Downey, E. C.: XEphem, http://www.clearskyinstitute.com/xephem (last access: September 2016), 2015.

Duriscoe, D., Luginbuhl, C., and Moore, C.: Measuring Night-Sky Brightness with a Wide-Field CCD Camera, Publ. Astron. Soc. Pacific, 119, 192–213, doi:10.1086/512069, 2007.

Fuller, V. R.: Auroral observations at the Alaska agricultural college and school of mines, 1930–31 – Concluded, Terr. Magnet. Atmos. Elect., 37, 150–166, 1931.

Galand, M., Baumgardner, J., Pallamraju, D., Chakrabarti, S., Lovhaug, U. P., Lummerzheim, D., Lanchester, B. S., and

Rees, M. H.: Spectral imaging of proton aurora and twilight at Tromsø, Norway, J. Geophys. Res., 109, 1–14, doi:10.1029/2003JA010033, 2004.

Gladstone, R., Mende, S., Frey, H., Geller, S., Immel, T., Lampton, M., Spann, J., Gerard, J.-C., Habraken, S., Renotte, E., Jamar, C., Rochus, P., and Lauche, H.: Stellar Calibration of the WIC and SI Imagers and the GEO Photometers on IMAGE/FUV, Trans. AGU, 81, F1034, 2000.

Griffin, R. and Lynas-Gray, A.: The effective temperature of Arcturus, Astron. J., 117, 2998–3006, 1999.

Grubbs, G. I., Michell, R., Samara, M., and Hampton, D.: A synthesis of star calibration techniques for ground-based narrow-band EMCCD imagres used in auroral photometry, J. Geophys. Res.-Space Phys., 121, 5991–6002, doi:10.1002/2015JA022186, 2016.

Hayes, D. S.: Stellar absolute fluxes and energy distributions from 0.32 to 4.0 um, in: Calibration of Fundamental Stellar Quantities, International Astronomical Union, Como, Italy, 225–252, 1985.

Hunten, D., Roach, F., and Chamberlain, J.: A photometric unit for the airglow and aurora, J. Atmos. Terr. Phys., 8, 345–346, doi:10.1016/0021-9169(56)90111-8, 1956.

Johnston, S.: Autonomous meridian scanning photometer for auroral observations, Opt. Eng., 28, 20–24, 1989.

Karkoschka, E.: Methane, Ammonia, and Temperature Measurements of the Jovian Planets and Titan from CCD Spectrophotometry, Icarus, 133, 134–146, doi:10.1006/icar.1998.5913, 1998.

Kieffer, H. H. and Stone, T. C.: The Spectral Irradiance of the Moon, Astron. J., 129, 2887–2901, doi:10.1086/430185, 2005.

Kinsey, J. H.: Auroral Parallactic Measurements at Byrd Station, Antarctica, during 1963, J. Geophys. Res., 70, 579–596, 1963.

Knudsen, D. J., Donovan, E. F., Cogger, L. L., Jackel, B. J., and Shaw, W. D.: Width and structure of mesoscale optical auroral arcs, Geophys. Res. Lett., 28, 705–708, doi:10.1029/2000GL011969, 2001.

Lyons, L. R., Nagai, T., Blanchard, G. T., Samson, J. C., Yamamoto, T., Mukai, T., Nishida, A., and Kokubun, S.: Association between Geotail plasma flows and auroral poleward boundary intensifications observed by CANOPUS photometers, J. Geophys. Res., 104, 4485–4500, doi:10.1029/1998JA900140, 1999.

Mallama, A.: The magnitude and albedo of Mars, Icarus, 192, 404–416, doi:10.1016/j.icarus.2007.07.011, 2007.

Mallama, A.: Improved luminosity model and albedo for Saturn, Icarus, 218, 56–59, doi:10.1016/j.icarus.2011.11.035, 2012.

Olmo, F. J., Cazorla, A., Alados-Arboledas, L., López-Alvarez, M. A., Hernández-Andrés, J., and Romero, J.: Retrieval of the optical depth using an all-sky CCD camera, Appl. Optics, 47, H182–H189, 2008.

Patat, F., Moehler, S., O'Brien, K., Pompei, E., Bensby, T., Carraro, G., de Ugarte Postigo, A., Fox, A., Gavignaud, I., James, G., Korhonen, H., Ledoux, C., Randall, S., Sana, H., Smoker, J., Stefl, S., and Szeifert, T.: Optical atmospheric extinction over Cerro Paranal, Astron. Astrophys., 527, A91, doi:10.1051/0004-6361/201015537, 2011.

Rees, M. H. and Luckey, D.: Auroral Electron Energy Derived From Ratio of Spectroscopic Emissions, 1. Model Computations, J. Geophys. Res., 79, 5181–5186, doi:10.1029/JA079i034p05181, 1974.

Román, R., Antón, M., Cazorla, A., de Miguel, A., Olmo, F. J., Bilbao, J., and Alados-Arboledas, L.: Calibration of an all-sky

camera for obtaining sky radiance at three wavelengths, Atmos. Meas. Tech., 5, 2013–2024, doi:10.5194/amt-5-2013-2012, 2012.

Rostoker, G., Samson, J. C., Creutzberg, F., Hughes, T. J., McDiarmid, D. R., McNamara, A. G., Jones, A. V., Wallis, D. D., and Cogger, L. L.: Canopus – A ground-based instrument array for remote sensing the high latitude ionosphere during the ISTP/GGS program, Space Sci. Rev., 71, 743–760, doi:10.1007/BF00751349, 1995.

Samson, J. C., Lyons, L. R., Newell, P. T., Creutzberg, F., and Xu, B.: Proton aurora and substorm intensifications, Geophys. Res. Lett., 19, 2167–2170, 1992.

Seidelmann, P. K.: The Explanatory Supplement to the Astronomical Almanac, 2nd Edn., University Science Books, Sausalito, California, 1992.

Seidelmann, P. K.: Explanatory Supplement to the Astronomical Almanac, 3rd Edn., University Science Books, Sausalito, California, 2005.

Sterken, C. and Manfroid, J.: Astronomical Photometry, Kluwer Academic Publishers, Dordrecht, Boston, 1992.

Stormer, C.: Preliminary report on the results of the aurora-polaris expedition to bossekop in the spring of 1913, Terr. Magnet. Atmos. Elect., 20, 1–12, 1915.

Strickland, D. J., Meier, R. R., Hecht, J. H., and Christensen, A. B.: Deducing Composition and Incident Electron Spectra From Ground-Based Auroral Optical Measurements: Theory and Model Results, J. Geophys. Res., 94, 13527–13539, doi:10.1029/JA094iA10p13527, 1989.

Thuillier, G., Hersé, M., Labs, D., Foujols, T., Peetermans, W., Gillotay, D., Simon, P. C., and Mandel, H.: The solar spectral irradiance from 200 to 2400 nm as measured by the SOLSPEC spectrometer from the ATLAS and EURECA missions, Solar Physics, 214, 1–22, 2003.

Tomasi, C. and Petkov, B.: Calculations of relative optical air masses for various aerosol types and minor gases in Arctic and Antarctic atmospheres, J. Geophys. Res.-Atmos., 119, 1363–1385, doi:10.1002/2013JD020600, 2014.

Vargas, M., Benítez, P., Bajo, F., and Vivas, A.: Measurements of atmospheric extinction at a ground level observatory, Astrophys. Space Sci., 279, 261–269, doi:10.1023/A:1015184127925, 2002.

Wang, Z.: Absolute Calibration of ALIS Cameras, Master's thesis, Lulea University of Technology, Lulea, 2011.

Wang, Z., Wu, B., and Sergienko, T.: A Non-linearity Correction Method for Calibration of Optical Sensor at Low Level Light, AsiaSim, 2012, 126–134, doi:10.1007/978-3-642-34384-1_16, 2012.

Whiter, D., Lanchester, B. S., Gustavsson, B., Ivchenko, N., and Dahlgren, H.: Using multispectral optical observations to identify the acceleration mechanism responsible for flickering aurora, J. Geophys. Res., 115, A12315, doi:10.1029/2010JA015805, 2010.

Zesta, E., Lyons, L. R., and Donovan, E. F.: The auroral signature of earthward flow bursts observed in the magnetotail, Geophys. Res. Lett., 27, 3241–3244, 2000.

Zhang, H.-W. H.-H., Liu, X.-W., Yuan, H.-B., Zhao, H.-B., Yao, J.-S., and Xiang, M.-S.: Atmospheric extinction coefficients and night sky brightness at the Xuyi Observation Station, Res. Astron. Astrophys., 13, 490–500, doi:10.1088/1674-4527/13/4/010, 2013.

# PERMISSIONS

# LIST OF CONTRIBUTORS

**Kaisa Lakkala, Hanne Suokanerva, Juha Matti Karhu, Tomi Karppinen, Markku Ahponen, Henna-Reetta Hannula, Anna Kontu and Esko Kyrö**
Finnish Meteorological Institute – Arctic Research Centre, Tähteläntie 62, 99600 Sodankylä, Finland

**Antti Aarva and Antti Poikonen**
Finnish Meteorological Institute, Observation Services, Helsinki, Finland

**Achim Morschhauser , Jürgen Haseloff, Oliver Bronkalla and Jürgen Matzka**
GFZ German Research Centre for Geosciences, Geomagnetism, Telegrafenberg, 14473 Potsdam, Germany

**Gabriel Brando Soares, José Protásio and Katia Pinheiro**
Geophysics Department, Observatório Nacional, Rio de Janeiro, CEP, 20921-400, Brazil

**Timo Sukuvaara, Kari Mäenpää, and Riika Ylitalo**
Finnish Meteorological Institute, P.O. Box 503, 00101 Helsinki, Finland

**Greg Kopp, Paul Smith, Chris Belting, Zach Castleman, Ginger Drake, Joey Espejo and Karl Heuerman**
Laboratory for Atmospheric and Space Physics, University of Colorado, Boulder, CO 80303, USA

**James Lanzi and David Stuchlik**
NASA Wallops Flight Facility, Wallops Island, VA 23337, USA

**Arvind Singh and Upendra Kumar Singh**
Department of Applied Geophysics, Indian Institute of Technology (Indian School of Mines), Dhanbad, Jharkhand 826004, India

**Janne Narkilahti, Riitta Hurskainen and Hanna Silvennoinen**
Sodankylä Geophysical Observatory, University of Oulu, POB 3000, Oulu, 90014, Finland

**Elena Kozlovskaya and Jouni Nevalainen**
Sodankylä Geophysical Observatory, University of Oulu, POB 3000, Oulu, 90014, Finland
Oulu Mining School, University of Oulu, POB 3000, Oulu, 90014, Finland

**David Brain**
University of Colorado, Boulder, USA

**Mohamed Elhag**
Department of Hydrology and Water Resources Management, Faculty of Meteorology, Environment & Arid Land Agriculture, King Abdulaziz University, Jeddah 21589, Saudi Arabia

**Haneen Alsubaie**
Biological Sciences Department, Faculty of Science, King Abdulaziz University, Jeddah 21589, Saudi Arabia

**Hanaa K. Galal**
Biological Sciences Department, Faculty of Science, King Abdulaziz University, Jeddah 21589, Saudi Arabia
Botany Department, Faculty of Science, Assiut University, Asyut, Egypt

**Richard Larsson1 and Yasuko Kasai1,**
National Institute of Information and Communications Technology, Tokyo, Japan

**Mathias Milz**
Luleå University of Technology, Kiruna, Sweden

**Patrick Eriksson and Jana Mendrok**
Chalmers University of Technology, Gothenburg, Sweden

**Stefan Alexander Buehler**
University of Hamburg, Hamburg, Germany

**Catherine Diéval**
Lancaster University, Lancaster, UK

**Paul Hartogh**
Max Planck Institute for Solar System Research, Göttingen, Germany

**Sergey Y. Khomutov and Oksana V. Mandrikova**
Institute of Cosmophysical Research and Radio Wave Propagation FEB RAS, Mirnaya str, 7, Paratunka 684034, Kamchatka, Russia

**Ekaterina A. Budilova**
Institute of Cosmophysical Research and Radio Wave Propagation FEB RAS, Mirnaya str, 7, Paratunka 684034, Kamchatka, Russia
Kamchatka State Technical University, Klyuchevskaya str, 35, Petropavlovsk-Kamchatsky 683003, Russia

**Kusumita Arora and Lingala Manjula**
CSIR – National Geophysical Research Institute, Uppal Road, Hyderabad-500007, Telangana, India

**Stefan Meyer, Marek Tulej and PeterWurz**
Physics Institute, Space Research and Planetary Sciences, University of Bern, Sidlerstrasse 5, 3012 Bern, Switzerland

**Andreas Georgiou and Dimitrios Skarlatos**
Civil Engineering & Geomatics Dept., Cyprus University of Technology, 30 Archbishop Kyprianou Str., 3036 Limassol, Cyprus

**Alexandre Gonsette, Jean Rasson, and François Humbled**
Centre de Physique du Globe, Royal Meteorological Institute, 5670 Dourbes, Belgium

**Ari-Matti Harri,Walter Schmidt, Jyri Heilimo, Mikhail Uspensky,Maria Genzer,Jouni Polkko, Mark Paton, Harri Haukka, Tero Siili, Osku Kemppinen and Jussi Leinonen**
Research Division, Finnish Meteorological Institute, Helsinki, Finland

**Konstantin Pichkadze, Sergey Alexashkin, Valery Finchenko, Vladimir Khovanskov, Viktor Vorontsov, Alexander Polyakov and Boris Ostesko**
Planetary Systems Department, Lavochkin Association, Moscow, Russia

**Lev Zeleny , Oleg Korablev, Vyacheslav Linkin and Alexander Lipatov**
Planetary Science Laboratory, Russian Space Research Center (IKI), Moscow, Russia

**Hector Guerrero, Ignacio Arruego and Marina Diaz-Michelena**
Microelectronics Department, Instituto Nacional de Tecnica Aeroespacial (INTA), Madrid, Spain

**Luis Vazquez, Pilar Romero and Francisco Valero**
Computational Mathematics Dept, Universidad Complutense de Madrid, Madrid, Spain

**Andrey Poroshin**
Dauria Ltd, Moscow, Russia

**Timo Siikonen and Matti Palin**
Finflo Ltd, Espoo, Finland

**Hannu Savijärvi**
Dept of Physics, University of Helsinki, Finland

**Maiju Linkosalmi, Mika Aurela, Juha-Pekka Tuovinen, Cemal M. Tanis, Ali N. Arslan, Tuula Aalto, Juuso Rainne, Juha Hatakka and Tuomas Laurila**
Finnish Meteorological Institute, Helsinki, Finland

**Mikko Peltoniemi**
Natural Resources Institute Finland (LUKE), Vantaa, Finland

**Pasi Kolari**
Faculty of Biosciences, University of Helsinki, Helsinki, Finland

**Kristin Böttcher**
Finnish Environment Institute (SYKE), Helsinki, Finland

**Ralf Srama**
Institute of Space Systems, University of Stuttgart, Pfaffenwaldring 29, 70569 Stuttgart, Germany

**Thomas Albin**
Institute of Space Systems, University of Stuttgart, Pfaffenwaldring 29, 70569 Stuttgart, Germany
Universitätssternwarte Oldenburg, Institute of Physics and Department of Medical Physics and Acoustics, Carl von Ossietzky University, 26129 Oldenburg, Germany

**Björn Poppe**
Universitätssternwarte Oldenburg, Institute of Physics and Department of Medical Physics and Acoustics, Carl von Ossietzky University, 26129 Oldenburg, Germany

**Detlef Koschny**
European Space Agency, ESA/ESTEC, Keplerlaan 1, 2201 AZ Noordwijk ZH, the Netherlands
Chair of Astronautics, Technical Univ. Munich, Boltzmannstraße 15, 85748 Garching, Germany

**Sirko Molau**
International Meteor Organisation, Abenstalstr. 13b, 84072 Seysdorf, Germany

**Brian J. Jackel, Craig Unick, Greg Baker, Eric Davis, Eric F. Donovan, CodyWilson, Jarrett Little, M. Greffen and Neil McGuffin**
Department of Physics and Astronomy, University of Calgary, Alberta, Canada

**Fokke Creutzberg**
Natural Resources Canada, Geological Survey, Geomagnetism Laboratory, Natural Resources Canada Geomagnetism Laboratory, Ottawa, Ontario, Canada

**Martin Connors**
Department of Physics and Astronomy, Athabasca University, Alberta, Canada

# Index